121 Springer Series in Solid-State Sciences

Edited by Peter Fulde

Springer
Berlin
Heidelberg
New York
Barcelona
Budapest
Hong Kong
London
Milan
Paris
Santa Clara
Singapore
Tokyo

Springer Series in Solid-State Sciences

Volumes 1–89 are listed at the end of the book

A. Okiji H. Kasai K. Makoshi (Eds.)

Elementary Processes in Excitations and Reactions on Solid Surfaces

Proceedings of the 18th Taniguchi Symposium
Kashikojima, Japan, January 22–27, 1996

With 106 Figures

Springer

Professor Ayao Okiji
Professor Hideaki Kasai

Department of Applied Physics, Osaka University
2-1 Yamadaoka, Suita, Osaka 565, Japan

Professor Kenji Makoshi

Faculty of Science, Himeji Institute of Technology
Kamigori-cho, Ako-gun, Hyogo 678-12, Japan

Library of Congress Cataloging-in-Publication Data

Elementary processes in excitations and reactions on solid surfaces : proceedings of the 18th Taniguchi Symposium, Kashikojima, Japan, January 22-27, 1996 / A. Okiji, K. Makoshi, H. Kasai (eds.).
p. cm. -- (Springer series in solid-state sciences ; 121)
"Taniguchi International Symposium on the Theory of Condensed Matter, held from 22nd to 26th January 1996, at the Shima-Kanko Hotel, Kashikojima, Mie, Japan."--Pref.
Includes bibliographical references and index.

1. Solids--Surfaces--Effect of radiation on--Congresses. 2. Charge transfer--Congresses. 3. Energy transfer--Congresses. 4. Exciton theory--Congresses. I. Okiji, A. (Ayao), 1934- . II. Makoshi, K. (Kenji), 1948- . III. Kasai, H. (Hideaki), 1952- IV. Taniguchi International Symposium on the Theory of Condensed Matter (18th : 1996 : Kashikojima, Japan) V. Series.
QC176.8.S8E44 1996
530.4'17--dc20 96-32198
CIP

ISBN-13: 978-3-642-64738-3 e-ISBN-13: 978-3-642-61185-8
DOI: 10.1007/978-3-642-61185-8

Softcover reprint of the hardcover 1st edition 1996

Typesetting: ASCO Trade Typesetting Limited, Hong Kong
SPIN: 10520743 54/3144 – 5 4 3 2 1 0 – Printed on acid-free paper

Preface

This volume contains the proceedings of the 18th Taniguchi International Symposium on the Theory of Condensed Matter, which was held from 22nd to 26th January 1996, at the Shima-Kanko Hotel, Kashikojima, Mie, Japan. The topic of the symposium was "Elementary Processes in Excitations and Reactions on Solid Surfaces".

In the last decade, our knowledge of static properties of solid surfaces has become precise and rich. Supported by the knowledge, the main stream of surface science is moving to explore the fundamental nature of dynamics in reactions at solid surfaces. Attempts are made to do first-principle calculations of quantum mechanical as well as classical molecular dynamics on the potential energy surfaces. These studies reveal various aspects of elementary processes in reactions on solid surfaces. Furthermore, pursuits of the time-development of electron systems during elementary processes in excitations on solid surfaces deepen the understanding of the mechanism involved.

We divide this volume into three frames, albeit crude because of the interdisciplinary nature of this field of science. Part I concentrates mainly on the rich reaction dynamics on the potential energy surfaces. Part II is devoted to the interplay of excitations and reactions with particular attention to the charge transfer as well as the energy transfer between well-characterized surfaces and beams. Part III introduces new and rapidly developing methods and aspects.

The symposium was fully supported financially by the Taniguchi Foundation established in 1929 by Mr. Toyosaburo Taniguchi from his personal funds, who regrettably passed away on 26th October 1994. The purpose of the Foundation is to promote research in the field of basic science thus contributing to the development of industry and the economy, and also to promote mutual understanding and international friendship especially among young researchers. The Taniguchi Foundation sponsors 18 international symposia on selected topics every year. The concept of these symposia is unique in that participants, both Japanese and from abroad, are limited in number to small discussion groups, and live together, although for a short period, as a close-knit community. Fruitful discussions held during the symposium have certainly encouraged participants, and will hopefully stimulate this field of science. On behalf of the participants, we would like to express our sincere thanks to the late Mr. Taniguchi and the Taniguchi Foundation for giving us an opportunity to have such a nice symposium. We would like to dedicate this volume to the late Mr. Taniguchi with our pleasant and unforgettable memory of the symposium.

Osaka
Osaka
Himeji
March 1996

Ayao Okiji
Hideaki Kasai
Kenji Makoshi

THE 18TH TANIGUCHI INTERNATIONAL SYMPOSIUM SHIMA KANKO HOTEL 1996.1.22~26

Participants in the Taniguchi International Symposium

January 22-26, 1996 (at Kashikojima, Japan)

Seated (left to right)
K. W. Jacobsen, M. Persson, H.-J. Freund, D.C. Langreth, W.J.R. Brenig, A. Okiji, A.W. Kleyn, D.M. Newns, A. Yoshimori, Y. Murata
Second row (left to right)
Ms. Jacobsen, Ms. Persson, Ms. Cho, Ms. Langreth, Ms. Brenig, Ms. Okiji, K. Makoshi, Ms. Makoshi, W. Hübner, Ms. Kleyn

Third row (left to right)
S. Tsuneyuki, K. Shobatake, K. Cho, H. Ishida, T. Uda, H. Nakatsuji, M. Kawai, Ms. Kasai, E. Torikai

Fourth row (left to right)
K. Fukutani, H. Nakanishi, H. Kasai, H.K.V. Lotsch

Contents

Part I Reaction Dynamics and Potential Energy Surfaces

Part II Beam-Surface Scattering and Charge and Energy Transfer

Part I

Reaction Dynamics and Potential Energy Surfaces

Reaction Dynamics of H_2/Si: A 5-D Model

W. Brenig[1], *A. Groß*[2], *U. Höfer*[3], *P. Kratzer*[1], and *R. Russ*[1]

[1]Physik-Department, Technische Universität München, D-85747 Garching, Germany
[2]Fritz-Haber-Institut der Max-Planck-Gesellschaft, D-14195 Berlin-Dahlem, Germany
[3]Max-Plank-Institut für Quantenoptik, D-85740 Garching, Germany

Several theoretical and experimental results on the dynamics of dissociative adsorption and recombinative desorption of hydrogen on silicon are reviewed.

Theoretically a 5-D model is presented taking lattice distortion, corrugation and molecular vibrations into account besides the translational motion of the molecule. The properties of this model are investigated by treating various 2-D and 3-D subspaces within the framework of coupled channel calculations.

While the temperature dependence of sticking is dominated by lattice distortion, the main effect of corrugation is a reduction of the pre-exponential factor by about one order of magnitude per lateral degree of freedom. Molecular vibrations have practically no effect on the adsorption/desorption dynamics itself but lead to vibrational heating in desorption with a strong isotope effect. *Ab initio* calculations for the H_2 interaction with the dimers of Si(100)2×1 show properties of the potential surface in qualitative agreement with the model, but its dynamics differs quantitatively from the experimental results.

Experimentally extremely small sticking probabilities in the range $10^{-9} - 10^{-5}$ could be measured using optical second-harmonic generation (SHG) for H_2 and D_2 on Si(111)7×7 and Si(100)2×1. Strong phonon-assisted sticking was observed for gases at 300 K and surface temperatures between 550 K and 1050 K. The absolute values as well as the temperature variation of the adsorption and desorption rates show surprisingly little isotope effect and differ only little between the two surfaces. These results indicate that tunnelling, molecular vibrations, and the structural details of the surface play only a minor role for the adsorption dynamics. Instead, they appear to be governed by the localized H–Si bonding and Si–Si lattice vibrations.

These results together with previously measured energy and angular distributions of desorbing molecules can be described very well with the theoretical model.

1. Introduction

The kinetics and dynamics of hydrogen interaction with silicon surfaces have recently become a topic of extensive experimental [1, 2, 3, 4, 5, 6, 7, 8, 9, 10, 11, 12, 13, 14, 15, 16] and theoretical efforts [17, 18, 19, 20, 21, 22, 23, 24, 25, 26, 27, 28, 29, 30, 31, 32, 33, 34, 35, 36, 37]. The great interest is, of course, partly due to the importance of H/Si in semiconductor chemistry [38, 39]. Equally important, however, hydrogen on silicon is a promising model system for studying general aspects of chemical reactions on covalent surfaces. It may be expected that the localized nature of bonding on semiconductor surfaces gives rise to a reaction behaviour that is in many respects qualitatively different from metal surfaces. One particularly far-reaching consequence of the localized nature of H-Si bonds, e.g., are the large diffusion barriers that were observed experimentally for H/Si(111)7×7 [5] and have been predicted theoretically for H/Si(100)2×1 [28, 19, 22].

Springer Series in Solid-State Sciences, Volume 121
Elementary Processes in Excitations and Reactions on Solid Surfaces
Eds.: A. Okiji, H. Kasai and K. Makoshi

Also, the finding that the desorption kinetics of H_2 from silicon depend on the details of the surface reconstruction [4, 6] is due to the local character of Si–H chemisorption.

One topic of the present paper is another, rather dramatic manifestation of the localized nature of hydrogen interaction with silicon surfaces: The occurrence of strong dynamic lattice distortions in the dissociative adsorption and recombinative desorption of molecular hydrogen on/from Si(111)7×7 and Si(100)2×1. Lattice relaxations provide the solution to the so-called 'barrier puzzle' resulting from recently performed state resolved experiments of hydrogen desorption from silicon surfaces [9, 24, 11]: It had been known for many years that the room temperature sticking coefficient of molecular H_2 on silicon surfaces is very low, less than $10^{-6} - 10^{-8}$ [40, 41, 42]. This suggests the presence of a high adsorption barrier of $V_{ads} > 0.5\,\mathrm{eV}$. However, the energetics of desorbing molecules, in particular the absence of translational heating [11], indicates the absence of a substantial potential drop during desorption.

Lattice relaxation is frequently associated with adsorption [43] and even for metal surfaces may influence adsorption/desorption barriers [44]. In the case of H/Si, theoretical investigations indicate that the diffusion [28, 19, 32] and desorption barriers [18] decrease substantially when the Si atoms relax. However, a simple *static* lattice distortion of Si between the two states 'hydrogen adsorbed' and 'hydrogen desorbed', would lead to a violation of *detailed balance* [11, 31]. This stimulated the first quantum mechanical coupled channel calculation of the adsorption/desorption reaction dynamics of H_2/Si. In the absence of a detailed ab initio potential surface (PES), a simple two dimensional model was proposed to study such reactions [31]. It contained – besides the center of mass distance of the molecule from the surface – only a single additional coordinate namely the amplitude of a 'representative surface oscillator', associated with lattice distortions.

The shape of the PES used in [31] was to a large extent dictated by the existing experimental facts. This implies that during desorption there should be little force exerted on the center of mass coordinate in order to avoid translational heating. The high adsorption

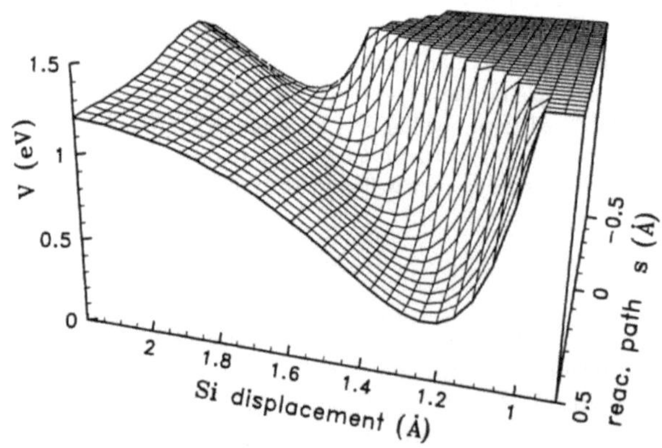

FIG. 1: Model potential for the interaction of molecular hydrogen with silicon surfaces. The three dimensional representation shows the projection of the PES on a characteristic Si–bulk spacing and the H_2–bulk distance, with the view from the gas phase region. The equilibrium Si–bulk distance of the clean surface is ~1.2 Å. With the lattice in this configuration, the incoming H_2 molecule experiences a large barrier to adsorption. In constrast to this, the desorption reaction proceeds via the upper minimum which has an increased Si–bulk spacing due to the presence of adsorbed hydrogen. Starting there, the potential energy surface has only a small gradient along the H_2–bulk direction.

barrier needed to explain the low sticking coefficient then has to originate mainly from a strong dependence of the PES on the lattice distortion. A plot of a slightly extended version of the PES proposed in Ref. [31] is depicted in Fig. 1. Such a potential obviously leads to a strong lattice excitation in desorption without significant translational heating. It somehow circumvents Polanyi's rules [45] for the usual 'curved' reaction paths which normally produce translational heating together with vibrational heating in exothermic processes. Application of time reversal to such a desorption process then leads to a large sticking coefficient for an excited lattice. Hence, such a model predicts strong *phonon-assisted sticking.*

Recently we have included more degrees of freedom in the model. We considered the effect of surface corrugation and molecular vibrations. In subsection 2.2 we show that the main effect of corrugation is a reduction of the sticking coefficient by about one order of magnitude per lateral degree of freedom (depending only slightly on temperature). Molecular vibrations have practically no effect on the adsorption/desorption dynamics itself but lead to vibrational heating in desorption with a strong isotope effect. Taken all together, our model allows for a satisfactory description of all currently available experimental results.

Ab initio quantum dynamical calculations for the interaction of H_2 with the Si–Si dimers of the Si(100)2×1 surface are found to agree qualitatively with the results obtained from the model PES, but underestimate the amount of coupling to the lattice. Possible reasons for the quantitative discrepancies and alternative reaction paths leading to stronger lattice distortions will be discussed.

Experimentally, the strong enhancement of the sticking probability was first observed for H_2 adsorption on Si(111)7×7 [7] and more recently also for Si(100)2×1 [8]. In these experiments sticking coefficients of the order of 10^{-9} could be measured with a technique based on optical second-harmonic generation (SHG). In section 3. of the present paper we summarize these adsorption experiments.

Specific microscopic reaction mechanisms will be considered in section 4. of the paper.

2. Theory

In this section we are going to present a 5-D model which can describe all experimental observations in a coherent way as the result of the dynamics on a suitably chosen potential energy surface. The most important degree of freedom, apart from the reaction coordinate s [46, 47, 48, 49] itself, is the motion of the substrate lattice during the adsorption/desorption process. We describe this part of the dynamics by a representative coordinate x. Additionally we will take the molecular vibrations (coordinate r) and the surface corrugation (coordinate $\boldsymbol{y}$) into account. In total our model covers five degrees of freedom.

We assume that the potential can be written as a sum of the following four terms:

$$V(x,\boldsymbol{y},r;s) = V_{\text{trans}}(s) + V_{\text{phon}}(x;s) + V_{\text{corr}}(\boldsymbol{y};s) + V_{\text{vib}}(r;s). \tag{1}$$

The first term $V_{\text{trans}}(s)$ then gives the value along the *minimum* of the multidimensional potential with respect to the coordinates orthogonal to s, i.e. along the line of steepest descent of the PES. The second term $V_{\text{phon}}(x;s)$ describes the coupling of adsorption and desorption to the substrate. The last two terms allow us to include the effect of a lateral one or two dimensional corrugation and the molecular vibrations. At present our model does not include orientational effects of the hydrogen molecule. This means first of all that we cannot describe the rotational cooling observed experimentally [9]. Concerning the temperature dependence of sticking coefficients we expect mainly a further reduction of the preexponential.

The organization of this section follows the different terms of equation (1). The results of the model including only the first two terms were reported in ref [31]. We briefly review them in subsection 2.1 in order to motivate the choice of shape and parameters of the PES. We then extend these calculations to three dimensions with different degrees of freedom included: In subsection 2.2 we additionally consider a 1-D surface corrugation. In subsection 2.3 we additionally study the vibrational dynamics of the molecules.

The experimental results are very similar for the Si(111) and Si(100) surface. This indicates that the basic physics is the same on both surfaces and the model considered here can be used to describe the adsorption and desorption dynamics in a general way. We will consider a specific microscopic mechanisms in subsection 2.4 and calculate the dynamical properties of an ab initio PES for the interaction of H_2 with the dimers of the Si(100)2×1 surface.

2.1 Basic 2-D model

The original 2-D model, considering only the first two terms of equation (1) (molecular translation and surface phonons) [31], is able to describe the quasi-thermal translational energy distribution in desorption [11] and the strong increase of the sticking coefficients with surface temperature (sec. 3., refs. [7, 8]).

For the potential $V_{\text{trans}}(s)$ along the *minimum* of the PES we use the following general parametrization:

$$V_{\text{trans}}(s) = (V_{\text{des}} - V_{\text{ad}})\frac{\tanh(\lambda s) - 1}{2} + \tfrac{1}{4}\Big(V_{\text{des}} + V_{\text{ad}} + 2\sqrt{V_{\text{des}}V_{\text{ad}}}\,\Big)\frac{1}{\cosh^2(\lambda s)} \tag{2}$$

The parameters of our model PES are then given by an effective adsorption barrier of $V_{\text{ad}} = 0.7\,\text{eV}$ for H_2, a desorption barrier of $V_{\text{des}} = 2.5\,\text{eV}$, and a potential decay length of $\lambda = 2.7\,\text{Å}^{-1}$. The desorption barrier V_{des} is known from experiment (sec. 3., refs. [2, 4, 6]) and λ was estimated. The results are not very sensitive to the latter two quantities, but only to the adsorption barrier V_{ad} which thus is the single essential parameter.

The phonon contribution $V_{\text{phon}}(x;s)$ we represent by a Morse potential:

$$V_{\text{phon}}(x;s) = \tfrac{1}{2}\,\hbar\omega_{\text{phon}}(s) \times d\Big(1 - \exp\big(-\alpha\gamma_{\text{phon}}(s)[x + \Delta x(s)]\big)\Big)^2. \tag{3}$$

The phonon potential depth d is given by the square root of the number of bound states which we chose to be $d^2 = 52$ for the model PES. The inverse oscillator length is $\gamma_{\text{phon}}(s) = \sqrt{M^*\omega_{\text{phon}}(s)/\hbar}$ with the frequency $\hbar\omega_{\text{phon}}(s) = E_1(s) - E_0(s)$ as the quantum of the first excited state and thus $\alpha = \sqrt{d}/(d-1)$. We estimate the effective phonon mass M^* to be the mass of one single Si atom, and for the frequency we use $\hbar\omega_{\text{phon,clean}} = 45\,\text{meV}$. The latter is a mean value of several optical modes [50, 51, 34]. The dynamics is insensitive to the exact value of this quantity due to the many phonons involved. The shift $\Delta x(s)$ in the lattice distortion ('phonon') coordinate is mainly dictated by our intention to avoid translational heating, i.e. not to have any potential drop in the translational coordinate between the barrier and the gas phase region. A contour plot of that PES for H_2 is shown in Fig. 2 a).

The results of the 2-D model are the following [31]:

(1) We find an approximate Maxwell distribution of translational energies in desorption shown in Fig. 3 as a function of final translational energy. The average energies are $\langle E_{\text{trans}}\rangle = 76\,\text{meV}$ at $T_s = 780\,\text{K}$ for H_2 and $\langle E_{\text{trans}}\rangle = 85\,\text{meV}$ for D_2 which agree well

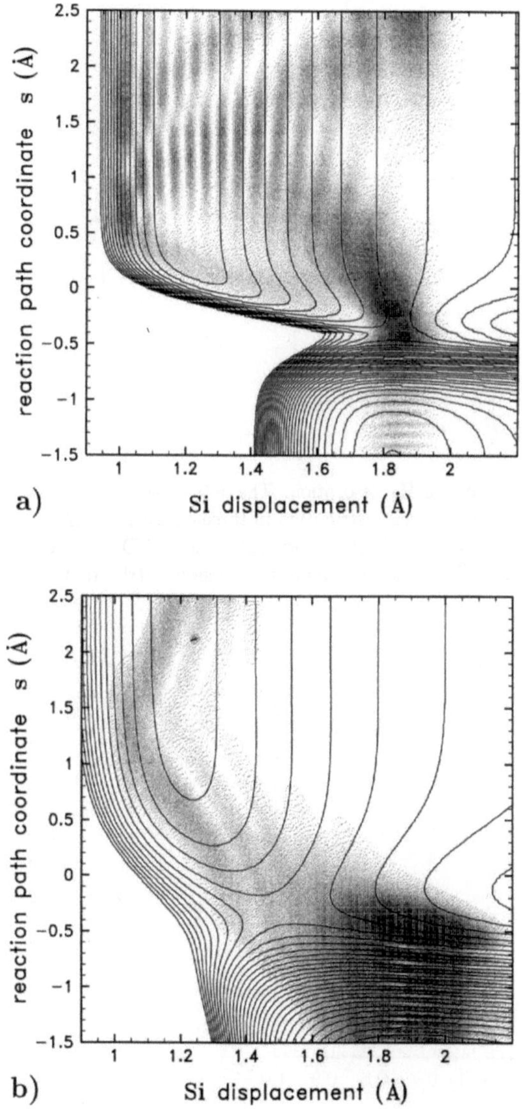

FIG. 2: Contour plots for H_2 of the projection of a) model PES and b) ab initio PES on the surface degree of freedom. The lattice distortion of the ab initio PES varies more slowly with the reaction path as compared to the model and only roughly 1/3 of the $\sim$0.6 Å shift takes place on the gas phase side of the barrier. We include dithered plots of desorption wave functions at $T_s = 1000\,\mathrm{K}$ (see discussion below and in subsection 2.4). The spacing between the contours is 100 meV.

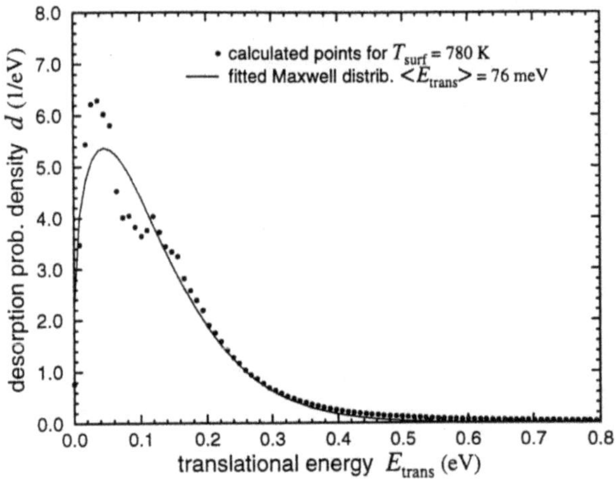

FIG. 3: The translational energy distribution for H_2 desorption. The dots show the calculated values and the solid line represents a fit of a Maxwell distribution in three dimensions. If it was simply thermal desorption one would expect a distribution corresponding to 1-D translation. The actual behavior is dictated by the Franck-Condon factors of the phonon distribution.

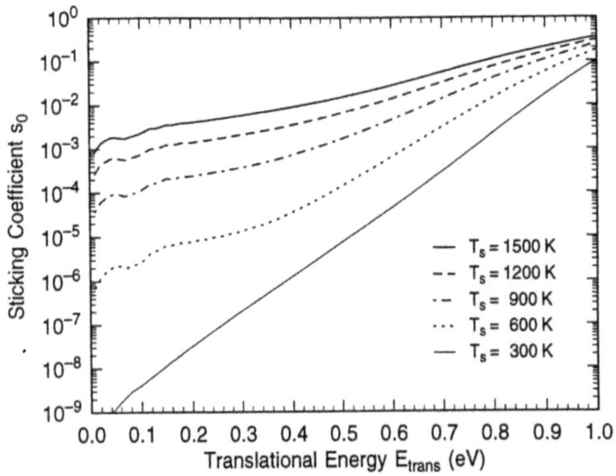

FIG. 4: Sticking coefficient of H_2 as a function of translational energy for five different surface temperatures, calculated from the basic 2-D model.

with the experiment [11]. In a qualitative way these results can be recognized from the contour plots of the wave function included in Fig. 2 a). They reveal a desorption process starting essentially from the ground state of the lattice oscillator and a translational energy of 1000 K. The large oscillation amplitude indicates that the potential drop is practically all converted into lattice energy. In the corresponding results of the ab initio PES (Fig. 2 b), a strong lattice oscillation is found as well, but a substantial portion of the potential drop is converted into translational energy.

(2) The sticking probability for a monoenergetic beam of H_2 is shown in Fig. 4 as a function of the translational energy for five different surface temperatures. As a consequence of the strong coupling of the model PES to lattice distortions one observes a pronounced dependence of the sticking coefficients on surface temperature. While the sticking at low surface temperatures is mainly based on translational energy as usual, it is dominated by phonon effects at high temperatures showing *phonon-assisted sticking.*

The sticking probability does not reach unity for a translational energy in the range of the minimum barrier of $V_0 = 0.7\,\mathrm{eV}$ as the molecules hit the much higher barrier straight ahead at the end of the entrance channel (see also Fig. 1). As a function of the inverse surface temperature we find an Arrhenius law with an Arrhenius energy of $E_{\mathrm{Arr}} = 650\,\mathrm{meV}$ for H_2 and $E_{\mathrm{Arr}} = 680\,\mathrm{meV}$ for D_2 which fits well to the experimental values reported in the following section.

For the later discussion of specific microscopic mechanisms we emphasize again at this point that the obtained results are mainly determined by the choice of the adsorption barrier V_{ad} and the displacement Δx. The detailed shape of the PES before and far after the transition state is of little importance for the adsorption/desorption dynamics. In particular it is not necessary that the lattice distortion is the result of a static Si displacement in the *adsorbed* state as we assume in this section for the purposes of explicit calculation. A transient lattice distortion around the transition state would result in an almost identical dynamical behaviour. Also the final dissipation of the local lattice excitation into thermal phonons can be expected to have little influence on the energy distribution of the desorbing molecules.

2.2 Effect of surface corrugation

The dominant effect of surface corrugation for molecules impinging under normal incidence on a surface is a reduction of the sticking coefficient [52, 53]. The variation of the barrier height across the surface leads to a 'keyhole effect', i.e. a more or less geometrical decrease of the effective surface area available for sticking which is independent of the energy at low energies. From our investigations of the situation for H_2/Cu [52] we know that this effect can cause a reduction of about two orders of magnitude. For non-normal incidence the role of surface corrugation is more complex. The additional parallel momentum can lead to an enhancement or a suppression of the sticking coefficient, depending on the type of corrugation and the energy regime [53, 54].

In this calculation we assume an 'energetic' corrugation of the potential where only the height but not the position of the barrier is varied along the surface. The corresponding term in equation (1) is:

$$V_{\mathrm{corr}}(\boldsymbol{y}; s) = \frac{V_0}{4\cosh^2(\lambda_{\mathrm{corr}} s)} \left\{ \left[1 - \cos\left(\frac{2\pi}{a} y_1\right)\right] + \left[1 - \cos\left(\frac{2\pi}{a} y_2\right)\right] \right\}. \tag{4}$$

Here $\boldsymbol{y}$ means the center of mass coordinate parallel to the surface with a periodicity length of $a = 2.4\,\text{Å}$ taken from the Si lattice spacing. The range of the corrugation perpendicular to the surface was taken to be $\lambda_{\mathrm{corr}} = 2.2\,\text{Å}^{-1}$ which we estimated from our experience with H_2/Cu [52]. The height V_0 is determined so as to lead to a lateral H_2 frequency of $\omega = 0.06$ eV, in agreement with electron energy loss spectroscopy experiments [55, 56]. This leads to the reasonable value of $V_0 = 0.25$ eV.

We will present a 3-D calculation where we consider a 1-D surface corrugation additionally to the basic model [31]. This reduces the absolute value of the sticking coefficient by about one order of magnitude. Another order of magnitude can be expected for a 2-D corrugation which in principal would have to be considered. However, one needs about 20 channels (reciprocal lattice vectors) to represent the scattering states in the lateral direction in these claculations, in addition to about 20 channels for the lattice vibrations. Thus, a 3-D model already requires 400 channels. A 4-D model with 2-D corrugation would require 8000 channels and presently lead to problems with computing time and memory space.

In Fig. 10 the calculated sticking coefficient for the flat and the corrugated surface are shown in comparison to experimental results. Indeed, the inclusion of 1-D surface corrugation leads to a reduction of the sticking coefficient by one order of magnitude. The sticking coefficients have been determined for a gas of hydrogen molecules with gas temperature $T_{gas} = 300\,K$. This means that molecules contribute to the sticking probability hitting the surface under all possible angles. A more detailed analysis reveals that for molecules under normal incidence the sticking probability is only reduced by a factor of three. The additional reduction of the sticking coefficient of the hydrogen gas comes from the fact that for an energetically corrugated surface the sticking probability of the molecules which hit the surface under non-normal incidence is suppressed by their additional parallel momentum [53, 54]. In a more realistic description also the variation of the barrier position, the so-called geometric corrugation, should be taken into account; the situation then becomes more complex [53, 54]. The effect of full 2-D corrugation is estimated in Fig. 10 by taking into account a further reduction factor corresponding to the reduction between the flat surface and the 1-D corrugated surface.

Besides the preexponential also the effective barrier is influenced by surface corrugation. First of all there is a contribution from the zero point energy (0.03 eV per lateral degree of freedom) and a further increase by about 0.05 eV due to thermal averaging of the potential (4). All together the two lateral degrees of freedom increase the barrier of the 2-D model by about 0.16 eV. To obtain agreement with experiment this minimum barrier has to be reduced accordingly.

In our calculations the rotational degrees of freedom of the hydrogen molecule are not considered. Since the barrier height also depends on the orientation of the molecule, taking these degrees of freedom into account would probably lower the sticking coefficient even more. On the other hand, the effective activation energy will also be increased since the corrugation in the polar and azimuthal coordinates of the H_2 molecule will have similiar effects as the lateral surface corrugation. Thus, finally, a smaller minimum barrier could be chosen to represent the experimental data. This may bring the barrier of the model closer to the barrier found in *ab initio* calculations.

Desorbing molecules mainly propagate along paths which pass the barrier region close to the minimum energy barrier position [53]. In order to check whether our model potential properly describes the experimental situation in the vicinity of the minimum energy path, we have also calculated the angular distribution of desorbing molecules. Experimentally it is not $\propto \cos\vartheta$, where ϑ is the angle from the surface normal, as one would expect from a thermal distribution but forward peaked $\propto \cos^n \vartheta$ with $3.9 \leq n \leq 5.2$ depending on the surface coverage [14]. Fig. 5 shows that there is a rather good agreement between theory and experiment. We also tested a surface corrugation corresponding to a lateral H_2 frequency of $\omega_s = 0.10$ eV which gave a narrower angular distribution in desorption than observed experimentally. Thus our dynamical simulation puts some constraints on the shape of realistic potential energy surfaces.

The non-cosine angular distribution in desorption is, indeed, an indication that the desorbing molecules do not really obey a thermal distribution. Even the *energy* distribution Fig. 3 in our model is not really thermal: it looks approximately $\propto E \exp(-E/kT_s)$ already in the original 2-D model, where a Boltzmann distribution actually should behave like $\propto \exp(-E/kT_s)$.

Our interpretation of this result is that the energy distribution – at least at low energies – is essentially a reflection of the Franck-Condon factors of the lattice vibrational

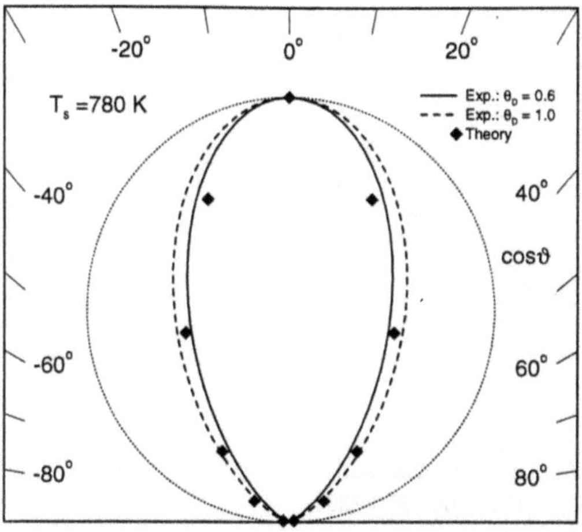

FIG. 5: Polar plot of the angular distribution of hydrogen molecules desorbing from Si(100). The experimental data for deuterium desorption have been determined for coverages of $\theta_D = 0.6$ (solid line) and $\theta_D = 1.0$ (dashed line) [14]. The theoretical results are indicated by diamonds. The $\cos\theta$ distribution is also shown as the dotted circle for comparison.

energy distributions in the distorted and undistorted lattice and *not* of an actual thermal equilibrium with the lattice. This is corroborated by the fact that the average kinetic energy of the desorbing particles at low energies increases less than kT_s (see Fig. 6) with increasing surface temperature as already mentioned in ref. [31]. In addition, Fig. 6 shows that by including surface corrugation the mean kinetic energy normal to the surface is slightly increased as compared to the non-corrugated surface due to the variation in barrier height, except for very low surface temperatures when almost all desorbing molecules follow a path through the minimum barrier position.

2.3 Molecular vibration dynamics

The molecular vibration dynamics leaves the adsorption/desorption dynamics of the model calculations presented above essentially unchanged. This is obvious from the small amount of energy transferred into the intra molecular degree of freedom. Although the population of the first excited state of the desorbing molecule at a surface temperature of 780 K is about a factor of 20 higher than for thermally equilibrated molecules, the difference in average vibrational energy is merely about 6 meV for H_2, 10 meV for HD and 25 meV for D_2.

Nevertheless the vibrational dynamics of the molecules including its isotope effect is interesting in itself. In this section we will present the study of a 3-D model considering the molecular vibrational degree of freedom in addition to the molecular translation and surface phonons which were already discussed above and in an earlier publication [31]. We apply the INTRA-LORE iteration scheme of the coupled-channel equations described in refs. [57, 48].

The vibrational dynamics of the molecule is included by considering a stretching of the intra molecular bond length leading to an elbow-like curved reaction path [46]. In order to apply the coupled channel method one has to transform to a locally orthogonal coordinate system spanned by the reaction path coordinate s and the oscillator coordinate

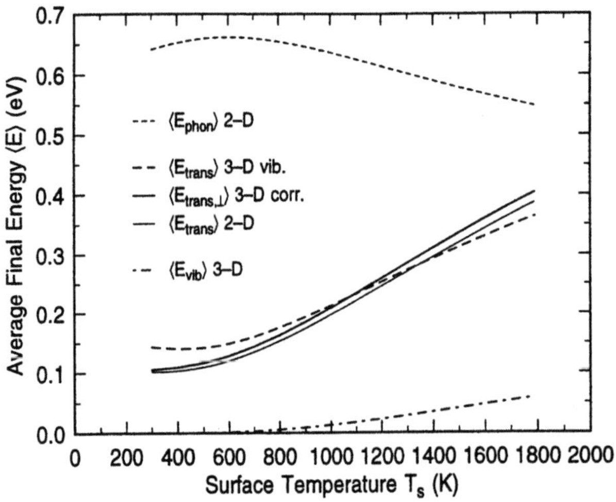

FIG. 6: Average final energies after desorption of translation $\langle E_{\rm trans}\rangle$, surface excitations $\langle E_{\rm phon}\rangle$, and molecular excitations $\langle E_{\rm vib}\rangle$ as a function of surface temperature. Results of the 2-D as well as of the two 3-D calculations are shown. For the exitation degrees of freedom the zero point energies have been subtracted.

r [58]. Thus molecular vibration is taken into account by adding a harmonic potential in equation (1):

$$V_{\rm vib}(r;s) \;=\; \tfrac{1}{2}\hbar\omega_{\rm vib}(s)\left(\gamma_{\rm vib}(s)r\right)^2. \tag{5}$$

The inverse oscillator length is $\gamma_{\rm vib}(s) = \sqrt{\mu\omega_{\rm vib}(s)/\hbar}$ with μ the reduced mass of the hydrogen molecule and $\hbar\omega_{\rm vib}(s)$ the frequency of the molecular vibration.

Vibrational dynamics has been the subject of various investigations of hydrogen desorbing from metal surfaces [59, 47, 60]. The vibrational heating is well understood and the properties of the PES are well known. Although the bonding of hydrogen to the silicon surface is quite different from metal surfaces, one can expect that the PES describing the *molecular* vibrational dynamics at semiconductor surfaces is similar apart from some details. The value of the curvature as well as a reduction of the vibrational frequency to about 50% for the adsorbed hydrogen are taken from the ab initio calculations (see subsec. 2.4).

Despite all similarities to metal surfaces, the energy distribution on the different degrees of freedom for this PES is quite different: The coupling to lattice distortions is strong and thus almost all of the potential drop is transferred to surface phonons while only a small portion ends up as additional molecular vibration.

This is documented by Fig. 6 where the state averaged final energies (above the ground state energy) of surface phonons, molecular translation, and molecular vibration are plotted as a function of surface temperature. The average phonon energy is dominant and remains nearly constant at 0.65 eV. The translational part increases with surface temperature due to the broadening of the wave function that makes it impossible to find an *optimal* way through the PES. The average vibrational energy is negligible on the whole temperature scale, although it rises for very high surface temperatures. The effects of taking additionally molecular vibrations into account is exhibited by comparison with the curves for the 2-D calculation shown as hairlines in Fig. 6. Though the average phonon

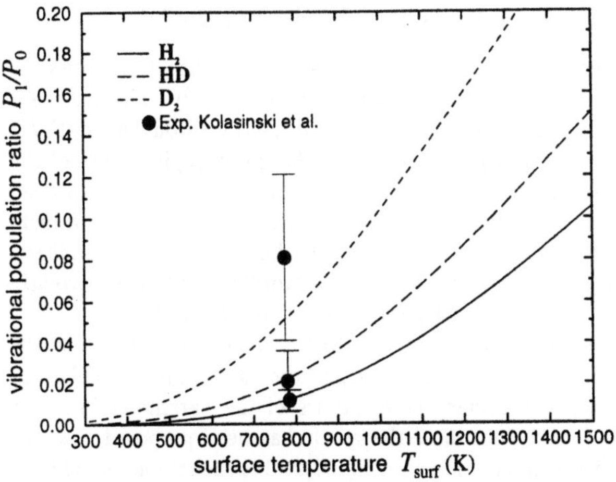

FIG. 7: Vibration population ratio as a function of surface temperature for H_2, HD and D_2. The experimental results [9] for a temperature of $T_s = 780\,\mathrm{K}$ coincide within their error bars.

energy is slightly reduced by a constant amount, the additional degree of freedom takes its energy from the translational part, i.e. it causes a reduction of translational heating at high surface temperatures.

The vibrational quantum of H_2 molecules in the gas phase is $\omega_{\text{vib}} = 516\,\mathrm{meV}$ and for HD and D_2 $\omega_{\text{vib}} = 450\,\mathrm{meV}$ and $\omega_{\text{vib}} = 371\,\mathrm{meV}$ respectively. The resulting vibrational population ratios are shown in Fig. 7. At the desorption temperature of $T_s = 780\,\mathrm{K}$ we find $P_1/P_0 = 0.012$, 0.022, and 0.052 for H_2, HD, and D_2. The ratio for the heavier molecules is larger due to their smaller vibrational quantum. These results are in good agreement with the vibrational state resolved measurements of Kolasinski et al. [9].

Another consequence of the different vibrational frequencies together with the reduction of these values for the adsorbed hydrogen is a difference in the effective potential barrier for the various isotopes. As D_2 has the lowest frequency, it has consequently the highest effective adsorption barrier. This results *i*) in a smaller sticking coefficient in adsorption. For a thermal gas at 300 K of which the calculated results are shown in Fig. 10 we find a difference in the sticking coefficient of a factor 1.5 to 2. This is also supported experimentally though less pronounced. *ii*) It causes a larger translational heating in desorption. The translational energy distribution is approximately Maxwellian and nearly in thermal equilibrium with the surface. For a surface temperature of $T_s = 780\,\mathrm{K}$ we calculated a mean translational energy of $\langle E_{\text{trans}} \rangle = 97\,\mathrm{meV}$ for H_2 and $\langle E_{\text{trans}} \rangle = 106\,\mathrm{meV}$ for D_2 . Unfortunately, up to now, experimental measurements of the translational energy distribution were only done for D_2 and therefore experimental knowledge to that consequence of the isotopic effect is not yet available. At least the thermal Maxwellian distribution is experimentally supported [11].

Another isotopic effect is caused by the stronger coupling to surface phonons for deuterium due to its heavier mass. As will be seen in Fig. 10 for the sticking coefficient of a thermal gas, this results in a higher value for the Arrhenius energy for D_2 compared with H_2. In the desorption reaction it leads to an additional increase of the translational energy for the heavier molecules.

An isotope effect based on quantum mechanical tunnelling is negligible because of the large barrier width.

2.4 Ab initio quantum dynamics for H_2/Si(100)2×1

In the previous section we have *proposed* a model and we have shown that it describes the experimental results correctly if its parameters are chosen appropriately. The dominant effect on the dynamics comes from strong lattice distortion in the transition state of H_2/Si as compared to the clean surface. We have not specified the microscopic nature of the relaxation process denoted by the coordinate x in eq. (1). It could simply be a Si–Si bond length or the distance of the reactive silicon atom to the underlying Si layer. In a more general way, this coordinate could also describe long ranged bulk or even non-adiabatic electronic relaxation processes. It is the exact nature of the coordinate x that is currently the missing link to arrive at a completely satisfactory microscopic description.

Recently, there have been several attempts to establish a microscopic picture of the substrate relaxations for hydrogen on Si(100)2×1 by performing ab initio calculations [33, 35, 30]. In this section we will discuss the dynamics of the parameter-free PES of ref. [33] which is based on density functional theory. It has been proposed that the most important process of hydrogen interaction with Si(100)2×1 is the interaction of both hydrogen atoms with a single Si–Si dimer on the Si(100)2×1 surface [2, 3, 6, 23]. Together with the pre-pairing of the hydrogen atoms on these dimers this mechanism provides the most natural explanation for the observed first-order desorption kinetics. The asymmetry between dissociative adsorption and recombinative desorption arises because the silicon dimers are found to be horizontally oriented when hydrogen is adsorbed [3, 61], whereas they are buckled on the clean surface [62, 63, 64]. Since the buckling is accompanied by charge transfer within the dimer and a re-hybridization of orbitals in the two silicon atoms [65, 62] there is a notable dependence of the H_2 adsorption barrier on the buckling angle [33, 35]. The minimum reaction path resulting from the calculations of ref. [33] is depicted in Fig. 8. It very clearly shows the displacement of two silicon atoms in the course of the reaction.

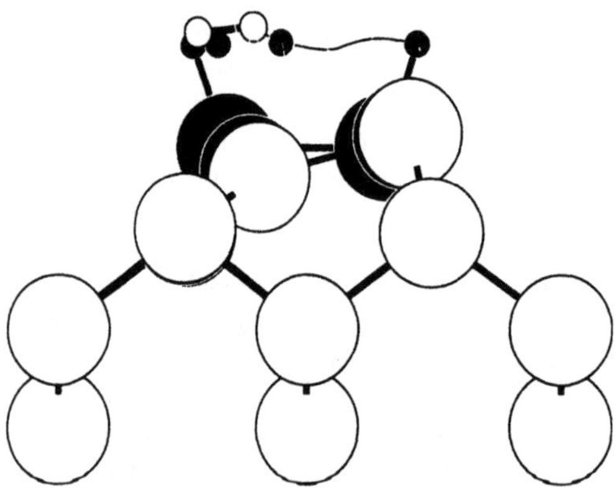

FIG. 8: Minimum energy path for H_2 dissociation at a buckled dimer of Si(100)2×1 resulting from density functional calculations [33]. In a ball-and-stick model, small balls indicate H atoms, while the Si lattice is shown as large white balls. Before the reaction the Si dimer is tilted by 18°. Since the "down" atom is less charged and thus less repulsive towards H_2 than the "up" atom, dissociation starts at the "down" atom. In the transition state the H–H distance is increased from 0.8 to 1 Å and the "down" atom exhibits an upward displacement of roughly 0.3 Å. (shown as light grey balls). At the end of the reaction the dimer is symmetric (dark grey balls). The total vertical Si displacements during the reaction are 0.5 Å for the "down" atom and 0.2 Å for the "up" atom.

The most difficult point for the dynamical calculation is to decide which phonon mode is the dominant one, because we want to restrict the calculation to one dimension for the translation, phonon and molecular vibration each (surface corrugation is not taken into account here). Since the Si–Si bond length changes only moderately upon hydrogen adsorption (from 2.32 Å on the clean surface to 2.44 Å in the monohydride [33]) we may, to a good approximation, represent the coupling to the lattice by the distance x of the lower dimer atom to the underlying Si layer. We note that the leaving out of further phonon modes yields a lower limit of the coupling strength.

The obtained projection of the PES on the phonon degree of freedom x is shown in Fig. 2 b). It is qualitatively similar to what one finds for the model PES but it differs quantitatively. We have used the same functional form as for the model potential, but with parameters adjusted to match the ab initio potential as closely as possible. For the details of the parametrization we refer the reader to ref. [34]. One can clearly recognize that the effect of lattice relaxation is not as strong as in the model PES. Although the total displacement between the asymptotic states is comparable, $\Delta x \simeq 0.6$ Å in both cases, the barrier of the ab initio PES occurs *late in desorption*. Therefore, the dynamically important shift of the barrier position in the x coordinate, measured relative to gas phase H_2 and the clean surface, is only $\Delta x \simeq 0.3$ Å (see Fig. 2), considerable less than in the model PES. Moreover, the range parameters λ in the two potential surfaces are different by a factor of about two. The effective minimum barrier heights including zero-point corrections are 0.38 eV for H_2 and 0.41 eV for D_2.

The dependence of the sticking probability on surface temperature resulting from the ab initio PES is not as pronounced as the corresponding one for the model PES, indicating a lack in coupling strength. This is indicated by the small Arrhenius energy of only about 250 meV for the sticking of a thermal gas of H_2 at 300 K, to be compared with $E_{\mathrm{Arr}} \approx$ 650 meV for the model PES. Due to the lower adsorption barrier the absolute value of the sticking coefficient at a common surface temperature is 10 to 100 times higher than the results of corresponding calculations in subsection 2.1. Furthermore, there is a substantial force along the translational coordinate leading to a strongly heated translational energy distribution after desorption. At T_s = 780 K we find a mean translational energy of $\langle E_{\mathrm{trans}} \rangle$ = 308 meV for H_2 and $\langle E_{\mathrm{trans}} \rangle$ = 336 meV for D_2, whereas the model PES yields 76 and 85 meV in agreement with experiment. In contrast to that, the molecular degree of freedom is quite well described, as can be seen from the desorption population ratio P_1/P_0 of the final vibrational states. The vibrational heating found in the experiment [9] is reproduced. In detail we get $P_1/P_0 = 0.007$ for H_2 and $P_1/P_0 = 0.074$ for D_2 at a surface temperature of T_s = 780 K, which is within the experimental error bars.

The PES we obtained from density functional theory calculations thus clearly shows that the substrate plays an important role for the dynamics of the reaction. Besides the experimental results this gives additional support for the qualitative correctness of our model discussed above and in ref. [31]. Furthermore, the vibrational population ratio, i.e. the molecular degree of freedom, is also quantitatively well described. Therefore, we have reasons to believe that the curvature of the reaction path, which we took over for the model PES, is reliable.

Nevertheless, the ab initio PES does not give quantitatively satisfying results for the surface effects. One possible reason for the discrepancies could be due to the fact that we have only considered the supposedly most important phonon mode and underestimate the coupling to the surface degrees of freedom. In addition, long range interactions which are not taken into account could be important. Probably even more important, the derivation of potential energy surfaces from density functional calculations, and in particular the use of gradient corrected density functionals, is a newly developing field and little experience is yet available to really know the degree of accuracy of these results [66]. The existence of several geometrically distinct transition states, however, requires an accurate knowlegde of the relative barrier heights to decide which of them is the dominant one. The symmetric transition state, for instance, was found to be only slightly higher in energy than the asymmetric one and involves a stronger effective lattice distortion. Therefore adsorption through this transition state leads to a somewhat stronger coupling to the lattice [67].

3. Experiment

Here we only give a brief account of the experimental data and refer to the original papers [9, 14, 11, 7, 8, 68] for details.

The recent investigations of adsorption-desorption dynamics of hydrogen molecules on silicon were stimulated a great deal by experiments of Kolasinski et al. [11] who measured the time-of-flight spectrum of D_2 desorbing from Si(100) and Si(111) surfaces after pulsed laser heating. The observed values of the mean translational energy $E_{\text{trans}}/2k$ were of the same magnitude as the peak desorption temperature. In total taking the small engergies involved in rotations and molecular vibrations into account the desorbing molecules have only littel excess energy above thermal in spite of a large barrier needed to explain the low sticking coefficient. This 'barrier puzzle' mentioned already in the introduction suggested the consideration of lattice relaxations and the *dynamics* of this relaxation in order to satisfy the principle of detailed balance. The experimental results concerning vibrational and translational state distributions have been discussed already in connection with the corresponding theoretical results in the foregoing sections.

A further rather dramatic result of the theory was the prediction of very strong *phonon assisted sticking*: The sticking coefficient, though still remaing small, varies strongly with surface temperature.

Recently extremely small sticking coefficients for dissociative adsorption of H_2/Si could be measured with optical second-harmonic generation (SHG) [7] with a sensitivity that exceeds that of most conventional techniques. In the past this high sensitivity has already been exploited for accurate measurements of H_2 desorption kinetics [4, 6] and surface diffusion of H/Si [5]. In addition one could take advantage of the compatibility of SHG with any gas pressure and measure the H coverage directly during exposure of the surface to a high flux of molecular H_2. This way it is possible to measure sticking coefficients for dissociative adsorption even for surface temperatures that exceed the H_2 desorption temperature. Furthermore, the coverage dependence of the sticking coefficient directly results from these experiments.

The Arrhenius plot for the desorption rates of H_2 and D_2 from Si(100)2×1 and Si(111)7×7 is shown in Fig. 9. The Si(100)2×1 surface exhibits higher desorption rates than Si(111) 7×7. However, the rates differ by less than a factor of five between the two surfaces in the investigated temperature range $700\,\text{K} < T_s < 850\,\text{K}$. Likewise, there is no large isotope effect; the rates of D_2 desorption are only $2-3$ times smaller than the corresponding rates for H_2. The resulting activation energies for desorption are all between $E_a = 2.3-2.4\,\text{eV}$ with an error of 0.2 eV, in good agreement to previous isothermal SHG [4, 6], laser-induced thermal desorption [2], and temperature programmed desorption (TPD) experiments [41].

The infered values for the initial, total sticking coefficient of H_2 and D_2 on Si(100)2×1 and Si(111)7×7 are displayed in Fig. 10 as a function of the inverse surface temperature. The smallest sticking coefficients could be observed for H_2/Si(111)7×7. They are in the range of 10^{-9} for a surface temperature of 600 K. The sticking coefficient s_0 increases by almost four orders of magnitudes to 10^{-5} when the temperature is raised to 1050 K. On Si(100)2×1 the overall sticking coefficients are by a factor of $2-10$ higher than

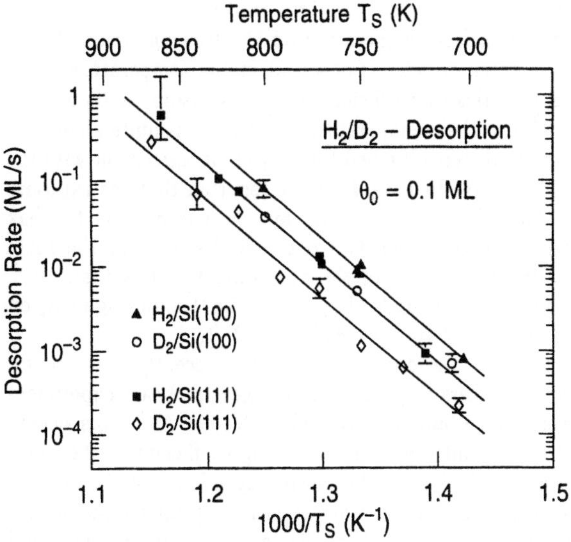

FIG. 9: Desorption rates of H_2 and D_2 from the monohydride states of Si(111)7×7 and Si(100)2×1 as deduced from the recovery of the SH signal after exposure to molecular H_2 or D_2.

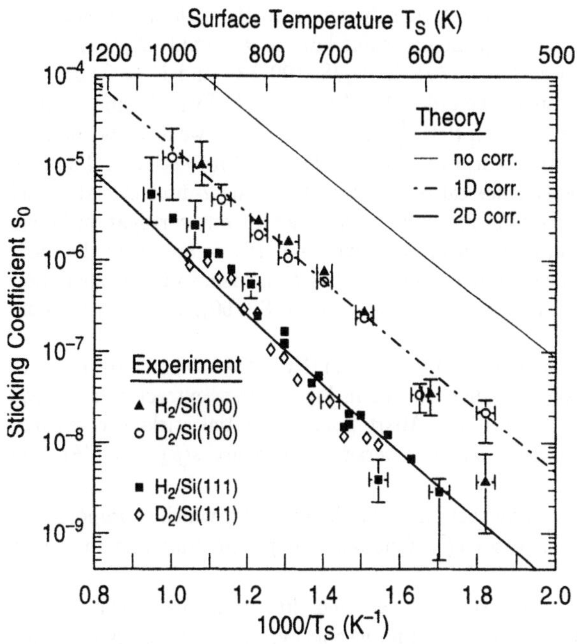

FIG. 10: Initial sticking coefficient of thermal H_2 and D_2 gases at $T_{gas} = 300\,\mathrm{K}$ on Si(111)7×7 and Si(100)2×1 surfaces as a function of their inverse temperature. The experimental values are denoted by symbols; lines are results of theoretical calculations described in section 2. (thin line: basic 2-D model without corrugation; dash-dotted line: 3-D calculation including corrugation in one dimension; thick line: exptrapolation to two-dimensional corrugation).

on Si(111)7×7 and the temperature dependence is slightly weaker. Nevertheless, also for H_2/Si(100)2×1 the effect of the surface temperature on the sticking probability is quite dramatic. An increase over three decades is observed in the temperature range 550 – 1000 K. For both surfaces the sticking coefficients of the two isotopes H_2 and D_2 are of a similar absolute value and also their temperature dependence is quite alike.

The small sticking probabilities that were observed this way are in agreement with the lower limits given earlier in the literature. In 1959, Law [40] reported that the exposure of single crystal Si filaments to large fluxes of molecular hydrogen leads only to weak flash desorption signals. Already from these old data one may deduce that the room temperature sticking coefficient on the terraces of the Si filaments is less than 10^{-8}. Later, Schulze and Henzler [41] gave upper bounds of $s_0 < 10^{-6}$ for the sticking of H_2 on Si(111)7×7 and Liehr *et al.* [42] found $s_0 < 10^{-8}$ for H_2/Si(100)2×1.

Much higher sticking coefficients, in the order of 10^{-5}, were recently reported for D_2/Si(100)2×1 by Kolasinki and coworkers [12] based on a molecular beam experiment with nozzle temperatures between 300 K and 1000 K. The effective Arrhenius energy describing the surface temperature dependence of their sticking coefficient in the range 300 K $< T_s <$ 650 K is less than 0.1 eV. This rather weak dependence of the sticking coefficient on surface temperature stands in marked contrast with the results of the present work (E_a = 0.7 eV). Although the reasons for these discrepancies are currently still unclear, it should be kept in mind that the beam experiment lacked the sensitivity to measure sticking coefficients below 5×10^{-6} [12]. For a more detailed comparison of the Si(100) data with the measurements of Kolasinski *et al.* see Ref. [8].

The strong effect of the surface temperature on the sticking probability that was observed, shows, independent of any particular theoretical model, that substrate excitations must play a decisive role in the reaction dynamics of H_2 with silicon surfaces. The behaviour may be approximated by an Arrhenius law, $s_0 = A\exp(-E_a/kT_s)$, with activation energies ranging between $E_a = 0.7\,\text{eV}$ for D_2/Si(100)2×1 and $E_a = 0.85\,\text{eV}$ for H_2/Si(111)7×7. The prefactors are of the order $A \simeq 10^{-2} - 10^{-1}$. Fits of some of these data to Arrhenius law have been shown in Ref. [7] and [8]. In Fig. 10 we directly compare the data with the results of the theoretical calculations described in section 2. The thin solid line is the temperature dependence calculated from the basic 2-dimensional model without corrugation. For the thick solid line and the dash-dotted line corrugation in one and two dimensions was taken into account in the calculation. Please note that the theoretical curves are *not fitted* to the experimental data. The coincidence of the calculation involving 1D/2D corrugation with the experimental data for Si(100)/Si(111) is to some extent accidental.

For comparison with theory the most interesting quantity is the sticking coefficient of the clean surface. For this reason, all values plotted in Fig. 10 were extracted from data with a total hydrogen coverage $\theta \leq 0.1$ ML. Apart from this initial sticking coefficient $s_0 = s(\theta = 0)$, also some experiments on the coverage dependence $s(\theta) = s_0 f(\theta)$ were performed.

For H_2/Si(111)7×7 the data clearly show that the slope $-d\theta/dt$ which is proportional to the sticking coefficient $s(\theta)$ increases as a function of exposure time and thus as a function of coverage.

The observed behaviour is in contrast to the expected effect from site-blocking which in this coverage range should already result in a noticeable decrease of the sticking coefficient. Obviously the data are also incompatible with a simple defect mediated adsorption process. If the low sticking coefficients were primarily caused by a low concentration of available adsorption sites then the sticking coefficient should decrease even more strongly with coverage, in complete contrast to the observed increase. The shape of measured $\theta(t)$

also deviates significantly from a linear increase of θ ($f(\theta) = 1$) which could be expected approximately for precursor mediated adsorption in the coverage range $\theta < 0.5$ ML.

The increase of the sticking coefficient with coverage is likely to be caused by a slight decrease of the adsorption barrier for sites in the vicinity of adsorbed hydrogen. This effect should in general depend on the specific local coordination of reacted and unreacted sites and is thus difficult to model quantitatively. To gain some insight, we have made a rather crude model and simply assume an average adsorption barrier that exhibits a weak negative coverage dependence

$$E_a = E_{a,0}(1 - \beta\theta). \tag{6}$$

If we include the usual site blocking then the sticking coefficient $s = A\exp(-E_a/kT)$ takes the simple form

$$s(\theta) = (1 - \theta)^2 s_0^{(1-\beta\theta)}. \tag{7}$$

Good fits of this functional form to the experimental data were obtained for $\beta \simeq 0.25$. The 5-fold increase of the sticking coefficient, as compared to the $(1 - \theta)^2$ behaviour, corresponds roughly to a 10% reduction of the adsorption barrier. Such a small reduction of the adsorption barrier could be caused by several mechanisms: For example, it is well known that the surface phonons of silicon exhibit some softening upon hydrogen adsorption [69]. According to our model, this should reduce the effective adsorption barrier at elevated surface temperature. In addition, also purely electronic effects, such as band-flattening or redistribution of charge between the dangling bonds of Si(111)7×7, could reduce the adsorption barrier with increasing hydrogen coverage. Which of these two effects is the more important one could be decided by investigating the coverage dependence for different gas and surface temperatures. For the Si(100)2×1 surface the coverage dependence of the sticking coefficient was found to be much weaker than for Si(111)7×7.

4. Discussion of microscopic mechanisms

The microscopic reaction mechanism considered in section 2.4 is not the only one that is conceivable. In fact, a number of models have been proposed based on ab initio theoretical work for H_2 adsorption and desorption on Si(100)2×1 [17, 18, 19, 20, 21, 22, 23, 24, 25, 26, 27, 28, 29, 30, 31, 32, 33, 34, 35, 36, 37]. Unfortunately, the results of the calculations are partially contradictory, and, despite the enormous efforts, no detailed picture has been established up to now. In this section we would like to discuss some of these models in the light of the experimental and theoretical results presented above. In this discussion we will implicitly assume that adsorption and desorption occur via the same mechanism. Peculiar non-equilibrium desorption mechanisms will not be considered.

Although no serious theoretical attempts have yet been devoted to H_2 interaction with Si(111)7×7 because of its large unit cell, from a qualitative point of view, the reaction mechanism currently appears to be even clearer for the Si(111) than for the Si(100) surface. The basic building blocks of the 7×7 reconstruction are fairly simple and well understood [70]. Moreover, the coverage dependence of sticking reported in section 3. indicates that defects should be unimportant. The reaction of H_2 with Si(111)7×7 should thus be dominated by the restatom and adatom dangling bonds (dbs) on the terraces. Since the restatom dbs are doubly occupied, whereas the adatom dbs are only partially filled [71], hydrogen dissociation (and recombination) is likely to have an asymmetry towards the adatoms. With 4 Å, the closest distance between a restatom and an adatom is large compared to the Si–H bond length of 1.6 Å and the H–H distance of 0.8 Å. Therefore, substantial distortion of the adatom bonding in the transition state is likely to occur, in complete agreement with the results of the present work. Such a picture of distorted adatoms is further supported by the fact that the activation energy for phonon-assisted sticking of 0.8 – 0.9 eV that we find for this surface agrees well with activation energy for adatom diffusion that has been estimated from high temperature STM measurements [72]. Whether the distortion of adatom bonding is large enough to lead to the formation of a di-hydride intermediate or whether desorption occurs in a direct 4-center reaction could probably already be decided in slab calculations with a 2×2 unit cell that contains one adatom and one restatom.

The presently discussed desorption models for H_2/Si(100) fall into two categories: Group one are direct one-step mechanisms. In the model already introduced above, two hydrogen atoms from doubly occupied dimers recombine and desorb through a symmetric [37] or more likely an asymmetric transition state [33, 35, 30, 36]. These models are in agreement with the desorption kinetics, and the slab-calculations based on density functional theory give the experimental activation energy for the desorption [2, 6]. However, the lattice distortion in the transition state is too weak at least by a factor two, as discussed in the last section for our own PES. Similar quantitative discrepancies between experimental and calculated values for translational heating and phonon assisted sticking may be expected for the PES of ref. [35, 30, 36]. In addition, the fact that the dimerization is a special feature of the Si(100) surface, whereas the energetics and dynamics of adsorption and desorption are very similar on Si(100) and Si(111), may be an argument against such a process.

Another one-step mechanism is the interdimer recombination of two hydrogen atoms from adjacent dimers on the same row [17, 30]. Vittadini and Selloni [30] find that the interdimer pathway is energetically only slightly less favourable than the intradimer desorption treated by the same theory (E_{des} = 2.7 eV vs. 2.5 eV). Moreover, the energy that remains in the substrate after desorption was found to be approximately equal to 0.8 eV which is in excellent agreement with the PES of Fig. 2 and with our experimental value for phonon-assisted sticking. The significantly larger amount of lattice distortion, as compared to the interdimer recombination, arises from the large distance between the two involved silicon atoms of 3.85 Å and its reduction by a concerted movement of Si atoms. (In case of the interdimer process a stretching of the dimer bond, which could in principle lower the adsorption barrier, instead leads to an enlargement of the distance of the reaction partners). A problem of this model is, that it does not readily explain the first-order desorption kinetics. However, clustering of adsorbed hydrogen would, e.g., lead to approximate first-order kinetics.

The second category of reaction mechanisms of H_2 with Si(100) surfaces are two-step processes with a di-hydride intermediate formed at steps or single-atom defects. These defect models have been advanced by several groups [27, 18, 21] because their ab initio cluster calculations for desorption from single dimers yielded activation energies for desorption between 3.4 and 4.1 eV [17, 18, 20, 21, 26] which are much higher than the 2.5 eV observed experimentally. While such an argumentation is certainly debatable in view of the much lower values for E_{des} that have recently been obtained by density functional slab [33, 35, 30] and cluster calculations imposing less geometrical constraints [36], defect mediated processes should not readily be ruled out for the Si(100) surface.

Based on the present results we can on the one hand safely exclude a very basic defect model that simply explains the small sticking coefficient by a low concentration of static defects. In such a model the observed surface temperature dependence of sticking would have to arise from diffusion of atomic hydrogen from the reactive sites across the terraces. However, the measured activation energy for Si(100), $E_a \simeq 0.7$ eV, is much smaller than any calculated diffusion barrier of 1.3 – 2.0 eV [28, 19, 22] and the measured value of 1.5 eV for Si(111)7×7[7].

On the other hand, if the defects as active sites are created transiently by a thermally activated process the mechanism becomes very similar to the model discussed in this work:

In the di-hydride state, one of the normally present three Si–Si bonds is broken or at least considerably weakened and the adsorption barrier will certainly strongly vary as a function of the Si–Si bond length x, like in our model. On the perfect terraces the energy required to bring the Si lattice into a di-hydride-like configuration is certainly too high. The isomerisation energy between monohydride and dihydride was calculated to be 1.9 eV [30] (compare also to the Si–Si dimer bond strength of $\sim$ 1.3 eV [18]) However, distorted dimers at steps might facilitate such excitations with energies in the order of the 0.7 eV we find for phonon-assisted sticking. High temperature STM measurements, e.g. show that above $T_s \simeq 600$ K the steps of Si(100) become rough and move rapidly across the surface [72]. The overall activation energy for desorption from such an activated defect agrees well with the experimental value of 2.5 eV because the activation energy for desorption from the di-hydride state is known to be slightly less than 2 eV [13].

5. Conclusions

We have presented the results from a comprehensive experimental and theoretical study on the reaction dynamics of molecular hydrogen with silicon surfaces. The sticking probabilities for dissociative adsorption of H_2 and D_2 on – as well as the rates for recombinative desorption from – Si(111)7×7 and Si(100)2×1 have been determined experimentally wit h optical second-harmonic generation. A rather similar behaviour is observed on both surfaces and for both isotopes.

At gas temperatures of 300 K, adsorption depends strongly on the surface temperature T_s and reveals phonon-assisted sticking. In the case of Si(111)7×7, the initial sticking probabilities increase dramatically from $s_0 = 2 \times 10^{-9}$ at $T_s = 580$ K to $s_0 = 5 \times 10^{-6}$ at $T_s = 1050$ K. For Si(100)2×1, dissociative adsorption is slightly more probable; in the temperature range $T_s = 550 - 1000$ K the sticking probabilities for H_2 and D_2 increase from 10^{-8} to 10^{-5}. The absolute value of desorption rates of H_2 and D_2 from both surfaces differ by less than a factor of five in the temperature range $T_s = 700 - 850$ K and reveal similar desorption barriers of 2.3–2.5 eV. On Si(111)7×7 the sticking coefficient *increases* by a factor of 3 between 0 and 0.4 ML hydrogen coverage.

From the fact that H_2 adsorption on Si(111)7×7 does not show signs of site blocking for coverages up to almost half a monolayer, we conclude, that the H_2 dissociation takes place on the terraces of this surface and is not mediated by step or defect sites. The relatively small isotope effect observed for the rates of desorption and adsorption of a thermal gas, suggests that tunnelling plays only a minor role in the H_2 interaction with silicon surfaces. The surprisingly similar quantitative and qualitative behaviour of Si(111)7×7 and Si(100)2×1 indicates that the structural details of the surface are of little importance for the energetics and dynamics of the reaction.

An effective five dimensional model containing, besides the translational coordinate, one substrate degree of freedom, the molecular vibrations, and surface corrugation is capable of describing *all* existing experimental data of molecular hydrogen adsorption and desorption from Si(111) and Si(100), namely: (1) the translational energy distribution of desorbing molecules (absence of translational heating), (2) the temperature dependence of the sticking coefficient as well as (3) its absolute value (preexponential), (4) the heating of molecular vibrations in desorption (including its isotope effect), and (5) the angular distribution of desorbing molecules. The model contains only very few parameters: The curvature was taken from ab initio calculations and three frequencies were obtained by fitting other experimental data. One essential parameter is the barrier height for adsorption determined so as to describe the low room temperature sticking coefficient.

The results of the dynamical calculations are to some extent independent of the microscopic interpretation of the *substrate degree of freedom*. To fix the attention we have considered an *effective surface oscillator*. One should keep in mind, however, that this oscillator has to describe several degrees of freedom in an average way and might even include non-adiabatic electronic processes. Several microscopic mechanisms that involve considerable lattice distortion in the transition state appear to be quite plausible. Attempts to establish the relationship to the dynamically relevant microscopic degrees of freedom by means of ab initio calculations for the H_2 interaction with the dimers of Si(100) have so far been only successful in a qualitative way. Reasons for the quantitative discrepancies to our model, as well as to the experiments, may come from inherent approximations of density functional theory, restrictions to small unit cells or clusters, or difficulties in finding the energetically most favourable reaction path on relatively complex surfaces.

Acknowledgement

The authors would like to thank J. Nørskov and M. Scheffler for valuable discussions. This research was supported by the Deutsche Forschungsgemeinschaft through Sonderforschungsbereich 338 (Projects C2 and C5), by the EEC Human Capital Programme (CHRX-CT93-0104) and by the Danish National Research Foundation.

References

[1] K. Sinniah, M. G. Sherman, L. B. Lewis, W. H. Weinberg, J. T. Yates, Jr., and K. C. Janda, Phys. Rev. Lett. **62**, 567 (1989); J. Chem. Phys. **92**, 5700 (1990).

[2] M. L. Wise, B. G. Koehler, P. Gupta, P. A. Coon, and S. M. George, Surf. Sci. **258**, 166 (1991).

[3] J. J. Boland, Phys. Rev. Lett. **67**, 1539 (1991); J. Vac. Sci. Technol. A **10**, 2458 (1992).

[4] G. A. Reider, U. Höfer, and T. F. Heinz, J. Chem. Phys. **94**, 4080 (1991).

[5] G. A. Reider, U. Höfer, and T. F. Heinz, Phys. Rev. Lett. **66**, 1994 (1991).

[6] U. Höfer, L. Li, and T. F. Heinz, Phys. Rev. B **45**, 9485 (1992).

[7] P. Bratu and U. Höfer, Phys. Rev. Lett. **74**, 1625 (1995).

[8] P. Bratu, K. L. Kompa, and U. Höfer, Chem. Phys. Lett. (in print).

[9] K. W. Kolasinski, S. F. Shane, and R. N. Zare, J. Chem. Phys. **96**, 3995 (1992). S. F. Shane, K. W. Kolasinski, and R. N. Zare, J. Chem. Phys. **97**, 3704 (1992);

[10] S. F. Shane, K. W. Kolasinski, and R. N. Zare, J. Chem. Phys. **97**, 1520 (1992).

[11] K. W. Kolasinski, W. Nessler, A. de Meijere, and E. Hasselbrink, Phys. Rev. Lett. **72**, 1356 (1994).

[12] K. W. Kolasinski, W. Nessler, K.-H. Bornscheuer, and E. Hasselbrink, J. Chem. Phys. **101**, 7082 (1994); Surf. Sci. **331-333**, 485 (1995).

[13] M. C. Flowers, N. B. H. Jonathan, Y. Liu, and A. Morris, J. Chem. Phys. **99**, 7038 (1993).

[14] Y-S. Park, J-S. Bang, and J. Lee, Surf. Sci. **283**, 209 (1992); Y.-S. Park, J.-Y. Kim, and J. Lee, J. Chem. Phys. **98**, 757 (1992).

[15] W. Widdra, S. I. Yi, R. Maboudian, G. A. D. Briggs, and W. H. Weinberg, Phys. Rev. Lett. **74**, 2074 (1995).

[16] Y. Morita, K. Miki, H. Tokumoto, Surf. Sci. **325**, 21 (1995).

[17] C. J. Wu and E. A. Carter, Chem. Phys. Lett. **185**, 172 (1991).

[18] C. J. Wu, I. V. Ionova, and E. A. Carter, Surf. Sci. **295**, 64 (1993).

[19] C. J. Wu, I. V. Ionova, and E. A. Carter, Phys. Rev. B **49**, 13488 (1994).

[20] P. Nachtigall, K. D. Jordan, and K. C. Janda, J. Chem. Phys. **96**, 852 (1991).

[21] P. Nachtigall, K. D. Jordan, and C. Sosa, J. Chem. Phys. **101**, 2648 (1994).

[22] P. Nachtigall and K. D. Jordan, J. Chem. Phys. **102**, 8349 (1995).

[23] M. P. D'Evelyn, Y. L. Yang, and L. F. Suctu, J. Chem. Phys. **96**, 852 (1992); Y. L. Yang and M. P. D'Evelyn, J. Vac. Sci. Technol. A **11**, 2200 (1993).

[24] J. Sheng, J. Z. H. Zhang, J. Chem. Phys. **97**, 596 (1992).

[25] Z. Jing and J. L. Whitten, Phys. Rev. B **46**, 9544 (1992).

[26] Z. Jing and J. L. Whitten, J. Chem. Phys. **98**, 7466 (1993); ibid **102**, 3867 (1995).

[27] Z. Jing, G. Lucovsky, and J. L. Whitten, Surf. Sci. Lett. **296**, L33 (1993).

[28] A. Vittadini, A. Selloni, and M. Casarin, Surf. Sci. **289**, L625 (1993).

[29] A. Vittadini, A. Selloni, and M. Casarin, Phys. Rev. B. **49**, 11191 (1994).

[30] A. Vittadini and A. Selloni, Chem. Phys. Lett. **235**, 334 (1995).

[31] W. Brenig, A. Groß, and R. Russ, Z. Phys. B **96**, 231 (1994).

[32] D. C. Sorescu, D. L. Thompson, and L. M. Raff, J. Chem. Phys. **101**, 1638 (1994).

[33] P. Kratzer, B. Hammer, and J. K. Nørskov, Chem. Phys. Lett. **229**, 645 (1994); Phys. Rev. B **51**, 13432 (1995).

[34] P. Kratzer, R. Russ, and W. Brenig, Surf. Sci., in print.

[35] E. Pehlke and M. Scheffler, Phys. Rev. Lett. **74**, 952 (1995).

[36] S. Pai and D. Doren, J. Chem. Phys. **103**, 1232 (1995).

[37] Guangwei Li, Yia-Chung Chang, R. Tsu, and J. E. Greene, Surf. Sci. **330**, 20 (1995).

[38] J. M. Jasinski, B. S. Meyerson, and B. A. Scott, Annu. Rev. Phys. Chem. **38**, 109 (1987).

[39] W. Kern, Semicond. Int., April, 94 (1984); E. Yablonovitch, D. L. Allara, C. C. Chang, T. Gmitter, and T. B. Bright, Phys. Rev. Lett. **57**, 249 (1986); G. S. Higashi, Y. J. Chabal, G. W. Trucks, and K. Raghavachari, Appl. Phys. Lett. **56**, 656 (1990).

[40] J. T. Law, J. Chem. Phys. **30**, 1568 (1959).

[41] G. Schulze and M. Henzler, Surf. Sci. **124**, 336 (1983).

[42] M. Liehr, C. M. Greenlief, M. Offenberg, and S. R. Kasi, J. Vac. Sci. Technol. A **8**, 2960 (1990).

[43] R. J. Behm and K. Heinz eds., *Surface Reconstruction: Structure and Dynamics*, Surf. Sci. **337**, 169-316 (1995).

[44] U. Bischler, P. Sandl, E. Bertel, T. Brunner, and W. Brenig, Phys. Rev. Lett. **70**, 3603 (1993).

[45] J. C. Polanyi and W. H. Wong, J. Chem. Phys. **51**, 1439 (1969).

[46] G. L. Hofacker, Z. Naturforschung **18a**, 607 (1963).

[47] W. Brenig and H. Kasai, Surf. Sci. **213**, 170 (1989).

[48] W. Brenig and R. Russ, Surf. Sci. **315**, 195 (1994).

[49] W. Brenig, A. Groß, and R. Russ, Z. Phys. B **97**, 311 (1995).

[50] D. C. Allan and E.J. Mele, Phys. Rev. Lett. **53**, 826 (1984).

[51] H. Bilz and W. Kress, *Phonon dispersion relations in insulators* (Springer, Berlin 197 8).

[52] A. Groß, B. Hammer, M. Scheffler, and W. Brenig, Phys. Rev. Lett. **73**, 3121 (1994).

[53] A. Groß, J. Chem. Phys. **102**, 5045 (1995).

[54] G. R. Darling and S. Holloway, Surf. Sci. **304**, L461 (1994).

[55] J. A. Schaefer, F. Stucki, J. A. Anderson, G. J. Lapeyre, and W. Göpel, Surf. Sci. **140**, 207 (1984).

[56] H. Froitzheim, U. Köhler, and H. Lammering, Surf. Sci. **149**, 537 (1985).

[57] W. Brenig, T. Brunner, A. Groß, and R. Russ, Z. Phys. B **93**, 91 (1993).

[58] C. C. Rankin and J. C. Light, J. Chem. Phys. **51**, 1701 (1969).

[59] J. Harris, S. Holloway, T. S. Rahman, and K. Yang, J. Chem. Phys. **89**, 4427 (1988).

[60] S. Küchenhoff, W. Brenig, and Y. Chiba, Surf. Sci. **245**, 389 (1991).

[61] Y. J. Chabal and K. Raghavachari, Phys. Rev. Lett. **53**, 282 (1984); Phys. Rev. Lett. **54**, 1055 (1985).

[62] J. Dabrowski and M. Scheffler, Appl. Surf. Sci. **56-58**, 15 (1992).

[63] R. A. Wolkow, Phys. Rev. Lett. **68**, 2636 (1994).

[64] E. L. Bullock, R. Gunella, L. Patthey, T. Abukawa, S. Kono. C. R. Natoli, and L. S. O. Johansson, Phys. Rev. Lett. **74**, 2756 (1995).

[65] P. Krüger, A. Mazur, J. Pollmann, and G. Wolfgarten, Phys. Rev. Lett. **57**, 1468 (1986).

[66] P. Nachtigall, K. D. Jordan, A. Smith, and H. Jónsson, preprint

[67] A. C. Luntz and P. Kratzer, to be published

[68] P. Bratu, W. Brenig, A. Gross, M. Hartmann, U. Höfer, P. Kratzer, and R. Russ, submitted to Phys. Rev. B (1995).

[69] Ch. Stuhlmann, G. Bogdáni, and H. Ibach, Phys. Rev. B **45**, 6786 (1992).

[70] K. Takayanagi, Y. Tanishiro, S. Takahashi, and M. Takahashi, Surf. Sci. **164**, 367 (1985).

[71] F. J. Himpsel, Surf. Sci. Rep. **12**, 1 (1990).

[72] H. Tokumoto and M. Iwatsuki, Jpn. J. Appl. Phys. **32**, 1368 (1993).

Eley-Rideal Dynamics of the Reaction of Gas-Phase H(D) with D(H) on Cu(111)

B. Jackson[1] *and M. Persson*[2]

[1]Department of Chemistry, University of Massachusetts, Amherst, MA 01003, USA

[2]Department of Applied Physics, Chalmers University of Technology and University of Göteborg, S-412 96 Göteborg, Sweden

Abstract. We present an overview of our theoretical studies of the Eley–Rideal dynamics of the reaction of gas-phase H(D) with D(H) adsorbed on Cu(111). The studies are based on a three-dimensional flat-surface model and an empirical potential energy surface. Results for reactive cross sections and rovibrational HD product state distributions are presented. The pronounced isotope effects on the vibrational state distribution are shown to have a simple dynamical origin. Our predictions for these product distributions are in semi-quantative agreement with experiments, whereas the calculated and measured reactive cross section disagree.

1. Introduction

Most gas-surface reactions that have been studied occur via the Langmuir-Hinschelwood (LH) mechanism, in which both reactants are adsorbed and thermally equilibrated with the surface. A much less common pathway, for which strong evidence has only recently appeared from state-resolved molecular beam experiments [1–7], is the Eley–Rideal (ER) mechanism where the gas-phase species reacts directly with an adsorbed species. Prime examples are provided by the reactions of gas-phase H atoms with Cl adsorbed on Au(111) [1,4,5] and of gas-phase H(D) atoms with adsorbed D(H) atoms on Cu(111) [3,6,7]. In these cases, the experimental evidence for the ER mechanism is based on two characteristic dynamical features; the sensitivity to the initial state of the incident atom, and the relatively large translational and internal energy of the product. The latter feature is due to the large exothermicity for the ER mechanism, in which only a single adsorbate-substrate bond has to be broken, compared to the LH mechanism where two such bonds are broken. However, many possible pathways lie between these limiting ER and LH mechanisms, such as the hot atom mechanism proposed by Harris and Kasemo [8]. To distinguish the ER mechanism from the LH mechanism is not a trivial matter because an hot atom that is introduced by the trapping of a gas-phase atom on a surface may retain a major part of its incident energy and lateral momentum.

Support for the ER reaction mechanism has also been produced by theoretical dynamics studies based on fully quantum-mechanical models [9–16]. Most

Springer Series in Solid-State Sciences, Volume 121
Elementary Processes in Excitations and Reactions on Solid Surfaces
Eds.: A. Okiji, H. Kasai and K. Makoshi

of these studies are based on a simple two-dimensional collinear geometry and treat the reactions of gas-phase hydrogen atoms with either hydrogen or chlorine atoms adsorbed on metal surfaces [9–11]. These studies have revealed several dynamical features that are in qualitative agreement with experiments, such as a large vibrational excitation of the product molecule [9,10], the importance of the relative masses of the reactants on the product state [10,11], and the strong dependence on the vibrational state of the adsorbed atom that leads to a low apparent thermal activation energy [11].

The collinear studies have been revealing but do not fully account for the dynamics, and can at best be viewed as qualitative. A more realistic description of the ER dynamics has been achieved by the application of a three-dimensional flat-surface model [13–15] to the reaction of H(D) with D(H) on Cu(111). The model is based on a flat-surface approximation for the potential energy surface, so it does not restrict the motion of the reactants and the product. This enables us to calculate reaction cross sections and detailed rovibrational product state distributions. The interaction of Hydrogen with Cu(111) is relatively well-characterized from theoretical and experimental studies of the dynamics of dissociative adsorption [16–17], helping us to construct a reasonable potential energy surface. Another appealing feature of this system is the possibility of probing the ER dynamics by studying the effects of isotopic substitution on the product state, and comparing with detailed state-resolved atomic beam experiments performed very recently by Rettner and Auerbach [6,7].

In this paper we present an overview of our studies, based mainly on the flat-surface model of the ER mechanism for the reaction of a gas-phase H(D) atom with an adsorbed D(H) atom on Cu(111). Results are presented for the isotope dependence of the reactive cross section and the vibrational and rotational state distributions. The model is assessed by a comparison of our predictions with experiment.

A detailed description of the model and the method of implementation is given in our paper [13], which henceforth will be referred to as PJ. We present the main assumptions and ingredients of the model in Section 2, the results are discussed in Section 3 and compared with experiment in Section 4, and we make some concluding remarks in Section 5. For easy reference, the initially adsorbed and incident atoms will be referred to as the targon and the incidon, respectively. Moreover, we shall use the terminology H-on-D, D-on-H and so on to describe the different isotopic combinations of the incidon and the targon.

2. Theory

The recombination of two atoms involves six degrees of freedom, assuming both electronic adiabaticity and a rigid substrate. The latter assumption is justified by the relatively small masses of H and D. At present, a full quantum-mechanical treatment including all these degrees of freedom is intractable, so it is necessary to make some physical approximations that reduce the high dimensionality of the problem. The small diffusion barriers of H atoms on a

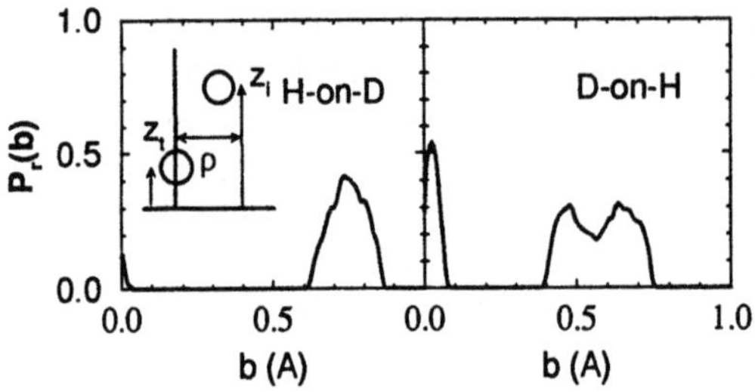

Fig. 1. Quasi-classical reaction probabilities, $P_r(b)$, as a function of impact parameter, b, for H-on-D and D-on-H. Same incidence conditions as in Table 1. The inset in the upper panel depicts the reactive scattering geometry in the flat-surface model together with the coordinates of the targon (t) and the incidon (i). The dotted line, $\rho = 0$, indicates the geometry used in the collinear model.

metal surface suggest a flat-surface appromixation, in which the lateral variation of the potential energy surface (PES) is neglected. The PES is then a function of only three variables ρ, z_i and z_t, which are defined in the inset of Fig. 1. This approximation reduces the dimensionality of the problem from six to three by introducing the following three constants of motion: $\mathbf{P}_{\|}$ the center-of-mass momentum parallel to the surface and J_z, the projection of the angular momentum normal to the surface. In this paper, we report results only for the case of the incidon at normal incidence, where $\mathbf{P}_{\|} = J_z = 0$. The case of off-normal incidence is a bit more complex, and was fully explored in PJ for H-on-H. We would like to stress that the flat-surface model gives a fully three-dimensional description of the free molecule.

The model is solved quantum-mechanically by propagating the wave function in time using well established pseudospectral techniques. The wavefunction is represented by its values on a three-dimensional grid of points. Initially, the wave function is a simple product of the ground state wave function for the targon-metal vibration, and a Gaussian wave packet for the incidon centered at a position far from the surface with an average momentum determined by the incident energy E_i. In this work we shall consider only $E_i = 75$ meV. The wavefunction is evolved numerically in time using a short time propagator algorithm based on the split-operator method. The product state distribution is obtained from the scattered wave function using a projection operator-correlation function approach.

We have elucidated the role of quantum effects on the dynamics by making a direct comparison with results from standard quasi-classical calculations. By using the term quasi-classical, we stress that it is important to account for the zero-point motion of the targon. The targon vibration, whose energy was set equal to its quantum-mechanical zero-point value, was sampled over its initial phase. Quantum numbers j and ν were assigned to the product HD molecule by setting $\hbar j$ and $\hbar(\nu + \frac{1}{2})$ equal to the classical angular momentum and the classical vibrational action variable, respectively. The quasi-classical reaction cross section is given by,

$$\sigma_r = 2\pi \int_0^\infty P_r(b) b db, \tag{1}$$

where the opacity function, $P_r(b)$, is determined by the relative number of reactive trajectories for each impact parameter, $b = \rho(t \to -\infty)$.

An appropriate model PES describing the rearrangement of chemical bonds between the H and D atoms and a metal surface like Cu(111) is provided by a London-Eyring-Polanyi-Sato (LEPS) potential. The two-body interactions describing the HD molecule and the H and D atoms adsorbed on Cu(111) are given by Morse potentials, whose parameters are, as discussed in PJ, readily determined from experimental and theoretical studies of H_2 and H on Cu(111). The relatively large difference between the H-D bond energy, D_m=4.745 eV, and the hydrogen chemisorption energy, D_a=2.45 eV, results in a large exothermicity $Q = D_m - D_a + E_i + \hbar\omega/2 \approx 2.44$ eV, where the isotopic differences in the zero-point energy $\hbar\omega/2$ are negligible. The Sato parameters, which determine the potential in the interaction region, are chosen to give values for the height and position of the activation barrier to dissociative adsorption that are consistent with values suggested by studies of this part of the PES [16,17]. These values, which correspond to the PES-I model in PJ, are consistent with the results from recent density functional calculations of the multi-dimensional PES for the dissociative adsorption of hydrogen [18]. This construction results in an attractive PES for the ER reaction, which means that a large part of Q is released in the entrance channel. As demonstrated for the H-on-H combination in PJ, this feature of the PES leads to a weak dependence on E_i and a highly vibrationally and rotationally excited H_2 product. In the case of the HD product, Q allows a maximum possible rotational excitation $j_{\max} = 21$ and vibrational excitation $\nu_{\max} = 5$.

3. Results and discussion

In Fig. 1, we show the calculated opacity functions, $P_r(b)$, for H-on-D and D-on-H. These results look qualitatively similiar with some exceptions. For both combinations there is a relatively narrow range of reactive b close to r_0, the equilibrium HD bond length. In contrast to D-on-H, there are almost no reactive collisions for H-on-D when b is close to zero, corresponding to a collinear arrangement. This behavior is consistent with an impulsive collision argument put forward in our collinear study of this reaction [10], but due to the area element in Eq. (1) these collisions contribute negligibly to the overall behaviour. Because the detailed structure of $P_r(b)$ is quite different for the two isotopic combinations, we expect that the underlying dynamics behind this difference should also affect the HD product distributions.

As demonstrated in Table 1, there are indeed strong isotope effects in the calculated reactive cross-sections and averages of the different HD product distributions. The quasi-classical (QC) results are in good agreement with the quantum mechanical (QM) results for the average internal energy, but

Table 1. Comparison between calculated and measured averages of internal energies, $\langle E_{\mathrm{int}} \rangle$, vibrational quantum numbers, $\langle \nu \rangle$, and rotational quantum number, $\langle j \rangle$ and reaction cross sections σ_r for H-on-D and D-on-H. The quasi-classical results are within the parenthesis. The calculated results are for normal incidence with $E_i = 0.075$ eV while the experimental data taken from Ref. [7] are recorded at $E_i = 0.07$ eV with an angle of incidence 10° from the surface normal.

	H-on-D		D-on-H	
	Theory	Expts	Theory	Expts
$\langle E_{\mathrm{int}} \rangle$ (eV)	1.48 (1.48)	1.2	1.67 (1.69)	1.3
$< \nu >$	1.05 (0.51)	1.4	2.13 (1.89)	1.6
$< j >$	11.2 (14.5)	7.3	8.8 (11.7)	7.0
σ_r (Å^2)	0.27 (0.31)	~5	0.39 (0.30)	~5

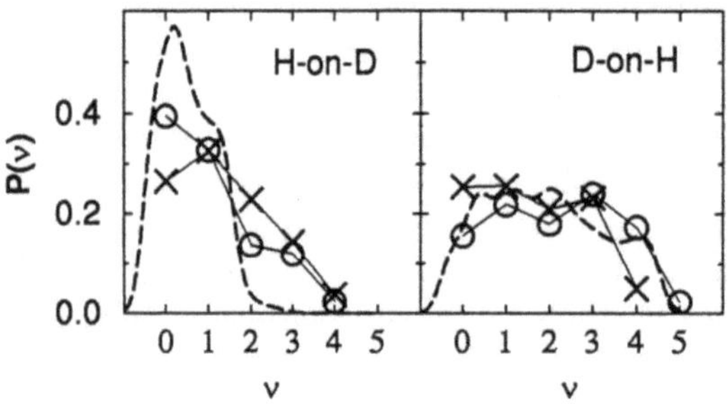

Fig. 2. Comparison between calculated and measured vibrational distributions, $P(\nu)$ for H-on-D and D-on-H. The quantum-mechanical and quasi-classical results are depicted by open symbols and dashed lines, respectively, while the experimental data taken from Ref. [6–7] are depicted by crosses. Same incidence conditions as in Table 1.

are less accurate for σ_r and the averages $\langle \nu \rangle$ and $\langle j \rangle$ of the HD rovibrational distributions. However, the trend of $\langle \nu \rangle$ and $\langle j \rangle$ with isotope is reproduced. As discussed in our paper [14], the impulsive collision argument fails with respect to the internal state distributions.

Strong isotope effects are also seen in the calculated detailed HD product distributions. Here we shall mainly scrutinize the HD product rovibrational state distributions. The results for the translational energy distribution have been reported and discussed in Ref. [14]. In Fig. 2, we compare the QM and QC HD product vibrational state distributions, $P(\nu)$ for the two isotopic combinations. The isotope effects on the calculated $P(\nu)$ are quite pronounced; D-on-H not only gives rise to a much larger $\langle \nu \rangle$ but also a much broader distribution than H-on-D. D-on-H gives even an appreciable population up to $\nu_{\max} = 5$.

The pronounced isotope effects in the calculated $P(\nu)$ are caused by changes in the relative time-scales for the motion of the reactants rather than differences

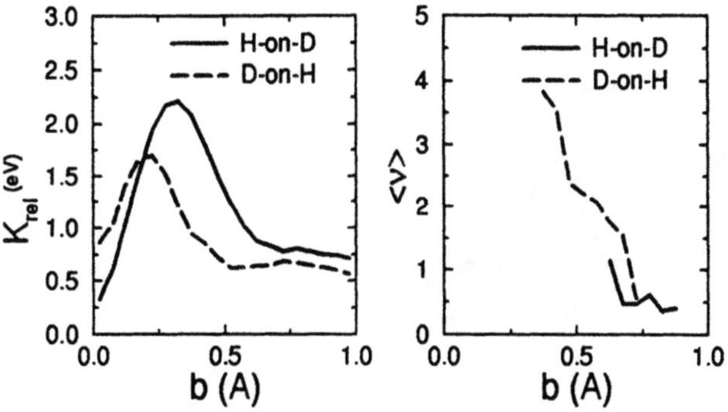

Fig. 3. Quasi-classical averages of the relative kinetic energy, K_{rel}, along the surface and vibrational quantum number, $\langle \nu \rangle$, as a function of impact parameter, b, for the two isotope combinations. Same incidence conditions as in Table 1.

in the zero-point energy of the targon. The latter effect is consistent with $P(\nu)$ being broadest for the combination D-on-H, since this gives the larger amplitude for the targon zero-point motion. However, this explanation was disproved in paper [14] by comparing results from QC calculations using two different zero-point energies for the targon; the proper value and the value corresponding to the other isotope. Here we shall instead show that the isotope effects in $P(\nu)$ are due to a combination of two dynamical effects: (1) different ranges of reactive b in $P_r(b)$ that result from different time-scales for the relative motion of the reactants, and (2) a decrease in vibrational excitation with increasing b. These two effects will now be demonstrated in turn.

In Fig. 3, effect (1) is illustrated by the calculated results for the kinetic energy, K_{rel}, of the relative lateral motion of the targon and the incidon for nonreactive classical trajectories as a function of b . For intermediate ranges of b, K_{rel} is so large that the two reactants separate too quickly to have time to form a molecule. By comparing with Fig. 1, this behavior appears to be responsible for the characteristic holes in $P_r(b)$. Furthermore, the larger values of the reactive b for H-on-D than for D-on-H is caused by a faster separation of H-on-D over a wider range of b than for D-on-H. Effect (2) is also demonstrated in Fig. 3, where QC results for the average vibrational excitation $\langle \nu \rangle$ are plotted as a function of b.

Finally, we show in Fig. 4 that the calculated rotational state probability distributions, $P(j)$, are also quite different for the two isotopic combinations. Both combinations give rise to a broad range of rotational excitations from $j = 0$ to $j_{\max} = 21$. H-on-D tends to populate higher j states than for D-on-H. This behavior is consistent with the variation of the QC reaction probability with b shown in Fig. 1, but a more detailed analysis of the QC calculation reveals that there is no simple mapping between the rotational excitations and b, as there is for the vibrational excitations. Once again we see that the QC distribution is too narrow for H-on-D but fairly good for D-on-H. In Ref. [14],

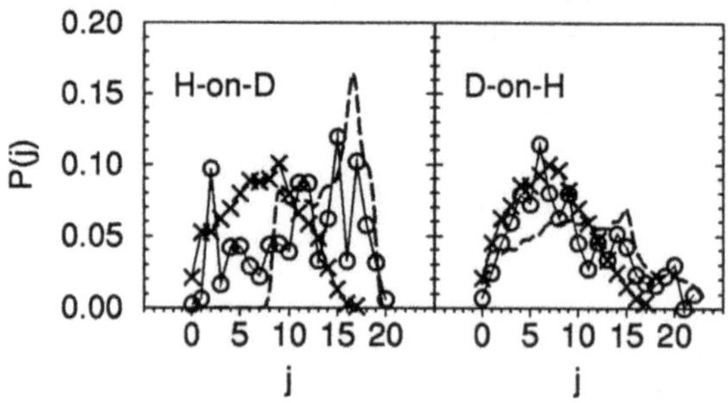

Fig. 4. Comparison between calculated and measured rotational state distributions, $P(j)$ for H-on-D and D-on-H. The quantum mechanical and quasi-classical results are depicted by open symbols and dashed lines, respectively, while the experimental data taken from Ref. [6–7] are depicted by crosses. Same incidence conditions as in Table 1.

we showed that these isotopic effects are not simply due to differences in the QC zero-point motion of the targon for the two isotopic combinations. In analogy with the results for H-on-H presented in PJ, the strong oscillations in the calculated $P(j)$ are probably due to interference effects, and are expected to be washed out at off-normal incidence.

4. Comparison with experiments

The recent and detailed atomic beam measurements of the internal state of the HD product for these reactions by Rettner and Auerbach [6–7] enable us to assess our flat-surface model for the reaction dynamics by making a direct comparison of our predictions with their data. We shall not attempt to make any exhaustive comparison here that will be made elsewhere.

As shown in Table 1, they observed a high rovibrational excitation of the HD product. Some of the observed rovibrational states had energies corresponding to the maximum possible energy released in this reaction. This observation provides strong experimental evidence for an ER-like pathway for the reaction. The overall magnitudes and the trend between the two isotopic combinations of the averages of the internal energy and the rovibrational state distributions are in good agreement with experiments, but the calculated isotope effects are more pronounced. This level of agreement suggests that our model is at least a reasonable starting point for the development of an understanding of the ER dynamics behind this reaction. This view is supported by the following comparison of the calculated vibrational and rotational state distributions with the experimental data.

We find a rather impressive agreement between the calculated and measured $P(\nu)$ for the two isotopic combinations, as shown in Fig. 2. The calculations reproduce semi-quantatitively the magnitudes of the observed $P(\nu)$ for the two cases. That the observed isotope effects of $P(\nu)$ are so well reproduced by our model suggests that the aforementioned dynamical effects behind the isotope effects in the calculated $P(\nu)$ make the latter effects rather insensitive to the details of the PES.

As shown in Fig. 4, the calculation also reproduces amazingly well the observed $P(j)$ for D-on-H, whereas the agreement for H-on-D is not as good. In the case of D-on-H, the main discrepancy shows up at high $j \sim$ 15–20 where the calculation gives larger values for $P(j)$ than the experiments, whereas in the case of H-on-D, the calculation does not reproduce the shape of the observed $P(j)$ at low j and gives too much rotational excitation at high j. The isotope effects in the observed $P(j)$ are smaller than in the calculations.

A most puzzling observation, in view of the rather impressive agreement between theory and experiment for the product internal state distributions, is the disagreement between the calculated and measured reactive cross-sections, which differ by an order of magnitude. A possible explanation for this disagreement is that, as discussed by Rettner and Auerbach [7], the incidon may react with the targon after making a few bounces with the surface. They argue that the dynamics for the reaction after a few bounces is still similar to the direct one; the large mass mismatch between the H(D) atoms and the Cu atoms ensures that the energy transfer to the Cu substrate is inefficient, so the incidon retains a major part of its incidence energy and momentum, even after a few bounces. Alternatively, the disagreement could result from a too crude PES or neglect of surface corrugation in the model. In the papers [13,15], we showed that it is possible to construct a PES, called PES-V, that can give rise to values for σ_r of the same order of magnitude as the observed values, but PES-V is qualitatively different from PES-I and gives, for example, a wrong isotope effect for $\langle \nu \rangle$.

5. Summary and concluding remarks

We have presented a brief overview of our theoretical studies of the dynamics underlying the Eley–Rideal reaction between a gas-phase H(D) atom and a D(H) atom adsorbed on a Cu(111) surface that are based on a flat-surface model and an empirical potential energy surface (PES). The final state distribution of the HD product and the reaction cross section have been calculated using both fully quantum and quasi-classical methods. The predictions of the model for these quantities have been directly compared with results from very recent state-resolved atomic-beam experiments performed by Rettner and Auerbach [6,7].

The results from the QC calculations are in general agreement with the quantum-mechanical results with respect to qualitative behaviour and trends.

The good agreement of the overall magnitude and the isotope effects of the rovibrational distributions with experiment suggests that the model is able to capture some essential ingredients of the reaction dynamics. The vibrational state distributions exhibit the most pronounced isotope effects. An analysis of the QC calculations reveal that these effects are caused by isotope differences in the relative timescales for the motion of the reactants rather than in the zero-point motions of the initially adsorbed atom. A most puzzling observation is the strong disagreement between the calculated and measured values of the reactive cross section. Two possible explanations of this disagreement are that the cross-section is enhanced by the gas-phase atom making a few bounces with the surface before it reacts or by steering introduced by the surface corrugation. Work is in progress to use density functional theory to construct a more accurate PES, which combined with QC calculations should be able to clarify this matter.

Acknowledgements

The work by MP has been supported by the Swedish Natural Science Research Council (NFR). Bret Jackson acknowledges support from the Division of Chemical Sciences, Office of Basic Energy Sciences, Office of Energy Research, US Department of Energy, under Grant # DE-FG02-87ER13744. Allocation of computer time by the Center for Parallel Computers (PDC) in Stockholm is also gratefully acknowledged.

References

1. K. R. Lykke and B. D. Kay, in: *Laser Photoionization and Desorption Surface Analysis Techniques*, vol. 1208, ed. N. S. Nogar, (SPIE, Bellingham, WA, 1990) p. 18.
2. E. W. Kuipers, A. Vardi, A. Danon, and A. Amirav, Phys. Rev. Lett. **66**, 116 (1991).
3. C. T. Rettner, Phys. Rev. Lett. **69**, 383 (1992).
4. C. T. Rettner and D. J. Auerbach, Science **263**,365 (1994).
5. C. T. Rettner, J. Chem. Phys. **101**, 1529 (1994).
6. C. T. Rettner and D. J. Auerbach, Phys. Rev. Lett. **74**, 4551 (1995).
7. C. T. Rettner and D. J. Auerbach, J. Chem. Phys. (in press)
8. J. Harris and B. Kasemo, Surf. Sci. **105**, L281 (1981).
9. P. Kratzer and W. Brenig, Surf. Sci. **254**, 275 (1991).
10. B. Jackson and M. Persson, J. Chem. Phys. **96**, 2378 (1992) ; Surf. Sci. **269/270**, 195 (1992) .
11. B. Jackson, M. Persson, and B. D. Kay, J. Chem. Phys. **100**, 7687 (1994).

12. D. D. Koleske, S. M. Gates, and B. Jackson, J. Chem. Phys. **101**, 3301 (1994).
13. M. Persson and B. Jackson, J. Chem. Phys. **102**, 1078 (1995).
14. M. Persson and B. Jackson, Chem. Phys. Lett. **237**, 468 (1995).
15. B. Jackson and M. Persson, J. Chem. Phys. **103**, 6257 (1995).
16. See the separate contributions by Michelsen *et al.*, B. Hayden, and S. Holloway in *Surface Reactions* ed. R.J. Madix (Springer, Berlin, 1993).
17. C. T. Rettner, H. A. Michelsen, and D. J. Auerbach, J. Chem. Phys., **102**, 4265 (1995).
18. B. Hammer, M. Scheffler, K. W. Jacobsen, and J. K. Nørskov, Phys. Rev. Lett. **73** (1994) 1400.

Dipped Adcluster Model Study of Surface Reactions

H. Nakatsuji

Department of Synthetic Chemistry and Biological Chemistry,
Faculty of Engineering, Kyoto University, Kyoto 606-01, Japan

Abstrat. Dipped adcluster model (DAM) for surface reactions involving electron transfer between admolecule and surface and the SAC/SAC-CI method for studying many different electronic states are explained and applied to harpooning, surface chemiluminescence, and electron emission in the halogen chemisorption on alkali-metal surfaces.

1. Introduction

For theoretically studying reactions on a metal surface, the first question is "how do we describe surface-molecule interaction in a tractable way ?". When admolecule-surface electron transfer is important, the cluster model (CM) may be inadequate. We have proposed dipped adcluster model (DAM) [1] in which the adcluster, which is a combined system of an admolecule and a cluster, is dipped onto the electron bath of the solid metal and an equilibrium is established for an exchange of electrons between the adcluster and the bulk metal. Electrostatic image force, characteristic to a metal surface, is also included [2]. Furthermore, for quantitative descriptions of molecule-surface interactions and reactions, electron correlations, electron transfers, and participations of lower exited states are very important. The SAC (symmetry adapted cluster)/SAC-CI method [3-5] is very suitable for such studies because it describes ground, excited, ionized, and electron attached states efficiently in a similar accuracy [3].

The DAM has been applied to oxygen chemisorptions on palladium [1,2] and silver [6] and halogen chemisorptions on alkali metal surfaces [7]. In the latter, the harpooning, surface chemiluminescence and surface electron emission were nicely explained in agreement with experiment. We further applied the DAM to silver-catalyzed partial oxydation of ethylene [8]. We give here a brief review of such studies.

Springer Series in Solid-State Sciences, Volume 121
Elementary Processes in Excitations and Reactions on Solid Surfaces
Eds.: A. Okiji, H. Kasai and K. Makoshi

2. Dipped Adculster Model (DAM)

For surface-molecule interacting systems in which electron transfer between surface and admolecule is important, the CM is insufficient, because the cluster itself must supply or accomodate electrons to or from the admolecule by strongly affecting the bonds of the cluster itself, if the cluster is small. In actual metal surfaces, a sufficient number of free electrons exist, so that the transfer of electron to or from the admolecule does not significantly affect the local bonding of the metal atoms of the cluster directly interacting with the admolecule.

There are many cases in which electron transfer seems to be very important; oxygen and halogen chemisorptions on a metal surface, the roles of alkali metals and halogens as promoters of catalytic reactions, and the activity of the electropositive metals for dissociative adsorptions of CO, N_2 etc. The dipped adcluster model (DAM) [5,6] has been proposed for dealing with such systems.

Fig. 1 illustrates the concept of DAM. We define "adcluster" as a combined system of admolecule and cluster. We dip it into the electron "bath" of the solid metal and let an equilibrium be established for the electron and/or spin transfer between them. The equilibrim would be established when the chemical potential of the adcluster becomes equal with the chemical potential of the surface, or more precisely when the adcluster reaches at the minimum of $E(n)$ in the range of

$$-\frac{\partial E(n)}{\partial n} \geq \mu \qquad (1)$$

where $E(n)$ is the energy of the adcluster, n is the number of electrons transferred from the bulk metal to the adcluster, and μ is the chemical potential of the electrons of the metal surface. Since the adcluster is a partial system, the number of the transferred electrons, n, is not necessarily an integer. In this model, the external effects such as those of promoters, cocatalysts, supports, temperature, electric potential, etc., are included through a variation in the chemical potential μ.

A typical behavior of the $E(n)$ curve is illustrated in Fig. 2. The $E(n)$ curve is a lower convex in case (a), but an upper convex in case (b). In case (a), electron flows

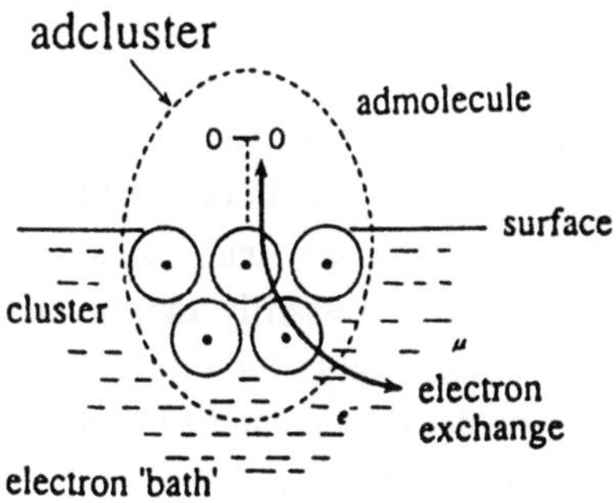

Fig. 1 Concept of the dipped adcluster model (DAM).

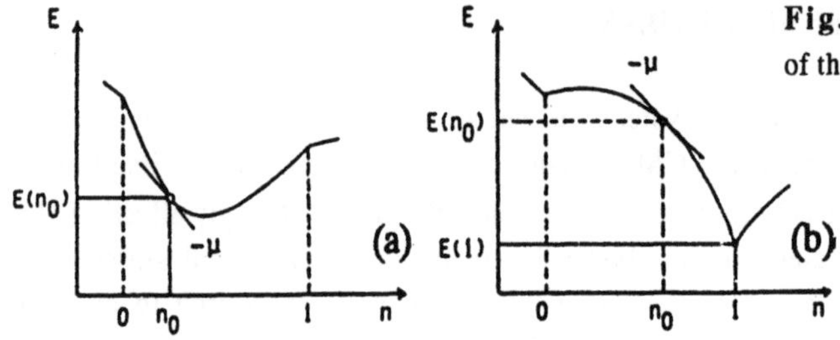

Fig. 2 Two examples of the $E(n)$ curve.

into the adcluster up to n_0, where the gradient of the $E(n)$ curve becomes equal to -μ, as eq(1) implies, since the potential of the adcluster is lower than that of the electron of the metal. The energy of the adcluster is $E(n_0)$. In case (b), there is a barrier to electron flow: electron flow does not occur until n_0 electrons "sink" into the adcluster through, say, tunneling or activation, but afterwards, the electron looks down a deep valley and flows into it up to i, as eq(1) indicates, i being an integer. At $n=i$, the system is most stable, and when n exceeds i, the potential suddenly becomes very high (even positive), so that electron flow ceases at $n=i$. The energy of the system is $E(i)$. It is interesting to note that in case (b), transfer of an integral number of electrons naturally results, so that the case (b) is, in some senses, quantum. When the gradient of the $E(n)$ curve does not become equal with -μ in some range of n, the electron flow does not occur. For more details, see ref. [1].

The energy of the adcluster $E(n)$ may be calculated by the molecular orbital (MO) method [1] or by the method including electron correlations [3-5]. The density-functional method [9] is also applicable to this purpose. In the MO method described in ref 1, we assumed that the adcluster exchanges electrons and spins with the solid through the active orbitals of the adcluster: e.g. highest occupied (HO) MO and the lowest unoccupied (LU) MO.

On a metal surface, the electrostatic force, called image force, is sometimes very important. This is especially so when the system involves a large amount of electron transfer. In the DAM, the image force effect is rather easily included [2] based on the theory of electricity.

3. SAC/SAC-CI Method

Electron correlations are sometmes very important for the surface phenomena. In particular, when the surface is a transition metal, the Hartree-Fock model is insufficient for describing the bonding between admolecule and surface. Furthermore, since the catalytically active states are not necessarilly the ground states of the system. Excited states are sometimes responsible for the reaction [6,10]. Therefore, we need a theory which can describe both ground and excited states in a same accuracy. The SAC/SAC-CI theory is such a theory.

The SAC theory [4] provides a method for calculating a ground state and the SAC-CI method [5] is the method for calculating correlated wave functions for excited, ionized, and electron attached states. The SAC/SAC-CI wave functions satisfy the correct relations among these states and are based on the approximate transferability of electron correlations among these states. The SAC/SAC-CI method is more rapidly convergent than ordinary CI expansion and furthermore it gives directly comparable results for the energies and the wave functions of many different states, different in energy and in the number of electrons. The theoretical simplicity, reliability, and usefulness have been shown in many publications in molecular spectroscopy [11] and excited state chemistry [12,13]. We believe that the method is simple and accurate enough to be useful. For more details, we refer to a recent review article [3]. The SAC/SAC-CI code [14] is available from the author on request.

4. Halogen Chemisorption on Alkali Metal Surface

Halogen chemisorption on an alkali metal surface involves interesting electron transfer processes from the surface to the halogen molecule, X_2. They are 'harpooning', surface chemiluminescence, and surface electron emission. When a halogen molecule approaches to some distance from the surface, an electron jumps into the molecule: this large distance electron transfer is called 'harpooning'. When the molecule receives an electron, it becomes an anion so that the molecule is elongated and at the same time, it is attracted to the surface. When it collides with the surface, the surface chemiluminescence and electron emission occur. They are the electron transfer processes from the surface to the molecule accompanied with the emissions of a photon and an electron, respectively. Up to three electrons are thus transferred from the surface to the molecule. We investigate these electron transfer processes by the DAM and the SAC/SAC-CI method [7].

4.1 Harpooning

Harpooning is a long-distance jump of an electron from a metal surface to a molecule. It is explained as an electron tunnelling from the alkali metal surface to the halogen molecule. However, in the conventional model, this process should be strongly vibronic, since the electron affinity of X_2^- is very much dependent on the X-X distance: a large electron affinity of X_2 is obtained only after an elongation of the distance.

We assume that Cl_2 molecule approaches the surface in the end-on form, since the σ^* MO of Cl_2 is the electron-accepting orbital. We think that the electron transfer occurs as a Franck-Condon process, so that the Cl-Cl distance is fixed to 2.0025 Å, theoretically optimized distance: the experimental equilibrium distance is 1.987 Å. We do not take into account molecular vibration effect. The alkali metal surface is

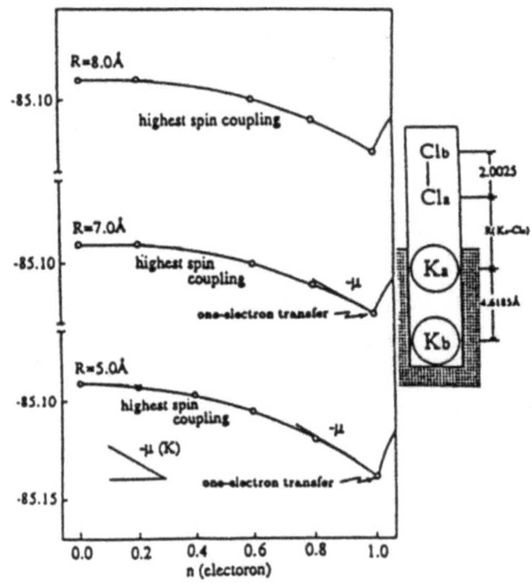

Fig. 3 *E(n)* curves calculated by the DAM for the Cl_2- potassium surface system at the Ka-Cla distances of 8, 7, and 5 Å.

represented by the two metal atoms dipped onto the electron bath of the alkali metal surface as illustrated in Fig. 3. We consider here only potassium surface.

We first apply molecular orbital model for the DAM [1]. The adcluster K_2 interacting with Cl_2 is dipped onto the electron bath of the potassium metal whose chemical potential μ is 2.30 eV. Fig. 3 shows the E(n) curves, calculated at different separations of Cl_2 from the surface. The curve is calculated by the highest spin coupling model [1] and the active orbital is chosen to be the σ* MO of Cl_2. The E(n) curve is an upper convex and its tangent can become larger (in absolute value) than -μ when the Cl_a-K_a distance is smaller than 7 Å. Since the E(n) curve is an upper convex, one electron (an integral number, one) is transferred from the surface to Cl_2. This one electron transfer is just the harpooning of an electron from the surface to Cl_2: this is the DAM picture of the harpooning. We note that we could explain the harpooning as a Franck-Condon process without resorting to the vibronic coupling. We also note that we have used the picture of 'tunneling' from E(0) to E(n) in the *n-E* space not in the *R-E* space. Similar results were also obtained for the sodium-Cl_2 and rubidium-Cl_2 systems, giving different harpooning distances [7].

4.2 Elongation and Chemisorption of Cl_2^-

The harpooning occurs as a Franck-Condon process in the DAM, and afterwards, the Cl-Cl distance is elongated to be 2.64 Å with much stabilization of about 48 kcal/mol [7]. The binding energy of Cl_2^- on the surface is calculated to be only 7 kcal/mol: very weak even in comparison with the experimental binding energy, 29 kcal/mol of a free Cl_2^-. Since the energy gain by the elongation is much larger than the binding energy, we expect that a neutral Cl atom is ejected out from the surface after the harpooning. Nevertheless, some Cl_2^- molecule would further adsorbed on the surface, with the excess energy being dissipated somewhere, say, to the alkali metal surface.

4.3 Surface Chemiluminescence Process

Next is the surface chemiluminescence process: the second electron is transferred from the metal surface to the halogen molecule with an emission of a photon. We here assume the reaction pathway of the adcluster K_2-Cl_2^- as shown in Fig. 4. The reaction is assumed to proceed from the end-on geometry on K_a, #1 and #2 positions, to the dissociated chemisorption structures, #8 and #9.

An interesting question here invoked is "from where the second electron originates?". Does it originate from the bulk metal or from the local K_2 site directly interacting with Cl_2? In order to answer this question, we calculated by the SAC/SAC-CI method the total energies of the adclusters with the formal charges of K_2-Cl_2^{2-} and K_2^{2+}-Cl_2^{2-}, the former being the product of the electron transfer from the bulk metal and the latter from the local K_2 site. As a result, we find that the latter is lower than the former. We therefore conclude that the second electron originates from the local K_2 site directly interacting with Cl_2^-.

We calculate the potential energy of the K_2-Cl_2^{2-} adcluster (initial state) and the K_2^+-Cl_2^{2-} adcluster (final state) by the SAC-CI method. Fig. 5 shows the result. The

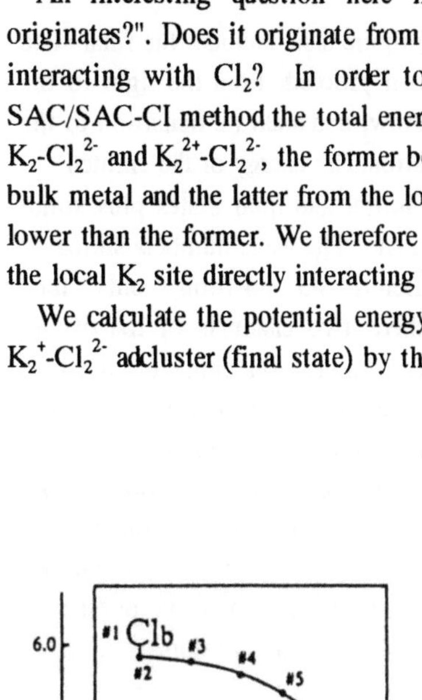

Fig. 4 Assumed reaction pathway for the surface chemiluminescence and electron emission processes.

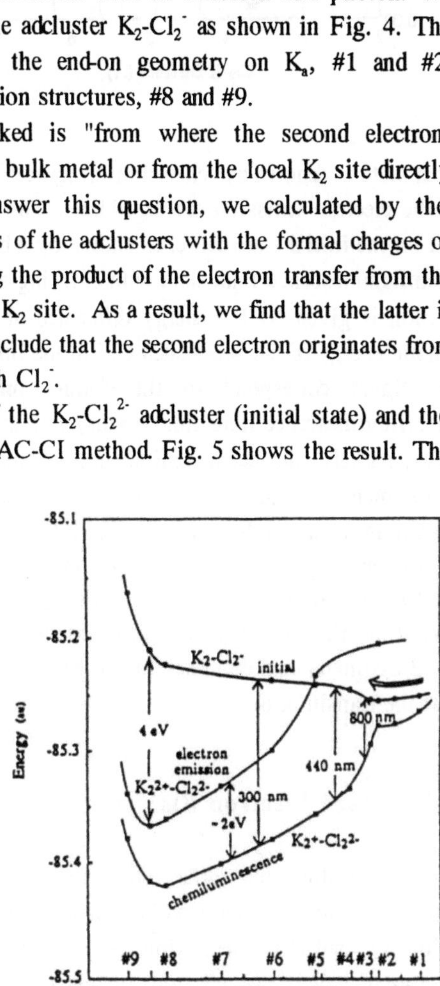

Fig. 5 Potential curves of the Cl_2-potassium surface system for the surface chemiluminescence and electron emission processes calculated by the SAC/SAC-CI method.

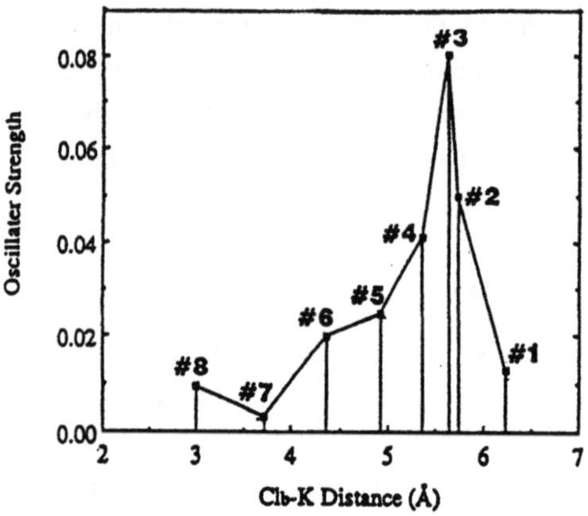

Fig. 6 Oscillator strength for the chemiluminescence along the reaction coordinate.

final state is denoted by 'chemiluminescence'. The figure also includes the final state of the electron emission discussed later. The system proceeds from the right to the left on the initial state curve, as shown by a big arrow, and makes a transition to the chemiluminescence final state with emitting a photon. The energy of the emitted photon is given by the energy difference of the initial and final states plus some kinetic energy of the accelerated Cl_2^- in the initial state. The wave numbers shown in the figure correspond to the former one. Before #5 position, only the chemiluminescence transition occurs, but after there, two channels of transitions exist, namely, the chemiluminescence channel and the electron emission channel. Experimentally, the spectrum spreads from about 800 nm to 300 nm with the peak maximum at about 440 nm [7]. The intensity rapidly decreases from about 400 nm to 300 nm. This rapid decrease is understood to be due to the transition to the electron emission channel: the crossing at about #5 geometry facilitates this transition. Fig. 6 shows the calculated probability of the transition: it is maximum at #3 geometry and decreases in the geometries in either side. Thus, we understand that the transition occurs in the course of the dissociative chemisorption.

4.4 Surface Electron Emission Process

Finally we study the surface electron emission process. This is the two-electron process in which one electron is transferred from the surface to the halogen molecule anion X_2^-, and at the same time, another electron is emitted out of the surface. We assume that these two electrons belong to the local K_2 site directly interacting with Cl_2^-, since they are so strongly correlated in this process. It is difficult to expect such a strong correlation if one or two of these electrons belong to the bulk metal. We use again the reaction path shown in Fig. 4.

Fig. 5 includes the potential curve for the final state of the surface electron emission calculated by the SAC-CI method. The initial state is common to both processes. The electron emission process occurs later to the chemiluminescence process, since the process proceeds from right to left on the initial-state curve. The emission occurs in the left of the geometry #5, where the initial and final state curves of the electron emission cross to each other. The excess (kinetic) energy of the emitted electron is about 4 eV at the turning point of the initial state curve, #8-#9, which is also the equilibrium geometry of the final state, namely KCl. We therefore expect a large transition probability at #8-#9 geometries from these features of the potential curves: the Franck-Condon factor should be large there. A large transition probability is also expected at the crossing point at about #5 geometry and may be calculated by the Landau-Zener model. Experimentally, the exoelectron energy distributions are measured for Cl_2 on yttrium and on Rb-dosed yttrium, and a maximum peak is seen at about 4 eV in accordance with the above expectation [7].

In Fig. 5, the difference between the final-state potential curves of the electron emission and the chemiluminescence is roughly about 2 eV at any geometry of the reaction pathway. This is because the difference corresponds to the work function of the potassium surface: the experimental value is 2.3 eV.

5. Conclusion

We have investigated various types of surface-molecule interactions by the methodology of our laboratory: the dipped adcluster model and the SAC/SAC-CI method. It is shown that this methodology is powerful for investigating surface-molecule interactions and reactions occurring on metal surfaces in thermal and photochemical situations. Our methodology is also useful for studying surface photochemistry [13].

References

[1] H. Nakatsuji, J. Chem. Phys., **87**, 4995 (1987).
[2] H. Nakatsuji, H. Nakai, and Y. Fukunishi, J. Chem. Phys. **95**, 640 (1991).
[3] H. Nakatsuji, Acta Chim. Hungarica, **129**, 719 (1992).
[4] H. Nakatsuji and K. Hirao, J. Chem. Phys., **68**, 2053 (1978).
[5] H. Nakatsuji, Chem. Phys. Letters, **59**, 362 (1978); **67**, 329, 334 (1979).
[6] H. Nakatsuji and H. Nakai, Chem. Phys Letters, **174**, 283 (1990); J. Chem. Phys. **98**, 2423 (1993); Can. J. Chem., **70**, 404 (1992).
[7] H. Nakatsuji, R. Kuwano, H. Morita, and H. Nakai, J. Mol. Catalysis, **82**, 211 (1993).
[8] H. Nakatsuji, H. Nakai, K. Ikeda, and Y. Yamamoto, submitted.
[9] R. G. Parr and W. Yang, *Density-Functional Theory of Atoms and Molecules* (Oxford Unversity, New York, 1989).
[10] H. Nakatsuji, Y. Matsuzaki, and T. Yonezawa, J. Chem. Phys., **88**, 5759 (1988).

[11] See for examople, H. Nakatsuji and M. Ehara, J. Chem. Phys., **102**, 6822 (1995); H. Nakatsuji, J. Hasegawa, and M. Hada, J. Chem. Phys., in press.
[12] M. Hada, Y. Imai, M. Hidaka, and H. Nakatsuji , J. Chem. Phys., **103**, 6993 (1995).
[13] H. Nakatsuji, H. Morita, H. Nakai, Y. Murata, and K. Fukutani, , J. Chem. Phys., **104**, 714 (1996).
[14] H. Nakatsuji, *Program System for SAC and SAC-CI Calculations, Program Library No. 146 (Y4/SAC)*, Data Processing Center of Kyoto University, 1985; H. Nakatsuji, *Program Library SAC85 (No. 1396)*, Computer Center of the Institute for Molecular Science, Okazaki, 1986.

Ab Initio Study on Photo-Induced Desorption of CO and NO from Pt(111)

H. Aizawa and S. Tsuneyuki

Institute for Solid State Physics, University of Tokyo,
7-22-1 Roppongi, Minato-ku, Tokyo 106, Japan

Abstract. Photo-induced desorption of CO and NO from Pt(111) is investigated by means of ab initio electronic structure calculations with the configuration interaction method for small clusters. An electronic excitation to an unoccupied level with some 5σ contributions is shown to lead to a repulsive potential energy surface, suggesting that an MGR-like mechanism is operative.

1. Introduction

Photo-induced desorption has been under active investigation recently not only because of basic scientific interest but also because of possibilities of its applications to technology. In particular, photodesorption of CO and NO molecules from a variety of surfaces has been extensively studied experimentally, partly because they are so active molecules as to play an important role in various chemical reactions of interest and partly because they can be state-selectively detected by means of either resonance-enhanced multiphoton ionization (REMPI) or laser-induced fluorescence (LIF) techniques.

Experiments on photo-induced desorption of CO [1] and NO [2] from Pt(111) have been carried out by Fukutani et al. at a surface temperature of 80 K. For both CO and NO, the on-top species have been found to desorb by laser irradiation, while the bridge species have not. The observed linear dependence of the desorption yields on laser fluence indicates that the desorption is a single-photon process. It has been found that the threshold energy for the desorption of NO is lower than 2.3 eV [3], while that of CO lies between 2.3 and 3.5 eV [1]. Experiments by Buntin et al. [4] for the NO/Pt(111) system annealed at 220 K have also shown the threshold energy for the desorption of NO to be rather low (between 0.65 and 1.17 eV), which is consistent with the above result.

In interpreting experimental results on photodesorption, photoemission and inverse photoemission data are usually exploited. The final state of a photoemission (inverse photoemission) process is, however, a state with an electron removed from (added to) the system, and energy

Springer Series in Solid-State Sciences, Volume 121
Elementary Processes in Excitations and Reactions on Solid Surfaces
Eds.: A. Okiji, H. Kasai and K. Makoshi

difference between an occupied level and an unoccupied one obtained from these data is not equal to excitation energy in photodesorption processes due to the electron-hole interaction. Furthermore, even if the final state of the photon absorption can be specified, one has to prove that an excitation to the state really results in desorption. In the so-called MGR mechanism [5], for instance, it has to be shown that a potential energy surface of the excited state is repulsive. Extracting information on shapes of potential energy surfaces from experimental data is, however, rather difficult with the present experimental techniques. By means of electronic structure calculations, on the other hand, shape of a potential energy surface of each excited state can be obtained as well as an excitation energy with the electron-hole interaction properly taken into account. Therefore they are expected to be of great help in understanding photodesorption processes. In particular, the configuration interaction (CI) method [6] can treat electronically excited states on firm ground in contrast to methods based upon the density functional theory. In the present paper, we perform electronic structure calculations with the CI method in order to elucidate the mechanism of photo-induced desorption of CO and NO from Pt(111).

2. CI Calculations with Small Clusters

The details of the calculations are described elsewhere [7]. Briefly, we perform the so-called Multi-Reference SDCI calculations using the HONDO program [8]. We make use of the effective core potentials and contracted Gaussian type orbitals (CGTO's) generated by Hay et al. [9] for Pt and by Stevens et al. [10] for C, N and O. Some of the CGTO's are split into two basis functions, forming a double-zeta basis set.

We perform CI calculations for small clusters composed of only three atoms, PtCO and PtNO. The three atoms are linearly placed with the C(N) atom bonded to the Pt atom so that they represent the on-top species, which adsorbs perpendicularly to the surface with the C(N)-end down.

In order to check the validity of the use of the three-atom clusters we also perform the Hartree-Fock (HF) calculation for a larger cluster with 10 Pt atoms , $Pt_{10}CO$, using a minimal basis set. The 10 Pt atoms are arranged so that they represent the (111) surface of a crystal with the face-centered cubic (f.c.c.) structure. The first layer of the surface is composed of 7 atoms, and the second layer of 3 atoms. We calculate the density of states weighted by the overlap population for each MO [11]. We call it the overlap population weighted density of states, or OPDOS for short. The overlap population proposed by Mulliken [12] is a quantity

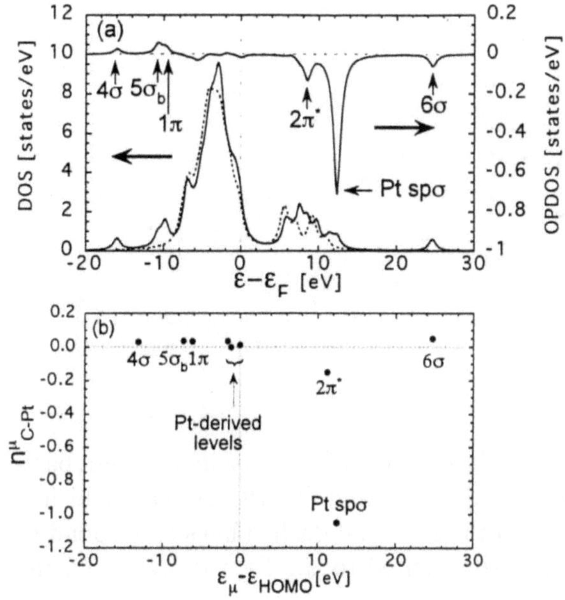

Fig. 1: (a) The overlap population weighted density of states (OPDOS) calculated with a minimal basis set for the $Pt_{10}CO$ cluster (upper solid line) together with the conventional density of states (DOS) for the $Pt_{10}CO$ (lower solid line) and Pt_{10} (dotted line) clusters. (b) The overlap population for each HF MO calculated with a minimal basis set for the three-atom cluster PtCO.

which indicates how each molecular orbital (MO) contributes to a bond. If it is positive, the corresponding MO has bonding character with respect to the bond, and if negative, it has anti-bonding character. The OPDOS with respect to the bond between the adsorbate and the surface is shown in Fig. 1(a) together with the conventional DOS for the $Pt_{10}CO$ and Pt_{10} clusters. Fig. 1(b) shows the overlap population for each HF MO calculated for the three-atom cluster PtCO using a minimal basis set. As is apparent, the characteristics of the larger cluster persist for the three-atom cluster. This suggests that the bond between the adsorbate and the surface has very localized nature, and that the use of the three-atom clusters is not so irrelevant as it seems.

3. Potential Energy Surfaces

The PES's for the three-atom cluster PtCO obtained by the CI calculations with a double-zeta basis set are shown in Fig. 2(a) as functions of

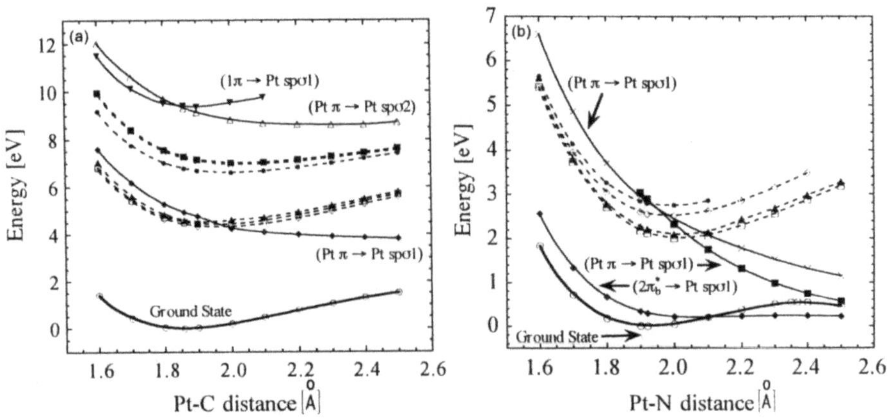

Fig. 2: The potential energy surfaces calculated with a double-zeta basis set for the three-atom clusters (a) PtCO and (b) PtNO. The bold line represents the ground state, and the solid and dashed lines correspond to the excited states with an electron excited to the Pt *spσ* levels and the $2\pi^*$-derived levels, respectively.

the Pt-C distance with the C-O distance fixed at the optimized value of 1.18 Å. In the figure, the bold line represents the CI ground state. The solid and dashed lines in the figure correspond to the states with an electron excited to the Pt *spσ* levels and the $2\pi^*$-derived levels, respectively. Each of the lines in the figure is depicted so that it joins the points which represent the CI states corresponding to the same electronic configuration, that is, the lines represent diabatic potential energy surfaces. The calculations with a double-zeta basis set yield two levels with the Pt *spσ* character (denoted by the Pt *spσ*1 and *spσ*2 levels), which are different from each other in the way the Pt *s* and *pσ* orbitals combine. The Pt *spσ* levels, which correspond to the Pt *spσ* level obtained with a minimal basis set (see Fig. 1), have some contributions from the CO 5σ orbital and have strong anti-bonding character with respect to the bond between the adsorbate and the surface. That is why an excitation to the Pt *spσ* levels leads to a repulsive PES. The state corresponding to the excitation $1\pi \rightarrow$ Pt *spσ*1 is the only exception, which has a bound-shaped PES due to the Coulomb attraction arising from the electron transfer from CO to Pt.

An excitation to the $2\pi^*$-derived levels does not result in a repulsive PES, though the minimum-energy point of the corresponding PES is displaced farther from the surface than the ground state PES. This can be simply understood from Fig. 1 (though it is obtained using a minimal basis set), where we can see that the $2\pi^*$-derived level has anti-bonding character which is not so strong as that of the Pt $sp\sigma$ level. In the HF calculations with a double-zeta basis set, the CO $2\pi^*$ level is actually split into the bonding $2\pi_b^*$ and anti-bonding $2\pi_a^*$ levels by the interaction with the metal $p\pi$ levels as has been suggested in the literature [13], but the obtained PES's imply that by taking electron correlations into account the $2\pi_b^*$ and $2\pi_a^*$ levels recombine into two levels, both of which have effectively anti-bonding character due to the effect of the interaction with the Pt $5d\pi$ level [7]. Strong Coulomb attraction seen in the case of the excitation from the 1π level does not come about in the case of the $2\pi^*$-related excitation, since the $2\pi^*$-derived levels are not so localized at the CO molecule as the 1π level, resulting in much less charge transfer.

Calculations for the three-atom cluster PtNO are performed similarly with the N-O distance fixed at the optimized value of 1.23 Å. The calculated PES's are shown in Fig. 2(b). The ground state of the PtNO has an electron occupying one of the $2\pi^*$-derived levels (the $2\pi_b^*$ level), which are completely empty in the ground state of the PtCO. The lines in Fig. 2(b) represent diabatic PES's except the bold line indicated as 'ground state' in the figure. It represents an adiabatic PES corresponding to the ground state at the equilibrium Pt-N distance [14], and has a barrier at around 2.37 Å. The state corresponding to the excitation $2\pi_b^* \rightarrow$ Pt $sp\sigma1$, which has different symmetry from that of the ground state at the equilibrium Pt-N distance, becomes the ground state at Pt-N distance larger than 2.1 Å. As in the case of the PtCO, an excitation to the Pt $sp\sigma$ levels results in a repulsive PES, while that to the $2\pi^*$-derived levels does not.

4. Desorption Mechanism

As shown in Figures 2(a) and 2(b), excitations to the Pt $sp\sigma$ levels result in a repulsive PES [15]. This suggests a mechanism in which the desorption is initiated by excitations to the Pt $sp\sigma$ levels and proceeds according to an MGR-like mechanism in the sense that a repulsive PES plays a crucial role. Inverse photoemission studies by Rangelov et al. [16] have suggested that one of the unoccupied bands of CO/Pt(110) contains substantial contributions from the $5\sigma_a$ level (which corresponds to the Pt $sp\sigma$ levels in the present calculations), while those of CO/Ni(110) and CO/Pd(110) do not. Therefore, the mechanism described above is quite consistent with the fact that no photodesorption of CO and NO from Ni or Pd surface has been observed [17] in contrast to that from Pt surface. The mechanism seems also consistent with the low threshold energy of NO as compared to that of CO, since the PtNO has a low-lying excited state corresponding to the excitation $2\pi_b^* \rightarrow$ Pt $sp\sigma1$, which is absent in the case of the PtCO. Note that the calculated threshold energy, namely the excitation energy of the lowest Pt $sp\sigma$-related excited state, is 4.9 eV for the PtCO, while the desorption of CO is experimentally observed at the photon energy of 3.5 eV, though the cross section is smaller than that at 5.0 and 6.4 eV [1]. The discrepancy is considered to be attributed to the use of the small cluster, because a molecular energy level can shift greatly due to the interaction with the extended surface.

Our calculations show, on the other hand, that any excitations to the $2\pi^*$-derived levels do not lead to a repulsive PES, ruling out the possibility that an MGR-like mechanism with the $2\pi^*$-related excitation is operative. The Antoniewicz mechanism [18], which has usually been discussed in connection with the hot electron mediated model [4, 19], is not supported either, because our calculations show that the $2\pi^*$-derived levels have anti-bonding character with respect to the bond between the adsorbate and the surface, suppressing an attractive force which is supposed to drive desorption in the Antoniewicz mechanism.

5. Conclusion

Ab initio electronic structure calculations using small clusters suggest a mechanism of photo-induced desorption of CO and NO from Pt(111) in which an excitation to the Pt $sp\sigma$ levels drives desorption in an MGR-like fashion. The mechanism is consistent with the substrate dependence of photodesorption as well as the observed low threshold energy for NO as compared to that for CO.

Acknowledgements

The authors wish to thank Y. Murata, K. Fukutani and M.-B. Song for valuable discussions.

References

[1] K. Fukutani, M.-B. Song and Y. Murata, J. Chem. Phys. **103**, 2221 (1995).

[2] K. Fukutani, Y. Murata, R. Schwarzwald and T.J. Chuang, Surf. Sci. **311**, 247 (1994); M.-B. Song, M. Suguri, K. Fukutani, F. Komori and Y. Murata, Appl. Surf. Sci. **79/80**, 25 (1994).

[3] K. Fukutani, Y. Murata, R. Schwarzwald and T.J. Chuang, private communication.

[4] S.A. Buntin, L.J. Richter, R.R. Cavanagh and D.S. King, Phys. Rev. Lett. **61**, 1321 (1988); S.A. Buntin, L.J. Richter, D.S. King and R.R. Cavanagh, J. Chem. Phys. **91**, 6429 (1989).
[5] D. Menzel and R. Gomer, J. Chem. Phys. **41**, 3311 (1964); P.A. Redhead, Can. J. Phys. **42**, 886 (1964).
[6] See, e.g., I. Shavitt, in: Modern Theoretical Chemistry, Vol. 3, edited by H.F. Schaefer (Plenum, New York, 1977)p.189.
[7] H. Aizawa and S. Tsuneyuki, to be published in Surf. Sci.
[8] Japan Chemistry Program Exchange, PROGRAM P048, HONDO v.7.01.
[9] P.J. Hay and W.R. Wadt, J. Chem. Phys. **82**, 299 (1985).
[10] W.J. Stevens, H. Basch and M. Krauss, J. Chem. Phys. **81**, 6026 (1984).
[11] R. Hoffmann, Rev. Mod. Phys. **60**, 601 (1988).
[12] R.S. Mulliken, J. Chem. Phys. **23**, 1833; **23**, 2343 (1955).
[13] Ph. Avouris, P.S. Bagus and C.J. Nelin, J. Electron Spectrosc. Relat. Phenom. **38**, 269 (1986).
[14] A diabatic PES corresponding to the main electronic configuration of the ground state at the equilibrium Pt-N distance is shown in Ref. [7].
[15] Similar results have been obtained in a recent calculation by Nakatsuji et al. (H. Nakatsuji, H. Morita, H. Nakai, Y. Murata and K. Fukutani, to be published in J. Chem. Phys.) for the Pt_2CO cluster with somewhat different atomic configuration from that in the present calculation.
[16] G. Rangelov, N. Memmel, E. Bertel and V. Dose, Surf. Sci. **251/252**, 965 (1991).
[17] F. Budde, A.V. Hamza, P.M. Ferm, G. Ertl, D. Weide, P. Andresen and H.-J. Freund, Phys. Rev. Lett. **60**, 1518 (1988); P.M. Ferm, F. Budde, A.V. Hamza, S. Jakubith, G. Ertl, D. Weide, P. Andresen and H.-J. Freund, Surf. Sci. **218**, 467 (1989); J. Yoshinobu, X. Guo and J.T. Yates,Jr., J. Chem. Phys. **92**, 7700 (1990); X. Guo, J. Yoshinobu and J.T. Yates,Jr., J. Chem. Phys. **92** 4320 (1990); J.A. Prybyla, T.F. Heinz, J.A. Misewich and M.M.T. Loy, Surf. Sci. **230**, L173 (1990).
[18] P.R. Antoniewicz, Phys. Rev. B **21**, 3811 (1980).
[19] J.W. Gadzuk, L.J. Richter, S.A. Buntin, D.S. King and R.R. Cavanagh, Surf. Sci. **235**, 317 (1990).

van der Waals Density Functional for Atoms and Surfaces

Y. Andersson[1], *E. Hult*[1], *D.C. Langreth*[2], *and B.I. Lundqvist*[1]

[1]Department of Applied Physics, Chalmers University of Technology and Göteborg University, S-412 96 Göteborg, Sweden

[2]Department of Physics and Astronomy, Rutgers University, Piscataway, NJ 08855, USA

Abstract. A general functional form for the exchange-correlation energy of the density-functional theory, valid for widely separated fragments of matter, has recently been proposed [1]. For many kinds of dimers it gives good van der Waals coefficients. Here we also show results for atoms outside surfaces, where it gives a z^{-3} behavior and reasonable van der Waals coefficients.

1. Formalism

Knowledge of weak and long-ranged atom-surface interactions is of key importance for interpreting, e.g., scattering experiments, atomic-force microscopy, and dynamics of surface reactions. Because of its non-local nature, the van der Waals interaction is not correctly described by the local or semi-local approximations that are commonly applied in density functional theory (DFT) calculations [2].

To restore the vdW interaction in DFT methods, we have used an approach which ties together several previous ideas where local or semi-local approximations are applied in various ways [3, 4, 5, 6]. The non-local interaction has been analyzed [6, 5, 7], and recently a general functional form for it has been proposed and shown to give promising results [1]. For the interaction between two small charge perturbations in a uniform electron gas the van der Waals interaction is described through the limiting behavior [4, 8, 7] of the linear response kernel K_{xc} which gives the exchange-correlation energy of a slightly non-uniform system [9],

$$E_{\mathrm{xc}} = \int d^3r_1 \int d^3r_2 K_{xc}(\mathbf{r}_1, \mathbf{r}_2) \delta n(\mathbf{r}_1) \delta n(\mathbf{r}_2) . \tag{1}$$

Rapcewicz and Ashcroft (RA)[4] analyzed the contribution from density inhomogeneities to K_{xc}, and obtained an expression for the effective long-range electron interaction which for large separations scales like r^{-6}, where r is the distance between the electrons. They described the electron gas through an effective density $n_{\mathrm{eff}} = \sqrt{n(\mathbf{r}_1)n(\mathbf{r}_2)}$.

If one wishes to create a density functional theory that is valid in both the uniform gas and separate atom limits one must have a form that is viable and physically motivated in both limits. We show here that such a situation exists

Springer Series in Solid-State Sciences, Volume 121
Elementary Processes in Excitations and Reactions on Solid Surfaces
Eds.: A. Okiji, H. Kasai and K. Makoshi

if we choose $n_{\text{eff}} = \left[\sqrt{n(\mathbf{r}_1)n(\mathbf{r}_2)}\left(\sqrt{n(\mathbf{r}_1)} + \sqrt{n(\mathbf{r}_2)}\right)\right]^{2/3}$.

In an analysis analogous to that of RA, we then get the effective long-range interaction between two electrons at $\mathbf{r}_1$ and $\mathbf{r}_2$,

$$\phi(\mathbf{r}_1, \mathbf{r}_2) \rightarrow -\frac{3e^4}{2m^2} \frac{1}{\omega_p(\mathbf{r}_1)\omega_p(\mathbf{r}_2)\left(\omega_p(\mathbf{r}_1) + \omega_p(\mathbf{r}_2)\right)|\mathbf{r}_1 - \mathbf{r}_2|^6}, \tag{2}$$

where $\omega_p(\mathbf{r}_i) = \sqrt{4\pi e^2 n_{\text{eff}}(\mathbf{r}_i)/m}$ is the local plasma frequency. The long-range interaction between two fragments of matter is obtained by multiplying Eq. (2) by the electronic densities of the fragments and integrating over their volumes. A cut-off is introduced to account for the fact that in regions of low electron density there will be no response to electromagnetic fields, and thus no contribution to the interaction.

2. Pairs of Atoms

For two atoms widely separated by a distance R, the integrated electron-electron interaction gives a long-range interaction energy which is on the van der Waals form, $E = -C_6/R^6$, where

$$C_6 = -\frac{3}{\pi}\int_0^\infty du\, \alpha_1(iu)\alpha_2(iu), \tag{3}$$

and

$$\alpha_i(\omega) = \int d^3r\, \chi_i(\omega, \mathbf{r}) \equiv \int d^3r\, \frac{1}{4\pi}\frac{\omega_p^2(\mathbf{r})}{\omega_p^2(\mathbf{r}) - \omega^2}. \tag{4}$$

Equation (3) is the standard expression [10] for C_6 in terms of the atomic polarizabilities α_i. The essence of the approximation is that the atomic polarizabilities are calculated in a local approximation.

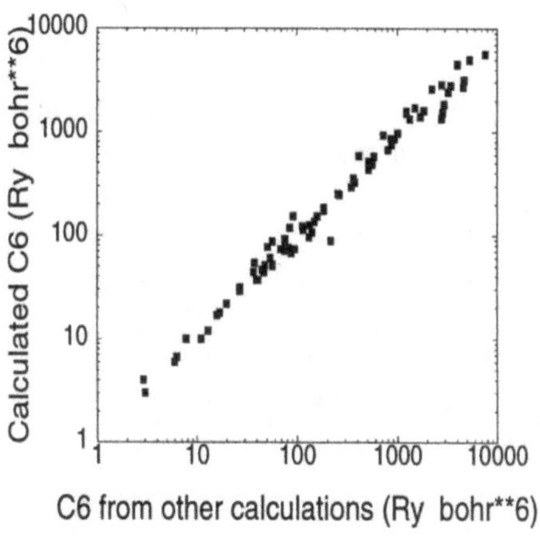

Figure 1. Van der Waals coefficients C_6 (in Ry atomic units) calculated from Eq.(3), plotted against corresponding values from first-principles calculations.

The vdW-constants C_6 for the interaction between a large number of different atoms and also for hydrogen molecules, calculated using available atomic wave functions [11, 12, 13], are displayed in Figure 1.

3. Atom Outside a Surface

The van der Waals interaction between an atom and a surface is calculated in the same way as in the atom-atom case, with one of the atomic electron densities replaced by the electron density of the surface. Also here the correct asymptotic behavior is obtained, $E_{\rm vdW} = -C_3/(z - Z_0)^3$, where z is the distance between the atom and the surface and Z_0 is the van der Waals plane.

Our expression for the van der Waals coefficient is,

$$C_3 = \frac{1}{4\pi} \int_0^\infty du\, \alpha(iu) \frac{\epsilon(iu) - 1}{2\epsilon(iu)}, \tag{5}$$

where $\epsilon(iu)$ is the bulk dielectric function. This expression differs from the exact [14] by a factor of $2\epsilon(iu)/(\epsilon(iu) + 1)$, since our functional is too crude to recognize the fact that outside a surface the true electric field is equal to the applied field multiplied by that factor. In addition, the image plane position, $[\epsilon(iu)/(\epsilon(iu)+1)]\, d(iu)$, is replaced by the centroid of the induced surface charge, $d(iu)$, in our expression for Z_0. Values of C_3 calculated using Eq.(5) are given in Table 1 for some systems where first-principles calculations are available. Since the results, in particular for He and H_2, are in reasonable agreement with the first-principles results, we believe that our expression for C_3 includes most of the physics of the interaction. The van der Waals plane position, Z_0, is up to a factor of two too large, which relates to its great sensitivity to the cut-off.

In the approximation described here we treat the response to the applied field as local. Presently we are improving the theory by treating the response to the true field as local, an approach which promises to give more accurate results [15].

Table 1: The van der Waals coefficient C_3 (Ry atomic units) for He, Be and H_2 outside jellium. $C_3^{\rm ref}$ are results from first-principles results.

	$r_s = 2$			$r_s = 4$		
	He	Be	H_2	He	Be	H_2
C_3	0.079	0.80	0.21	0.041	0.56	0.12
$C_3^{\rm ref}$	0.10[a]	1.5[b]	0.32[a]	0.045[a]	0.97[b]	0.16[a]

[a] Ref. [16], [b] Ref. [17]

4. Conclusions

A general functional form for the exchange-correlation energy of the DFT is presented and applied to interacting atoms and atoms outside surfaces. The suggested functional captures the basic physics of the van der Waals interaction, although it has been recognized that the theory needs some improvements, e.g., a better treatment of the local field and a fine-tuning of the cut-off. The results compare fairly well with first-principles calculations, giving promise for use in, e.g., simulations of soft condensed matter, where van der Waals interactions are crucial.

Acknowledgments

Work at Rutgers supported in part by NSF Grant No. DMR 94-07055. Financial support from the Swedish Natural Science Research Council and the Swedish National Board for Industrial and Technical Development is also acknowledged.

References

[1] Y. Andersson, D. C. Langreth, and B. I. Lundqvist, Phys. Rev. Lett. **76**, 102 (1996).
[2] R. O. Jones and O. Gunnarsson, Rev. Mod. Phys. **61**, 689 (1989), and references therein.
[3] D. C. Langreth and S. H. Vosko, Phys. Rev. Lett. **59**, 497 (1987).
[4] K. Rapcewicz and N. W. Ashcroft, Phys. Rev. B **44**, 4032 (1991).
[5] W. Kohn and W. Hanke (unpublished).
[6] J. F. Dobson, in *Topics in Condensed Matter Physics*, edited by M. P. Das (Nova, N. Y., 1994).
[7] B. I. Lundqvist *et al.*, Int. J. Quantum. Chem **56**, (1995).
[8] A. C. Maggs and N. W. Ashcroft, Phys. Rev. Lett. **359**, 113 (1987).
[9] W. Kohn and L. J. Sham, Phys. Rev. **140**, A1133 (1965).
[10] H. Margenau and N. R. Kestner, *Theory of Intermolecular Forces* (Pergamon Press, Oxford, 1969).
[11] E. Clementi and C. Roetti, At. Data Nucl. Data Tables. **14**, 177 (1974).
[12] C. F. Bunge, J. A. Barrientos, and A. V. Bunge, At. Data Nucl. Data Tables. **53**, 113 (1993).
[13] A. D. McLean and R. S. McLean, At. Data Nucl. Data Tables. **26**, 197 (1981).
[14] E. Zaremba and W. Kohn, Phys. Rev. B **13**, 2270 (1976).
[15] E. Hult, Y. Andersson, D. C. Langreth, and B. I. Lundqvist, to be submitted.
[16] B. N. J. Persson and E. Zaremba, Phys. Rev. B **30**, 5660 (1984).
[17] C. Holmberg, Ph.D. thesis, Chalmers University of Technology, 1986.

Lifetime of the Intermediate Excited State in Laser-Induced Desorption of NO from Pt

Y. Murata[1] and *K. Fukutani*[2]

[1]Institute for Solid State Physics, The University of Tokyo, 7-22-1 Roppongi, Minato-ku, Tokyo 106, Japan
[2]Institute of Industrial Science, The University of Tokyo, 7-22-1 Roppongi, Minato-ku, Tokyo 106, Japan

Abstract. The rotational energy distribution $N(J)$ of NO desorbed from a Pt(111) surface induced by valence electron excitation is analyzed by a simple impulse model. The $\ln N(J)$ versus $\sqrt{E_r}$ plot shows a linear relation, where E_r is the rotational energy. Both the lifetime and the critical residence time in the intermediate excited state can be estimated to be ~10 fs from the gradient and the onset of the linear relation, respectively, although some ambiguity remains in the estimation of angular frequency for frustrated rotation of adsorbed NO at the intermediate excited state.

1 Introduction

Ultraviolet and visible laser-induced dynamical processes of molecules on solid surfaces accompanied by valence-electron excitation, such as desorption, dissociation, and reaction, have recently attracted much attention in the field of surface physics and chemistry. Desorption of molecules via a nonthermal single-photon process is the simplest surface reaction. These studies of nonthermal processes from a quantum-mechanical point of view are finally expected to give us essential information on catalysis of metals. However, desorption of molecules chemisorbed on metal surfaces is scarcely induced by valence electron excitation as a single-photon process, because the electronically excited intermediates are rapidly deexcited to the ground state owing to strong interaction between the adsorbate and the substrate metal.

State-selective detection of desorbed molecules is very important for detailed analysis of dynamical behavior. Only NO and CO desorbed from Pt(001) and Pt(111) surfaces have been state-selectively observed as the strong-chemisorption systems [1-5]. NO on Ni(001) and Ni(111) is not desorbed but decomposed [6, 7], and NO on Pd(111) is neither desorbed nor decomposed [8]. The rotational energy distributions of NO desorbed from Pt(001) and Pt(111) have been precisely analyzed using the (1+1)-resonance-enhanced multiphoton ionization (REMPI) [2, 3], while CO desorption from Pt(001) and Pt(111) was observed using the (2+1)-REMPI method [4, 5]. Laser-induced fluorescence was also applied to observation of the rotational energy distribution of CO desorbed from NiO(111) [9]. Recently, the rotational energy distribution of desorbed CO from $Cr_2O_3(111)$ was observed by (1+1′)-REMPI method [10]. A similar observation is desired on CO from Pt surfaces.

The rotational energy distributions of NO from as-adsorbed Pt(111) and Pt(001) surfaces induced by 3.5~6.4 eV laser irradiation are represented by the Boltzmann distribution [2, 3]. This result is very curious, because the rotational quantum number is not defined in the initial state of the desorption process and the desorption occurs impulsively. Furthermore, the desorption is clearly induced by a nonthermal process, since the observed translational, rotational, and vibrational temperatures of desorbed molecules are much higher than the specimen temperature of 80 K [2, 3].

The desorption mechanism has been discussed using the translational, rotational, and vibrational energy distributions of desorbed molecules. Gadzuk *et al.* showed that the vibrational excitation of desorbed molecules is closely related to the lifetime of the negative ionic state [11]. Hasselbrink showed that the translational energy obtained from the

Springer Series in Solid-State Sciences, Volume 121
Elementary Processes in Excitations and Reactions on Solid Surfaces
Eds.: A. Okiji, H. Kasai and K. Makoshi

velocity distribution increases with increasing rotational energy of the desorbed molecules [12]. In both calculations, the assumed form of the potential energy surface (PES) in the excited state plays an important role. Mull *et al.* calculated the bimodal translational energy distribution of desorbed molecules by a classical scheme [13]. Zimmermann and Ho discussed rotational excitation in photodesorption using a simple impulse model [14]. In particular, rotational energy distributions for two spin-orbit states of NO desorbed from oxidized Ni(001) and Pt(111) are analyzed. Their treatment is quite different from the present one. The former considers only the ground state, while the latter is dominated by the intermediate excited state. Zimmermann and Ho also discussed velocity distributions [15] and rotational-translational correlations [16]. Although the latter discusses the same problem as that described in the present paper, the approach is also quite different.

In the present paper, we propose a novel simple model using the impulse scheme, in which the significant effects of the potential energy surface and of the transition process are confined within the momentum transferred from the substrate to an adsorbate atom. A tentative result has been already reported [17]. Although the intermediate excited state plays an important role in desorption induced by electronic transition, this property has not been elucidated. This simple model sheds light on the property of the intermediate excited state. As a result, the lifetime of the intermediate excited state is estimated from analysis of the rotational energy distribution, although some ambiguity remains in the estimated value of frustrated rotational frequency for adsorbed NO of the excited state. The lifetime of the intermediate excited state in laser-induced desorption and the critical residence time in the excited state leading to desorption are important to discuss the desorption mechanism [18].

2 Impulse model for the desorption process

The desorption process of NO from Pt(111) is tentatively described by the Menzel-Gomer-Redhead (MGR) model [19, 20], although the actual desorption process in the present system might be described by a substrate-mediated excitation model, in which the intermediate excited state is related to the unoccupied $2\pi_a$ state of adsorbate NO [17, 21]. Recently, Aizawa and Tsuneyuki calculated the potential energy surface of the excited state for the Pt-NO system [22]. Their result shows that the repulsive potential is effective to the desorption process. Hence, we discuss simply using the MGR model shown in Fig. 1. The adsorbate-substrate system is electronically excited by laser irradiation via the Franck-Condon process. Since the excited state is represented by the repulsive

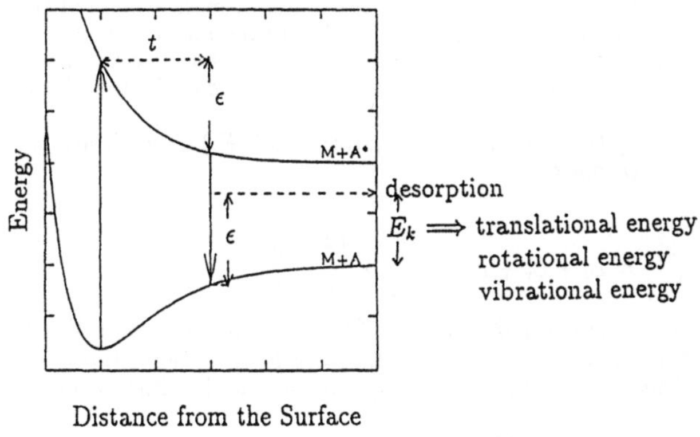

Fig. 1. Schematic energy diagram and transition illustrating the deexcitation and desorption processes. Electronic excitation due to laser irradiation occurs via the Franck-Condon transition. After a residence time t at the excited state, deexcitation occurs with an excess energy E_k surpassing the surface barrier for desorption. This energy is trasferred to the desorbed NO molecule and partitioned to translational, rotational, and vibrational energies of the NO molecule. The value of E_k depends strongly on t.

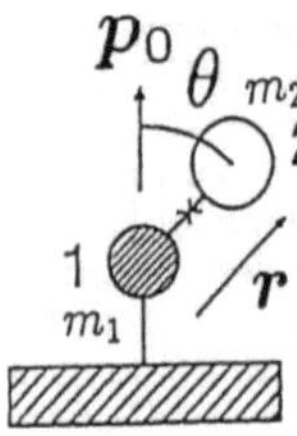

Fig. 2. Schematic representation of the impulse model. Bond breaking occurs between atom 1 and the substrate.

potential, the abscissa of the metal-adsorbate distance can be replaced by the reaction coordinate and then by the time that elapsed after excitation.

The electronic energy is gradually transferred to the nuclear motion energy, which is confined in the expansion of the metal-adsorbate distance for the case of Fig. 1. After a residence time in the intermediate excited state, vertical deexcitation occurs. At that time, the nuclear motion energy ϵ contributes to the desorption. When the residence time t is shorter than the critical time t_c, the energy ϵ is not enough to surpass the dissociation barrier resulting in recapture. On the other hand, the deexcitation into the ground state leads to desorption of NO for $t > t_c$. The energy E_k shown in Fig. 1 is the excess energy surpassing the dissociation barrier of the metal-adsorbate bond after the deexcitation to the ground state. This energy is partitioned by the translational, rotational, and vibrational energies of the desorbed molecule and these values are constantly held after desorption. Therefore, the internal energy distributions have much knowledge on the intermediate excited state in the desorption process induced by electronic transition. Rotational and translational energies appear after desorption, while the vibrational energy of the desorbed NO molecule includes the effect of the NO stretching vibration for the adsorbate NO as well as the excitation effect in the deexcitation process, as discussed by Gadzuk *et al.* [11].

It is assumed that the desorbing NO molecule sits at the on-top site. In the impulse model, the momentum $\boldsymbol{p}_0$ and the energy $E_k(= p_0^2/2m_1)$ are given to only the N atom bound to a Pt atom in the direction parallel to this bond at the time of metal-adsorbate bond breaking, where m_1 is mass of the N atom (Fig. 2). Recoil of the substrate atom can be ignored, because mass of the Pt atom is much heavier than that of the N atom. After desorption, the coordination system is changed to the center-of-mass system of the desorbed NO molecule. Hence, the momentum $\boldsymbol{p}_0$ is converted to the linear momentum for the center of mass, $\boldsymbol{P}=\boldsymbol{p}_0$, and the linear momentum for the internal coordinate, $\boldsymbol{p}=(m_2/M)\boldsymbol{p}_0$, is generated, where M and m_2 are mass of the NO molecule and the O atom, respectively.

The energy E_k is partitioned by the translational (E_t), rotational (E_r), and vibrational (E_v) energies of the desorbed NO molecule,

$$E_k = \frac{p_0^2}{2m_1} = E_t + E_r + E_v = \frac{P^2}{2M} + \frac{L^2}{2I} + \frac{p_r^2}{2\mu}, \tag{1}$$

where I and μ are moment of inertia and reduced mass of the NO molecule, respectively, the rotational angular momentum $\boldsymbol{L}=\boldsymbol{r}\times\boldsymbol{p}=pr\sin\theta$, and $p_r = p\cos\theta$. Here, the angle θ is the tilt angle of the NO molecular axis from the $\boldsymbol{p}_0$ direction at the time of metal-adsorbate bond breaking.

The momentum $\boldsymbol{p}_0$ depends on the PES of the intermadiate excited state, on the residence time at the excited state, and on the deexcitation due to diabatic of the excited state. That is, the PES of the excited state is not the adiabatic potential but accompanied with the electronic excitation near the Fermi level of Pt. Fortunately, p_0 can be eliminated from both relations of $E_t = P^2/2M = p_0^2/2M$ and $E_r = L^2/2I = (m_2/M)^2 p_0^2 \sin^2\theta/2\mu$. The relation between E_t and E_r is given by

$$E_r = (m_2/m_1)E_t \sin^2\theta. \quad (2)$$

The rotational energy is represented by both the tilt angle of the molecular axis from the surface normal and the translational energy. That is, the translational and rotational energies state-selectively observed are related by Eq. (2). The relation beteeen E_r and E_t is observed by two methods. One is the REMPI spectrum, in which the rotational energy distribution is observed at a mean translational energy $\langle E_t\rangle$ determined from the experimental condition. The other is the rotational quantum number dependence of the time-of-flight (TOF) spectrum. The translational energy distribution is observed as a function of the rotational quantum number [3].

If the tilt angle θ is proportional to the residence time t, and if the adsorbate NO stands upright on the surface in the ground state, Equation (2) is approximated by

$$E_r(t) \simeq (m_2/m_1)E_t\omega_e^2t^2, \quad (3)$$

where ω_e is the angular frequency for frustrated rotation of the adsorbate NO on the surface at the intermediate excited state. As the residence time increases, the rotational energy increases. The residence time is proportional to the square root of the rotational energy E_r. On the other hand, the lifetime τ of the intermediate excited state is defined by the relation of

$$N = N_0 \exp(-t/\tau). \quad (4)$$

From Eqs. (3) and (4), the $\ln N(J)$ versus $\sqrt{E_r}$ curve obtained from the REMPI apectrum is expected to be a linear relation of

$$\ln N(J) = -\frac{1}{\tau}\frac{\sqrt{m_1}}{\sqrt{m_2}\omega_e\sqrt{\langle E_t\rangle}}\sqrt{E_r} + \ln N_0 = A\sqrt{E_r} + C, \quad (5)$$

when the impulse model is correct and the tilt angle is proportional to the residence time. Here, $N(J)$ is the population in the rotational state J observed by the REMPI spectrum. The lifetime of the intermediate excited state can be obtained from the gradient A by the relation of

$$\tau = \frac{\sqrt{m_1}}{\sqrt{m_2}A\omega_e\sqrt{\langle E_t\rangle}}. \quad (6)$$

3 Results and Discussion

Figure 3 shows the rotational energy distributions in the Boltzmann plot for NO from Pt(111) induced by ArF (λ =193 nm) and XeF (λ =352 nm) excimer lasers [3]. According to Eq. (5), $\ln N(J)$ versus $\sqrt{E_r}$ curves are plotted. Figure 4 shows the rotational energy distributions of this plot for a spin-orbit state of Ω=1/2 shown in Fig. 3. The linear relation in the $\ln N(J)$ versus $\sqrt{E_r}$ curve is satisfied in the large $\sqrt{E_r}$ region, although the Boltzmann plot shows some deviation from the linear relation in the larger E_r region. This result shows that the adsorbate NO of the intermediate excited state rotates frustratingly with a constant angular frequency. On the other hand, the deviation from the linear relation for the $\ln N(J)$ versus $\sqrt{E_r}$ curve in the small $\sqrt{E_r}$ region appears and is considered to be caused by the effect of the critical residence time t_c. Since the desorption does not occur in the regidence time shorter than t_c, the desorption yield is reduced. The observed values of A obtained from the $\ln N(J)$ versus $\sqrt{E_r}$ curve in the large $\sqrt{E_r}$ region are listed in Table I. The mean values of E_t, $\langle E_t\rangle$, are determined from the experimental condition, as listed in Table I [3].

Next, we try to estimate the angular frequency ω_e for obtaining the lifetime. At first, we discuss the J dependence of the TOF spectra for this purpose. However, the angular frequency cannot be obtained from these observed data, because the potential energy surface for the frustrated rotation of NO at the excited state is concluded to be nearly flat, as described below. In analysis using Eq. (2) of the J dependence of the TOF spectra, the rotational energy is given by $E_r = BJ(J+1)$, where B is the rotational constant of the

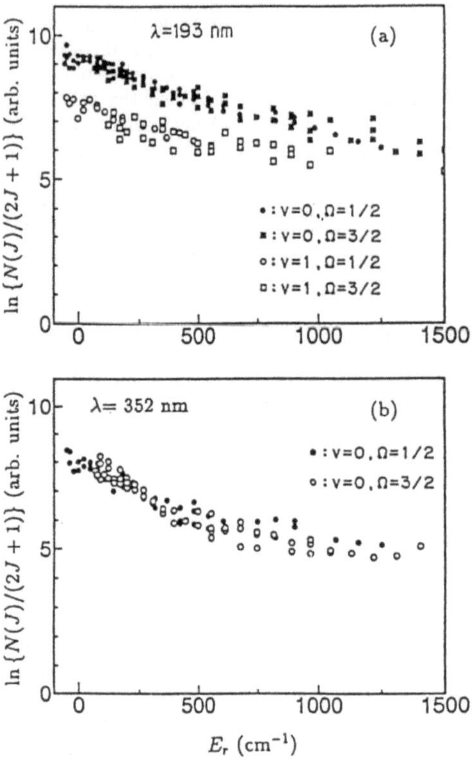

Fig. 3. Rotational energy distributions of NO from Pt(111) in the Boltzmann plot. (a) λ=193 nm. Solid circle and square are Ω=1/2 and 3/2 in the v=0 state, respectively. Open circle and square are Ω=1/2 and 3/2 in the v=1 state, respectively. (b) λ=352 nm. Solid and open circles are Ω=1/2 and 3/2 in the v=0 state, respectively.

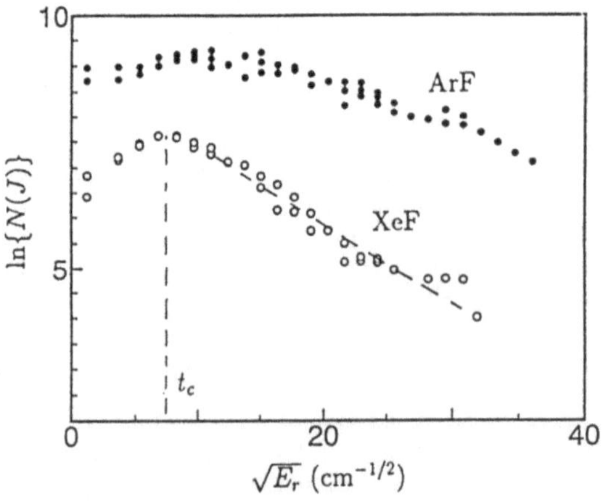

Fig. 4. Rotational energy distributions in the form of ln $N(J)$ versus $\sqrt{E_r}$ for NO (v=0, Ω=1/2) from Pt(111) with 193 nm (solid circle) and 352 nm (open circle) pump lasers.

Table I. Estimated lifetime (τ) and critical residence time (t_c) in the intermediate excited state for NO desorption from Pt(111) and Pt(001).

	Pt(111)		Pt(001)		
λ/nm	193	352	193	248	352
$A/K^{-1/2}$	0.069	0.138	0.222	0.221	0.181
$\langle E_t \rangle$/K	2700	1800	1000	1000	1000
τ/fs	15	9	7	7	9
t_c/fs	4	4	12	12	9
T_v/K	2900	1000	1200	1100	1200

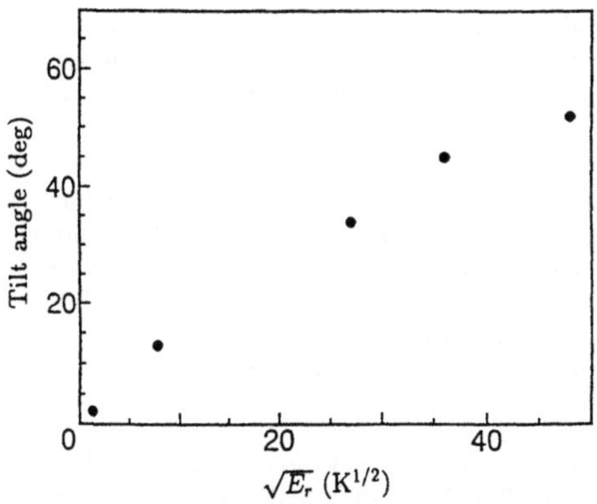

Fig. 5. Tilt angle at the time of desorption as a function of $\sqrt{E_r}$, which is proportional to the residence time in the excited state for NO from Pt(111) with 193 nm pump laser.

NO molecule. The mean translational energy is fitted by a modified Maxwellian distribution curve [23]. The tilt angle θ at the time of NO desorption is represented as a function of $\sqrt{E_r}$, which is assumed to be proportional to the residence time. The result is shown in Fig. 5 [24]. The turning point of the bending vibration of NO on Pt(111) is larger than 55°. On the other hand, the mean amplitude of molecular vibration is calculated from the vibrational frequency [25] and the value of the root-mean-square amplitude $\sqrt{\langle \Delta\theta^2 \rangle_T}$ at 80 K is 3.7° [24], which is estimated from a bending frequency of $\tilde{\nu} = \omega/2\pi c$=380 cm^{-1} in the ground state observed by high-resolution electron energy loss spectroscopy [26]. On the other hand, the vibrational amplitude of the bending vibration of CO adsorbed on Ni(110) was observed by the polar angle distribution of C 1s photoelectron intensity with X-ray photoelectron diffraction [27]. The root-mean-square amplitude at 120 K is 9° for [1$\bar{1}$0] azimuth and 12° for [001] azimuth. These values are larger than the above calculated value, but much smaller than the observed vibrational amplitude of the frustrated rotation at the electronically excited state; $\geq$ 55°. It is concluded that the potential energy surface of the bending vibration at the excited state is nearly flat.

We try to estimate ω_e by the other method, i.e. from the zero point vibration. The angular momentum of nuclear motion should be conserved in the electronic transition. Therefore, the angular frequency is considered to be given by the angular momentum derived from the zero point vibration of NO bending vibration in the ground state. Since only the kinetic-energy part of the vibrational energy corresponds to the angular momentum in this process, the averaged angular momentum can be derived from a relation of $\hbar\omega_e = \hbar\omega/4$ given by the virial theorem. Here, ω is the bending frequency in the ground state ($\tilde{\nu}$ = 380 cm^{-1} [26]). Thus, the apparent frequency corresponding to the angular frequency is given by $\tilde{\nu}_e$ = 95 cm^{-1}. The lifetimes calculated from Eq. (6) using $\tilde{\nu}_e$ = 95

cm^{-1} are also listed in Table I. The value of $\omega_e\tau$ can be roughly obtained from another experimental result and corresponds to the tilt angle obtained from the J dependence of the TOF spectra at the maximum $N(J)$ position; ~ 20°, which is in fairly good agreement with the value given from the gradient A, 15°.

We discuss the effects of the critical residence time. For molecules adsorbed in the configuration upright to the surface at t=0, the $\ln N(J)$ versus $\sqrt{E_r}$ curve is expected to be a triangle shape which rises at $\sqrt{E_r}$ corresponding to t_c and decreases linearly with a gradient of A, as $\sqrt{E_r}$ increases. However, this triangle shape is modified by the time-dependent PES of the excited state coupled with excitation of electrons near the Fermi level. The shape is rounded around the top of the triangle and has a tail toward the lower rotational energy side. On the other hand, the zero point vibration generates various bent configurations of the adsorbed molecules at $t = 0$ due to the Franck-Condon transition. That is, at the critical residence time, various tilt angles of $(\omega_e t_c \pm \theta_0)$ appear, where θ_0 is a tilt angle due to the zero-point vibration and populated by the Gaussian distribution with a width of mean vibrational amplitude. Although the rotational energy distribution for the $\ln N(J)$ versus $\sqrt{E_r}$ curve is given by the convolution of the modified triangle shape and the Gaussian form, the convolution curve gives the gradient A in the large $\sqrt{E_r}$ region.

The deviation from the linear relation in the $\ln N(J)$ versus $\sqrt{E_r}$ curve appearing in the small $\sqrt{E_r}$ region gives us information on the critical residence time t_c, which is more important than the lifetime τ from the quantum-mechanical standpoint. Although this estimation depends on the modified triangle shape, the critical residence time is roughly estimated from the $\sqrt{E_r}$ value at the onset of deviation from the linear relation, as indicated in Fig. 4. In this estimation, the value of 95 cm^{-1} was used for ω_e. The results are listed in Table I and similar values to the lifetime of the intermediate excited state are obtained. However, the critical residence time has larger error than the lifetime due to ambiguity in determination of the onset position.

In the above discussion, it is assumed that a desorbed NO molecule sits at the on-top site. This assumption is doubted in NO on Pt(111). Dynamical low-energy electron diffraction (LEED) analyses show that adsorbed NO molecules sit on the threefold hollow-site [28]. The present argument is acceptable, even if the NO molecule sits on the threefold hollow site. The momentum is considered to be also transferred to the surface normal, since the N-Pt bond is formed by the σ bond between electrons of the 5σ state of NO and the valence state of Pt atoms. Furthermeor, the LEED results show that adsorbed NO molecules tilt from the surface normal by 10°. However, the surface examined in the LEED analyses was annealed at 250 K [28] and is clearly different from the as-adsorbed surface used in the present study. Infrared-spectroscopic measurement shows that a wavenumber of the NO stretching vibration on the annealed surface is higher than that on the as-adsorbed surface by 5 cm^{-1} and corresponds with that on oxygen coadsorption surface [29]. Furthermore, REMPI spectra from both annealed and as-adsorbed surfaces are quite different each other [3]. The non-Boltzmann distribution observed on the annealed surface is interpreted by tilted NO molecules, being analyzed by the same impulse model as that described in the present manuscript [29]. This tilt structure is in good agreement with that obtained by LEED analysis.

It is possible to estimate the lifetime and the critical residence time for NO desorption from Pt(001). The results are listed in Table I. This case is not simple, since the flat region in the $\ln N(J)$ versus $\sqrt{E_r}$ curve is wide. This result may be caused by a bent equilibrium configuration of adsorbed NO species active to the laser-induced desorption. A wide flat region in the $\ln N(J)$ versus $\sqrt{E_r}$ curve due to the bent equilibrium configuration appears in NO desorption from the annealed Pt(111) surface [17, 28]. The adsorbed NO species active to the laser-induced desorption from Pt(001) are located in the step sites between 1×1 and hexagonal areas of a partially de-reconstructed Pt(001) surface [30]. Therefore, the direction of $\boldsymbol{p}_0$ for the desorbing species may be slightly tilted from the surface normal.

The vibrational excitation in the N-O stretching of the desorbed molecules is discussed to be higher, as the lifetime at the excited state is elongated [3, 11, 17]. The results obtained in this study are consistent with these discussions, since the vibrational temperature T_v derived from the population in the v=1 state has a nice correspondence with the lifetime, as seen in Table I.

References

[1] S.A. Buntin, L.J. Richter, R.R. Cavanagh, and D.S. King, J. Chem. Phys. **91**, 6429 (1989).

[2] K. Fukutani, A. Peremans, K. Mase, and Y. Murata, Phys. Rev. B **47**, 4007 (1993).

[3] K. Fukutani, Y. Murata, R. Schwarzwald, and T.J. Chuang, Surf. Sci. **311**, 247 (1994).

[4] A. Peremans, K. Fukutani, K. Mase, and Y. Murata, Phys. Rev. B **47**, 4135 (1993).

[5] K. Fukutani, M.-B. Song, and Y. Murata, J. Chem. Phys. **103**, 2221 (1995).

[6] F. Budde, A.V. Hamza, P.M. Ferm, G. Ertl, D. Weide, P. Andresen, and H.-J. Freund, Phys. Rev. Lett. **60**, 1518 (1988).

[7] T.T. Magkoev, M.-B. Song, K. Fukutani, and Y. Murata, Surf. Sci. **330**, L669 (1995).

[8] J.A. Prybyla, T.F. Heinz, J.A. Misewich, and M.M.T. Loy, Surf. Sci. Lett. **230**, L173 (1990).

[9] M. Asscher, F.M. Zimmermann, L.L. Springsteen, P.L. Houston, and W. Ho, J. Chem. Phys. **96**, 4808 (1992).

[10] K. Al-Shamery, I. Beauport, H.-J. Freund, and H. Zacharias, Chem. Phys. Lett. **222** 107 (1994).

[11] J.W. Gadzuk, L.J. Richter, S.A. Buntin, D.S. King, and R.R. Cavanagh, Surf. Sci. **235**, 317 (1990).

[12] E. Hasselbrink, Chem. Phys. Lett. **170**, 329 (1990).

[13] Th. Mull, B. Baumeister, M. Menges, H.-J. Freund, D. Weide, C. Fischer, and P. Andresen, J. Chem. Phys. **96**, 7108 (1992).

[14] F.M. Zimmermann and W. Ho, Phys. Rev. Lett. **72**, 1295 (1994).

[15] F.M. Zimmermann and W. Ho, J. Chem. Phys. **100**, 7700 (1994).

[16] F.M. Zimmermann and W. Ho, J. Chem. Phys. **101**, 5313 (1994).

[17] Y. Murata and K. Fukutani, J. Elect. Spect. Rel. Phenom. **64/65**, 533 (1993).

[18] Ph. Avouris and R.E. Walkup, Annu. Rev. Phys. Chem. **40**, 173 (1989).

[19] D. Menzel and R. Gomer, J. Chem. Phys. **41**, 3311 (1964).

[20] P.A. Redhead, Can. J. Phys. **42**, 886 (1964).

[21] K. Fukutani, M.-B. Song, and Y. Murata, Faraday Discuss. Chem. Soc. **96**, 105 (1993).

[22] H. Aizawa and S. Tsuneyuki, Surf. Sci. in press.

[23] Y. Murata and K. Fukutani, in *Laser Spectroscopy and Photochemistry on Metal Surfaces*, ed. by H.-L. Dai and W. Ho (World Scientific, Singapore, 1995) p.

[24] Y. Murata and K. Fukutani, J. Mol. Structure, **352/353**, 519 (1995).

[25] Y. Morino, J. Chem. Phys. **18**, 395 (1950).

[26] G. Pirug, H.P. Bonzel, H. Hopster, and H. Ibach, J. Chem. Phys. **71**, 593 (1979).

[27] D.A. Wesner, F.P. Coenen and H.P. Bonzel, Surf. Sci. **199**, L419 (1988).

[28] N. Materer, A. Barbieri, D. Gardin, U. Starke, J.D. Batteas, M.A. Van Hove, and G.A. Somorjai, Phys. Rev. B **48**, 2859 (1993).

[29] M.-B. Song, M. Suguri, K. Fukutani, F. Komori, and Y. Murata, Appl. Surf. Sci. **79/80**, 25 (1994).

[30] M.-B. Song, S. Mizuno, K. Fukutani, and Y. Murata, Chem. Phys. Lett. **196**, 559 (1992).

Electronic Excitations at Oxide Surfaces

H. Kuhlenbeck, M. Bäumer, M. Bender, D. Cappus, B. Dillmann, D. Ehrlich, F. Rohr, M. Schönnenbeck, O. Seiferth, and H.-J. Freund

Physical Chemistry 1, Ruhr-Universität Bochum,
D-44780 Bochum, Germany

The present paper summarizes the possibilities to use electron energy loss spectroscopy (EELS) to investigate the electronic properties of systems with highly localized electronic states, such as oxide surfaces. Surface excitations may be clearly distinguished from bulk excitations in EELS, and eventually a ligand field spectroscopy may be developed on this basis. This ligand field spectroscopy is used to study surface phase transitions at oxide surfaces. Adsorbates influence the surface properties of oxides dramatically. Examples are discussed.

1. Introduction

It is well known from the investigation of metal surfaces, that within the gaps of the projected bulk band structure surface states exist which may be detected via photoemission if they are occupied or via inverse photoemission if they are unoccupied [1,2]. It is characteristic of such surface states that their energy exhibits no wave vector dependence with respect to a direction perpendicular to the surface plane. Also, surface states may be quenched or energetically shift upon interaction with adsorbates. The reason for the appearance of surface states, of course, is the change of the electronic structure at the surface with respect to the bulk. The atoms in the surface feel a different potential because their coordination number is reduced and consequently the energy levels change. Very recently, Bertel and coworkers [3] have developed a very appealing picture of how these surface states on metals influence the chemistry, i.e. adsorption at metal surfaces including aspects of coadsorption in the sense of poisoning and promoting effects. For oxide surfaces, photoemission has, so far, not been sucessfully applied to detect intrinsic surface states [5]. Neither in valence photoelectron spectroscopy nor in core-electron spectroscopy has a clear detection of surface states on oxide surfaces been possible so far. This is somewhat surprising, but it is probably connected with the presence of strong many particle effects which are caused by localization phenomena in oxides, documented by intense satellite structure in the photoelectron spectra [4].

We shall discuss in the following how electron energy loss spectroscopy (EELS) can be used to clearly differentiate between surface and bulk excitations [6]. The examples are grouped in two parts. In the first part we shall discuss transition metal oxides with rock salt structure, namely NiO and related materials. In the second part we include corrundum type structures, such as Cr_2O_3, where the surface provides enhanced structural flexibility, in our discussion.

Springer Series in Solid-State Sciences, Volume 121
Elementary Processes in Excitations and Reactions on Solid Surfaces
Eds.: A. Okiji, H. Kasai and K. Makoshi

2. Experimental Aspects

The experiments are performed in ultrahigh vacuum chambers which are equipped with preparation sections. Several techniques are being used to prepare the oxide surfaces [6]. Oxide surfaces may be prepared by cleavage of bulk samples or via the creation of epitaxially grown thin oxide films on top of metallic substrates. While with the former preparation method non-polar surfaces with equivalent number of cations and anions in the surface are often well accessible, polar surfaces of highly ionic surfaces cannot be easily prepared via cleavage. The latter type of surfaces may be prepared via epitaxial growth. In addition, via the control of the match or mismatch of film and substrate lattice constants the defect density of the oxide film may be controlled. The geometrical structure of the films may be characterized by electron diffraction (LEED) and by scanning tunnelling microscopy (STM). We employ electron spectroscopic methods such as UPS, XPS, AES, and particularly EELS and HREELS for the present study. We also use photon based methods such as IRAS and laser induced desorption [7].

3. Results and Discussion

The idea is to excite electronic transitions of a transition metal ion which are optically forbidden for the free ion. These transitions are characterized by relatively small line widths, and they are energetically often located within the optical band gap of the material. Several groups have used EELS to identify such transitions for various materials [8-15]. A comprehensive collection of spectra accompanied by the appropriate interpretation can be found in the book by P.A. Cox [16]. Briefly, the situation relevant for the interpretation of the data is shown in Fig.1. A NiO(100) surface has been chosen as the example and shows the coordination of metal and oxygen ions in this case. We differentiate between the coordination in the second and deeper bulk layers of the rock salt structure and the one in the surface plane. Schematically the 3d orbital energies of the transition metal ion in the two surroundings are given in Fig.1. The depicted situation is representative for a single transition metal ion, but it is valid to a good approximation because the d-levels, especially of the late transition metals, are highly localized in space and the direct metal-metal coupling is negligible. Let us consider the bulk situation first: An octahedral field splits the d-orbitals in a threefold t_{2g}- and a twofold degenerate e_g set. Ni^{2+} carries eight electrons, six occupying the t_{2g}-set and two unpaired electrons the e_g-set [17]. If the spin couplings are properly taken into account, this leads to a $^3A_{2g}$ ground state. The two lowest lying excited states in this octahedral field are of $^3T_{2g}$ and $^3T_{1g}$ type and are situated at 1eV and 1.81eV respectively. If we reduce the ligand field by taking one of the ligands off, the symmetry as well as the repulsion is reduced [8]. This leads to splitting of the t_{2g} sub set into an e- and a b_2-component, and of the e_g sub set into a a_1- and a b_1-component. At the same time the splitting in the d-levels is reduced and the excited states shift to lower energy, and their number increases. The calculated excitation energies for both coordinations in NiO are plotted as an energy level diagram in Fig.2. This figure also contains the calculated excitation energies for Co ions in CoO which is isostructural to NiO. Co in CoO, however, is in a d^7 configuration. The excited

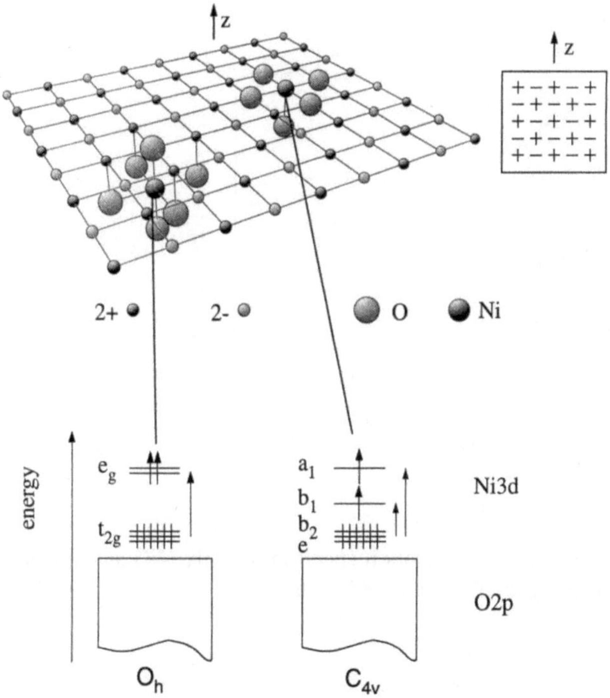

Fig. 1 Schematic representation of coordination at NiO(100) and in the bulk. Orbital energy diagrams are included.

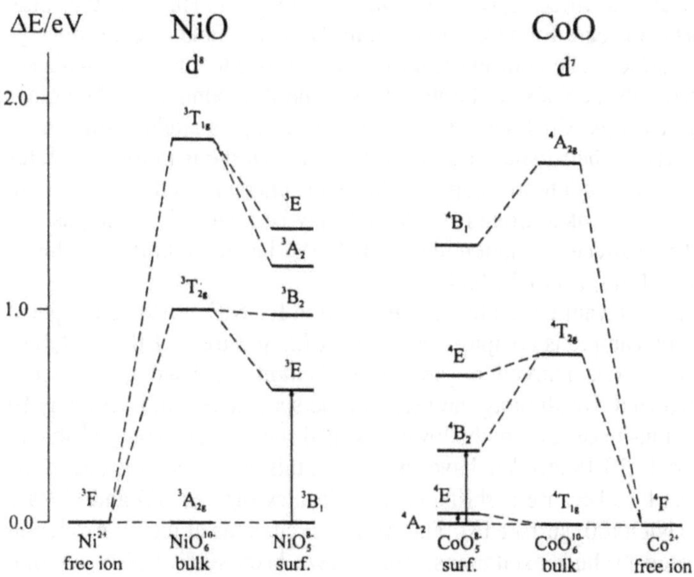

Fig. 2 Total energy diagrams of the lowest d-d excitations in NiO and CoO at the surface and in the bulk.

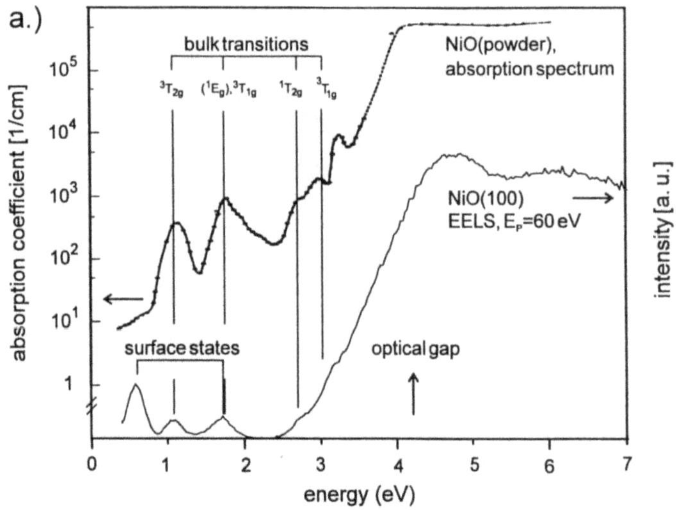

Fig. 3 Comparison of the optical absorption spectrum of bulk NiO as compared with the EELS spectrum of a NiO(100) surface.

states are, of course, comparable to NiO, but there is one very low lying state in the case of the surface coordination [18], because the ground state of a d^7 system in O_h symmetry is a spatially degenerate $^4T_{1g}$ state. The important point for the interpretation of the experimental data is to recognize that this spatial degeneracy of the ground state is lifted upon lowering the O_h symmetry of the bulk coordination. This is the case for a Co^{2+} ion within the ligand field of C_{4v} symmetry on the CoO(100) surface. The splitting resulting from this effect is very small, i.e. the E state lies only 55 meV above the non-degenerate A ground state. The other excited states occur on the CoO surface at excitation energies slightly smaller as compared to NiO(100). The reason for this similarity is that the manifolds of states in a d^7 and a d^8 configuration are identical, but their energetic order is inverted. The bulk excitation is situated around 0.9eV and may be assigned on the basis of literature data. The surface excitation lies at 0.45eV, i.e. it is shifted to lower excitation energy with respect to the bulk.

Fig.3 contains the relevant experimental information for NiO [8,19]. The optical spectrum of the bulk sample is compared with an EELS spectrum of the NiO(100) surface. The band gap is marked by an arrow. Within the band gap the d-d excitations are localized. By aligning the peaks in the spectra several peaks may be identified as bulk transitions, except the lowest excited state which is only visible in the EELS spectrum [9,13,19,20]. We have shown that this peak clearly is due to the excitation at the surface because it shifts when molecules such as NO and CO are adsorbed onto the NiO(100) surface [8]. The surface excitations at higher excitation energies are covered by bulk excitations. Those have been studied also via spin polarized ELS measurements [11]. Fig.4 collects HREELS data for CoO(100) [18] in comparison with NiO(100) within the range of the very intense Fuchs-Kliewer phonon excitations. The predicted feature due to excitation of the E state at 55 meV

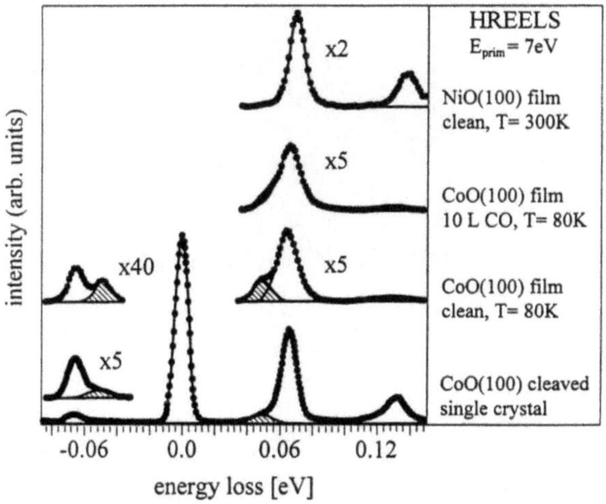

Fig. 4 EELS spectra of CoO in the range of the Fuchs-Kliewer phonon excitations in comparison with NiO.

is found in both the loss as well as the gain part of the HREELS spectrum. On the surface prepared by cleavage of a bulk sample the relative intensity of the peak close to the intense phonon peak is very small. The reason is the large sample volume, i.e. many oxide layers, contributing to the phonon intensity, while the surface state is only due to excitation in the first layer. If we, however, limit the thickness of the oxide by preparation of a thin, well structured oxide film we can decrease the relative intensity of the Fuchs-Kliewer phonons with respect to the surface excitation. This effect is seen in the second experimental trace shown in comparison with the cleaved bulk sample. Upon exposure of the surface to CO, for example, the surface state is quenched as indicated in the third trace from the bottom [18]. For comparative purposes we show in the topmost spectrum the corresponding range of Fuchs-Kliewer excitations in a NiO(100) thin film of comparable thickness. It is quite obvious that the low lying surface excitations are not present on a NiO(100) surface as predicted by the calculations [18].

Very similar studies have been undertaken for polar rock salt surfaces. NiO(111), CoO(111), and FeO(111) have been investigated [7,21-25]. These surfaces when prepared via oxidation of the corresponding metal single crystals are covered with a layer of hydroxyl groups, which stabilizes the polar surface by reduction of charge in the first layer. Upon removal of the hydroxyl layer the polar surface is destabilized and it restructures via the so called octopolar reconstruction. In this latter situation, microfacets of (100) orientation are formed on the surface. In these microfacets the vast majority of transition metal ions are in similar local environments as on the NiO(100) surface. EELS allows us to spectroscopically follow the changes in coordination upon reconstruction [21].

The above discussed structural rearrangement in NiO(111) involves several layers of the oxide. The mass transport involved leads to a relatively disordered surface indicated by diffuse LEED superstructure spots. There are other polar oxide surfaces, however, where the creation of the most stable surface does not involve

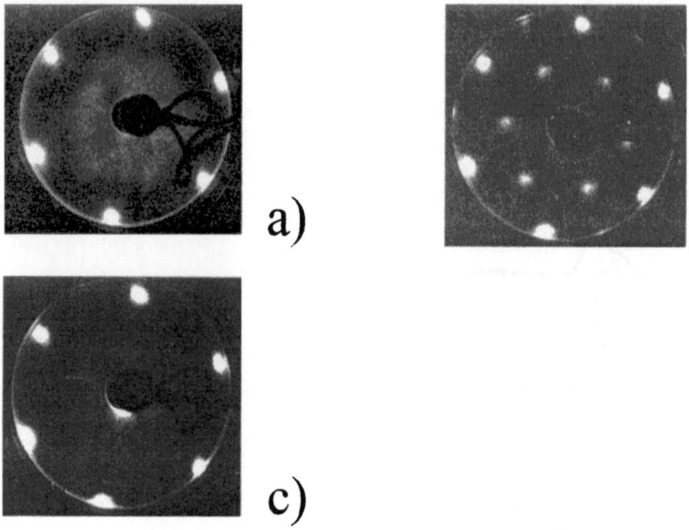

Fig. 5 LEED pattern of $Cr_2O_3(0001)$ as a function of surface temperature a) 100K, b) 150K, c) 300K

more than the top layer. The (0001) surface of Cr_2O_3 crystallizing according to the corrundum type has been recently studied in our laboratory in some detail [26-32]. The structural quality of a film prepared via oxidation of a Cr(110) surface is rather high as revealed by X-ray diffraction data [33]. LEED data [30,32] of a surface at low temperature (100K) exhibit the typical hexagonal pattern of $Cr_2O_3(0001)$ surface (see Fig.5). As the temperature of the surface is changed the LEED pattern of the surface changes. Fig.5a-c shows the sequence of LEED patterns observed. At 150K a $(\sqrt{3}x\sqrt{3})30^{o}$ superstructure is observed, which becomes diffuse as the temperature reaches room temperature [30,32]. We can follow this transition by investigating the intensity of the first order superstructure spot as a function of temperature. This is shown in Fig.6a. It appears that there are two phase transitions involved in the transformation, namely an order-to-order transition from low to intermediate temperature, and a order-to-disorder transition from intermediate to room temperature. In order to see this more clearly, we look at the I/V curves of the superstructure spot in Fig.6b. Two temperatures are compared. At 150K where the spot is very sharp and at 300K where its intensity is rather diffuse. Obviously, the I/V behaviour is very similar except that the amplitude of the maximum is strongly decreased. This is another strong indication that the surface sites involved in the low temperature and in the high temperature transition are identical. The comparison of the temperature dependence of the main structure spot intensity (I/V-curve in Fig. 6c) with the superstructure spot intensity in Fig.6a corroborates this conclusion. It is near at hand to speculate that the observed phase transitions are connected with the positions of the Cr ions in the surface layer. ELS spectra can now be used to identify the sites involved in the process [30]. Fig.7 shows EELS spectra taken from the clean, flashed oxide surface at two temperatures together with the results of the calculations. At 90K the

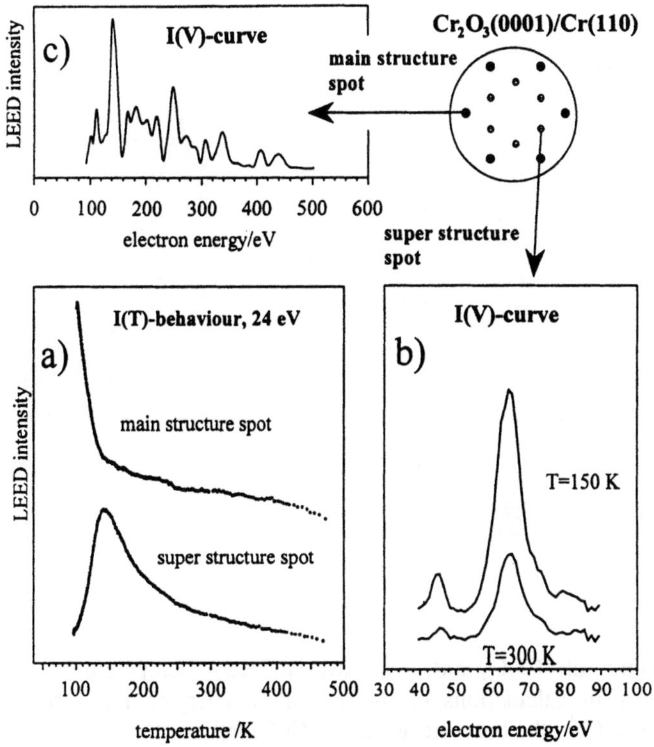

Fig. 6 LEED intensities for Cr_2O_3(0001)

a) as a function of temperature

b) I/V-curves of the superstructure spot for two surface temperatures

c) I/V-curve of the main structure spot for 100K.

spectrum exhibits two signals at loss energies of 1.2 eV and 1.4 eV, denoted by A and B respectively. They are accompanied by a broad feature C at about 1.8eV. As the temperature rises to 300K, signals A and B loose relative intensity whereas the relative intensity of feature C increases. In order to proove the surface nature of the features we studied the influence of adsorbates onto the spectral features (not shown here). The spectra in the considered energy range change dramatically indicating that all three features are due to surface excitations see below). The surface feature C also overlaps with some bulk excitations as revealed by comparison with information from bulk chromia samples [34] and ruby crystals [35]. The central idea for the interpretation of the data is that the various surface excitations are connected with Cr ions in different sites at the surface. The population of the sites changes as a function of temperature, and this gives rise for the temperature dependent changes in the EELS spectra, and concommitantly causes the observed phase transitions. Quantum chemical ab-initio cluster calculations by V. Staemmler and his group [30] may be used to corroborate these ideas. The clusters used consist of one Cr^{3+} ion surrounded by either six O^{2-} ions which form a slightly distorted octahedron (bulk Cr_2O_3) or by three O^{2-} ions which represent the next neighbours of

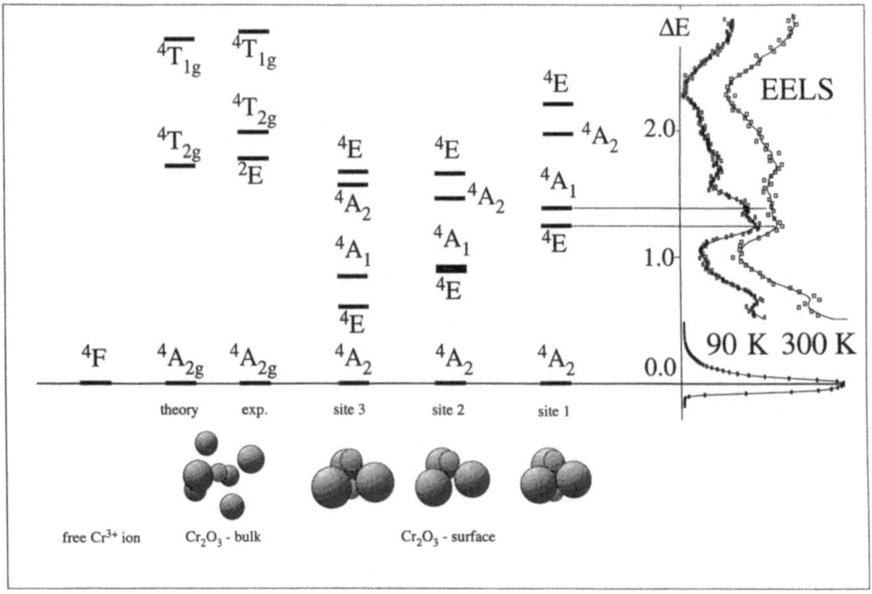

Fig. 7 EELS spectra of $Cr_2O_3(0001)$ at two different temperatures in comparison with calculations of d-d excitation energies for various surface sites of Cr^{3+}, the bulk site and a free Cr^{3+} ion.

a Cr^{3+} ion at the $Cr_2O_3(0001)$ surface. This cluster is embedded in an quasi infinite or quasi semiinfinite field of point charges +3e and -2e which simulate the Madelung field of the chromiumoxide ionic crystal, assumed to have the bulk geometric structure. Cr^{3+} is a d^3 system and its electronic state in the gas phase is a ^4F state. This ^4F state is split within a regular octahedral field into a $^4A_{2g}$ ground state, a first excited $^4T_{2g}$ and a second excited T_{1g} state. The threefold degenerate states are further split by 0.02eV by the slight distortion of the octrahedral symmetry in Cr_2O_3. The lowest excitation energies are calculated at the CASSCF level to be 1.73 eV ($^4T_{2g}$) and 2.73 eV ($^4T_{1g}$). At the oxide surface the Cr^{3+} ions are exposed to a weaker field due to the removal of half of the oxygen coordination sphere. The symmetry is reduced to below C_{3v} symmetry. Therefore one may expect on the basis of the experience with the rock salt oxides that the energy differences between the components of ^4F are smaller at the surface than in the bulk, and that quasidegeneracies of the T_{2g} and the T_{1g} levels are split further. The next step in our analysis is to consider the structure at the Cr_2O_3 surface. Fig. 8 shows schematic drawings of the surface in the stable chromium terminated configuration, i.e. with half the Cr ions removed. There are three possible sites for the metal ions to reside on the surface, denoted by 1,2, and 3. These three sites correspond to three different ligand fields at the surface, and consequently to three different surface excitation spectra. Fig. 7 compares the calculated excitation energies with the experimental spectra at lowest and at highest temperatures. The clusters used for the calculations are schematically represented. For energetic reasons site 2 is unfavourable. The

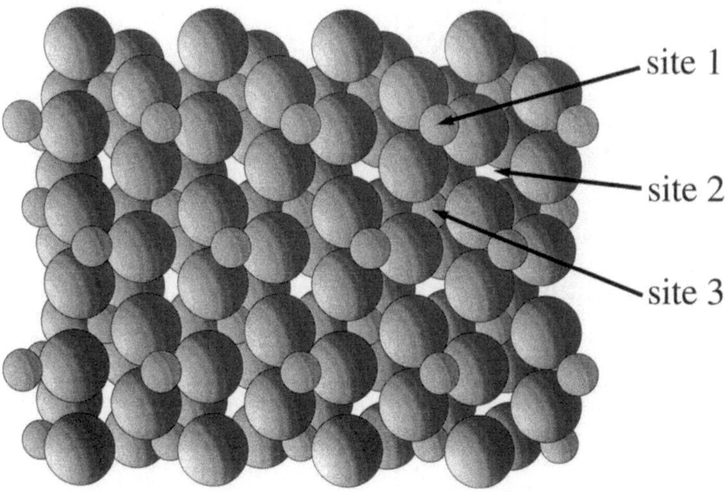

Fig. 8 Schematic representation of a $Cr_2O_3(0001)$ surface.

above qualitative arguments are confirmed by the calculations: For all three positions both the $^4T_{2g}$ and the $^4T_{1g}$ states are lowered considerably with respect to the $^4A_{2g}$ ground state and are slightly split ($^4T_{2g}$ into $^4A_1$ and ^4E, and $^4T_{1g}$ into $^4A_2$ and ^4E). Despite of these similarities, the d-d spectra of the Cr^{3+} ions at the three nonequivalent sites differ to some extent. Due to the weaker ligand fields, the Cr^{3+} ions at sites 2 and 3 have considerably different excitation energies from those at site 1. The interpretation of the EELS spectra on the basis of the CASSCF calculations is rather straightforward. The high energy part of the broad signal C between 1.7 eV and 2.2 eV contains the $^4T_{2g}$ state of bulk Cr_2O_3 and the ^4E surface component of the $^4T_{1g}$ state of Cr^{3+} in site 1. The low energy part of signal C which is to a large extent quenched upon adsorption is attributed mainly to the surface component of the $^4T_{1g}$ state of the same ions and probably also to the ^2E bulk state which is not included in the calculations. Finally, the signals A and B are probably the surface excitations correlating with the first excited $^4T_{2g}$ state of bulk Cr_2O_3. It is the split into the $^4A_1$ and ^4E component that gives rise to the observed doublet. As a function of temperature, the EELS intensity moves from the positions A and B towards C concommitant with the increasing population of site 3 with increasing temperature. This temperature dependent occupation of sites allow us to explain the observed phase transitions: At low temperature the Cr ions all occupy site 1 on the surface. At 150K the appearence of a $(\sqrt{3}x\sqrt{3})R30^\circ$ superstructure indicates the formation of a larger unit cell. This could be accompanied by the correlated motion of half of the Cr ions from site 1 into site 3. A driving force for such a rearrangement could be the magnetic coupling between the Cr ions in the first and the second transition metal ion layers as discussed elsewhere. At higher temperatures the two different Cr^{3+} sites are statistically occupied, giving rise to a unit mesh typical for a lattice gas. This again leads to a p(1x1) structure with very diffuse intensity in the range of the superstructure positions.

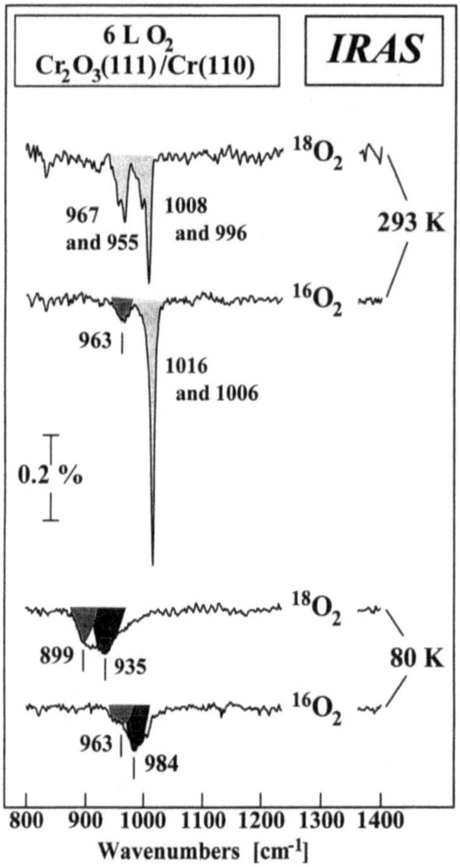

Fig. 9 Infrared-absorption -reflection spectra of $^{16}O_2$ and $^{18}O_2$ adsorbed at two surface temperatures.

Interestingly enough, the above described phase transitions may be strongly influenced by the presence of adsorbates. There are adsorbates that quench the phase transition, like for example molecular oxygen, but there are also adsorbates that seem to stabilize the phase transition at 150K. An example is CO [27]. We know that CO lies flat on the surface. Our assumption is that it interacts with two Cr^{3+} ions on the surface which, of course have to have the right interatomic separation to allow for optimal bonding. The idea is that only when Cr ions are in neighbouring sites 1 and 3 simultaneously, the CO molecule can be stabilized on the surface thus stabilizing the phase transition. In order to investigate how molecular oxygen on the other hand destabilizes the phase transition [28] we have performed IRAS experiments [30,36] employing a combined HREELS/IRAS apparatus described in detail elsewhere [37]. Fig.9 shows a series of FTIR spectra recorded after the Cr_2O_3 surface has been dosed with oxygen at 80K until saturation is reached, and heated in steps to 300K afterwards. Three features are observed.

We find absorptions between 963 cm^{-1} and 984 cm^{-1}, as well as at 1011cm^{-1}, and at 1024 cm^{-1}, due to adsorbed oxygen. As has been discussed before [28,36] the absorptions are due to the existence of distorted molecular oxygen at lower temperatures and atomically bound oxygen at higher temperatures. The atomically bound oxygen forms double bonds towards the Cr ions which are identified by comparison with IR spectra of chromyl species [38] that exhibit vibrational frequencies very close to those observed in the present data. There is even the splitting in the chromyl absorptions visible in the spectra. This is due to the in-phase and out-of phase combinations of two chromium-oxygen double bonds in a geminal configuration. Most importantly the resolution of the IRAS experiments is high enough to allow for isotopic labelling experiments. Some data are included in Fig.9. They are fully consistent with the existence of molecular oxygen at low temperature and atomic oxygen at high temperature. At high temperature, in the range of atomic oxygen on the surface we realize that although we have dosed the surface only with $^{18}O_2$, the formed chromyl bonds involve ^{18}O as well as ^{16}O, indicating that exchange with the lattice oxygen occurs. The temperature sequence also shows that this starts at rather low temperature. Returning at this point to the quenching of the phase transition by adsorption of molecular oxygen, it is quite clear how, via the formation of the chromyl species that the mobility of the chromium ions is blocked so as not to allow for the necessary site hopping to undergo the phase transition.

4. Conclusions

We have shown that the electronic structure at oxide surface may be studied via electron energy loss studies in the regime of low lying electronic excitations. A ligand field spectroscopy may be established that allows us to differentiate between sites of different coordination in the surface. Applying this spectrocopy it is possible to follow the rearrangements within surface phase transitions occuring at rather low temperatures on the Cr_2O_3 corrundum (0001) surface. Adsorbates may be employed to support the surface nature of the above discussed excitations by using the so called "crud test", but they may also be used to either stabilize or destabilize surface phase transitions. It is shown that IR absorption spectrocopy may be successfully applied to the study of thin film oxides, opening up the field of more catalytically oriented studies. For Cr_2O_3 surface states associated with charge-transfer excitations are identified.

5. Acknowledgements

Without the continous exchange of ideas and help from the theoretical chemistry group headed by Volker Staemmler the progress would have been much slower. We have profited from grants of the following agencies: Deutsche Forschungsgemeinschaft, Bundesministerium für Bildung und Forschung des Landes NRW, European Communities (BRITE-EURAM, COST), the German-Israeli Foundation, as well as the Fonds der Chemischen Industrie.

6. References

[1] S.D. Kevan (Ed.) "Angle Resolved Photoemission" Vol. 74, "Studies in surface Science and Catalysis", Elsevier 1992, Amsterdam, Chapters 4, 5

[2] E. Bertel, M. Donath (Eds.) „Electronic Surface and Interface States on Metallic Systems", World Scientific (1995)

[3] a) N. Memmel, E. Bertel, Phys. Rev. Lett. 75, 485 (1995)
b) E. Bertel, P. Sandl, K.D. Rendulic, M. Bente, submitted

[4] S. Hüfner „Photoelectron Spectroscopy", Series in Solid State Scienes Vol. 82, (1995) Springer Verlag Heidelberg

[5] V.E. Henrich, P.A. Cox "The Surface Science of Metal Oxides", Cambridge Univ. Press (1994)

[6] H.-J. Freund, H. Kuhlenbeck, V. Staemmler, Rep. Progr. Phys., in press

[7] H.-J.Freund, Ber. Bunsenges. Phys. Chem. 99, 1261 (1995)

[8] A. Freitag, V. Staemmler, D. Cappus, C.A. Ventrice, K. Al-Shamery, H. Kuhlenbeck, H.-J. Freund, Chem. Phys. Lett. 210, 10 (1993)

[9] P. A. Cox, A. A. Williams, Surf. Sci. 152-3, 791 (1985)

[10] J. P. Kemp, S. T. P. Davies, P. A. Cox, J. Phys. Condens. Matter 1, 5313 (1989)

[11] B. Fromme, M. Schmitt, E. Kisker, A. Gorschlüter, H. Merz, Phys. Rev. B50, 1874 (1994)

[12] M.-C. Wu, P. J. Møller, Surf. Sci. 235, 228 (1990)

[13] A. Gorschlüter, H. Merz, Phys. Rev. B 49, 17293 (1994)

[14] M. H. Mohamed, H. R. Sadeghi, V. E. Henrich, Phys. Rev. B37, 8417 (1988)

[15] W. Göpel, J. A. Anderson, D. Frankel, M. Jaehnig, K. Phillips, J. A. Schäfer, G. Rocker, Surf. Sci. 139, 333 (1984)

[16] P.A. Cox "Transition Metal Oxides. An Introduction to their electronic structure and properties", Clarendon Press, Oxford (1992).

[17] A.B.P. Lever „Inorganic Electronic Spectroscopy", Studies in Physical and Theoretical Chemistry", Vol. 33 Elsevier, Amsterdam (1984)

[18] M. Haßel, H. Kuhlenbeck, H.-J. Freund, S. Shi, A. Freitag, V. Staemmler, S. Lütkehoff, M. Neumann, Chem. Phys. Lett., 240, 205 (1995)
S. Shi, V. Staemmler, Phys. Rev. B., B 52, 12345 (1995).

[19] R. Newman, R. M. Chrenko, Phys. Rev. 114, 1507 (1959)
R. J. Powell, Stanford Electronics Laboratory Report No. 5220-1, 1967 (unpublished)

[20] V. Propach, D. Reinen, H. Drenkhahn, H. Müller-Buschbaum, Z. Naturforsch. 33B, 619 (1978)

[21] D. Cappus, M. Haßel, E. Neuhaus, F. Rohr, H.-J. Freund, Surf. Sci., 337, 268 (1995).

[22] H.-J. Freund, Phys. Stat. Sol., (b) 192, 407 (1995)

[23] C. A. Ventrice Jr, H. Hannemann, A. Brodde, H. Neddermeyer, Phys. Rev. B 49 5773 (1994)

[24] M. Schönnenbeck, D.Cappus, J. Klinkmann, H.-J. Freund, L.G.M. Pettersson, P.S. Bagus, Surf. Sci., in press

[25] D. Cappus, C. Xu, D. Ehrlich, B. Dillmann, C. A. Ventrice, K. Al-Shamery, H. Kuhlenbeck, H.-J. Freund, Chem. Phys. 177, 533 (1993)

[26] C. Xu, M. Haßel, H. Kuhlenbeck, H.-J. Freund, Surf. Sci. 258, 23 (1991)
[27] C. Xu, B. Dillmann, H. Kuhlenbeck, H.-J. Freund, Phys. Rev. Lett. 67, 3551 (1991)
[28] H. Kuhlenbeck, C. Xu, B. Dillmann, M. Haßel, B. Adam, D. Ehrlich, S. Wohlrab, H.-J. Freund, U.A. Ditzinger, H. Neddermeyer, M. Neuber, M. Neumann, Ber. Bunsenges. Phys. Chem. 96, 15 (1992)
[29] C.A. Ventrice, D. Ehrlich, E.L. Garfunkel, B. Dillmann, D. Heskett, H.-J. Freund, Phys. Rev. B 46, 12892 (1992)
[30] M. Bender, D. Ehrlich, I.N. Yakovkin, F. Rohr, M. Bäumer, H. Kuhlenbeck, H.-J. Freund, V. Staemmler, J. Phys.: Condens. Matter 7, 5289 (1995)
[31] M. Haßel, I. Hemmerich, H. Kuhlenbeck, H.-J. Freund, Surface Sci. Spectra, submitted
[32] F. Rohr, M. Bäumer, H.-J. Freund, to be published
[33] A. Stierle, P. Bödeker, H. Zabel, Surf. Sci. 327, 9 (1995)
[34] T.I.Y. Allos, R.P. Briss, M.R. Parker, E. Ellis, D.W. Johnson, Sol. State Comm. 24, 129 (1977)
[35] A.L. Schawlow, D.L. Wood, A.M. Clogston, Phys. Rev. Lett. 3, 271 (1959)
[36] H.-J. Freund, B. Dillmann, O. Seiferth, G. Klivenyi, M. Bender, D. Ehrlich, I. Hemmerich, D. Cappus, Catalysis Today, in press
[37] B. Dillmann, Thesis, Ruhr-Universität Bochum, 1995
[38] F.A. Miller, G.L. Carlton, W.B. White, Spectrochimica Acta 15, 709 (1959)
W.E. Habbs, J. Chem. Phys. 28, 1220 (1958)

Site Occupation of Adsorbed CO on Ni(100)

M. Kawai and J. Yoshinobu

The Institute of Physical and Chemical Research, 2-1 Hirosawa, Wako-shi, Saitama 351-01, Japan

Abstract. Adsorption of CO on Ni(100) surface is presented. At the thermodynamic equilibrium, the site occupation between the terminal and the bridged sites are described with the free energy of the system, including the vibrational entropy. Adsorption of CO onto a cold surface as 20 K has also been studied by infrared reflection absorption spectroscopy (IRAS). The occupation ratio of *bridged CO to terminal CO species* on Ni(100) at 20 K ranges from ~2.8 to 0.7 at the total coverage from 0.003 to 0.15 ML. Such strong coverage dependence of the occupation ratio even at small coverages suggests that the interaction between CO molecules operates at relatively long range (>10 Å). The isotope experiments suggest that there is substantial interaction between preadsorbed (accommodated) CO species and incoming (mobile) CO species.

1. Introduction

Carbon monoxide (CO) on transition metal surfaces is one of the prototype systems in the studies of thermodynamic properties, adsorbed states, adsorption/desorption kinetics, and dynamics [1]. A microscopic picture of adsorption process of CO on a metal surface may be summarized as follows. When gaseous CO approaches the surface, the molecule interacts with adsorption potential. The potential depth (adsorption energy, E_{ad}) of chemisorbed CO on transition metal surfaces is typically about 0.5-2 eV [2]. When the CO molecule collides with the surface, although a small part of molecules are going apart from the surface, the kinetic energy (adsorption energy + kinetic energy of gaseous CO) is transferred into various kinds of internal excitations (vibrational, rotational, electronic, phonon, etc.). Usually the molecule does not stick on the surface by the first collision, and it travels on the surface ("mobile precursor states"), dissipating its kinetic energy further [3,4]. Finally it is trapped at one of the potential minima (i.e., a certain adsorption site). After an adsorbed molecule is thermalized at the surface temperature, thermal motions of the adsorbed molecule can be excited by the heat of substrate. The statistical thermodynamic equilibrium of total adsorption system may be reached through the surface migration (hopping) of adsorbed molecules.

Springer Series in Solid-State Sciences, Volume 121
Elementary Processes in Excitations and Reactions on Solid Surfaces
Eds.: A. Okiji, H. Kasai and K. Makoshi

There are several possible adsorption sites for CO on transition metal surfaces, i.e., a terminal site, a bridged site, and hollow sites. Under thermal equilibrium, the occupation probability of each site is controlled by the free energy of adsorption (adsorption energy and entropy of the system at a certain temperature) [5-9]. If the surface temperature is too low to activate the surface migration, non-equilibrium adsorbed states, where molecules are initially trapped at a certain local potential minimum, may be observed. Namely, on a low temperature surface, the adsorption site occupancy is determined by the elementary adsorption dynamics and not by thermodynamic equilibrium[10]. On the other hand, when we observe the adsorption of CO at higher temperature (e.g., room temperature) thermodynamic equilibrium between the adsorption sites is quickly achieved. Therefore, the knowledge about the initial adsorption sites at low temperature would give some new information about the probability of adsorption at a certain site.

2. Experimental

The experiments were carried out in an ultrahigh vacuum chamber, which was equipped with a three-grid retarding-field analyzer for low energy electron diffraction and Auger electron spectroscopy, a quadrupole mass spectrometer for thermal desorption spectroscopy (TDS), and gas delivering systems. The base pressure was $<4x10^{-11}$ Torr. The details of the apparatus has been published elsewhere [11]. The clean surfaces of Ni(100) were prepared by Ar-ion bombardment, annealing, oxidation and flashing cycles. The sample crystal was cooled down to 20 K [Ni(100)] by a cryogenic refrigerator and was heated up to 1400 K by electron bombardment from the rear. Gaseous CO was introduced onto the sample through a pulsed gas doser or a continuous gas doser, where the gas temperature was room temperature. The coverage is determined from the integrated area of TDS peaks, assuming that the coverage of c(2x2)-CO on Ni(100) is 0.5 ML [1 ML = $1.6x10^{15}$ molecules cm^{-2} on Ni(100)]. The IRAS spectra were taken with a Fourier transform infrared spectrometer (Mattson RS-1) using a mercury-cadmium-telluride detector. The clean surface is served as a background reference for an absorption spectrum.

3. Site occupation at thermodynamic equilibrium

Terminal CO and bridged CO species are observed on Pt(111) and Ni(100) surfaces, and some parameters to describe adsorption potential energy surface (PES) have been obtained experimentally [5-9, 12-14]. On Pt(111) at ~100 to 300 K, only terminal species are observed at the coverage θ below ~0.15 monolayer (ML) and both species are observed at $\theta > 0.15$ ML [5]. In the case of CO on Ni(100), both species are

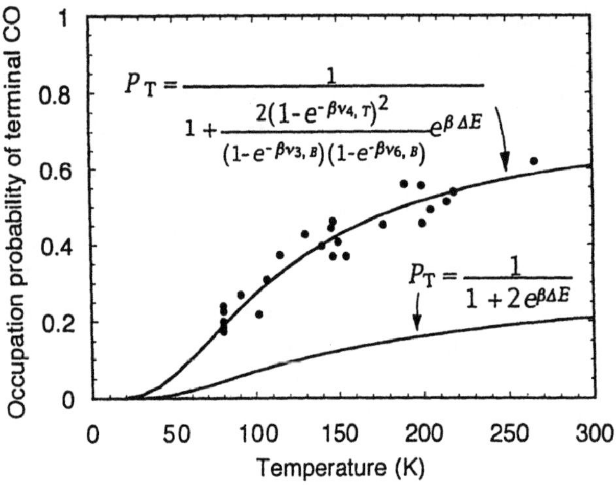

Fig. 1
Calculated occupation probability of the terminal CO on Ni(100) as a function of temperature with ΔE = 11 meV. Filled circles are the occupation probability from the experimental data for θ_{total} = 0.04 ML. Here, ΔE is the binding-energy difference between the terminal and bridged sites and $\beta = 1/k_BT$. For the details see ref [6].

observed from low coverage regions, where the occupation ratio strongly depends on the surface temperature [6-9].

At thermodynamic equilibrium, the occupation probability of each site is given by minimizing the free energy of the system. When the binding-energy difference between the possible sites is large, the energetically stable site is predominantly occupied. Namely, the binding energy E controls the free energy $F = E - TS$ in the system. However, when the difference is small the entropy factor plays an important role. Figure 1 represents the occupation probability of CO at terminal site on Ni(100) at different temperatures. Here the total coverage of CO was kept at θ = 0.04 ML [7]. The experimental data (shown in solid circle) are reasonably well fitted by considering the vibrational entropy of the adsorbed system. Especially, the excitation of low-energy hindered translational modes $\nu_{4,T}$ (4.0 meV) of the terminal CO, $\nu_{6,B}$ (4.2 meV) and $\nu_{3,B}$ (27.6 meV) of the bridged CO became dominant in this temperature range [6]. From the curve fit in Fig.1, the binding-energy difference between the terminal and bridged sites is determined to be 11 meV, where the bridged site is more stable in its ground state. Because of the C_{4v} symmetry of the adsorption system at the terminal site, the lowest hindered translational mode $\nu_{4,T}$ is degenerated. Thus the terminal site has higher density of states (DOS) than the bridged site.

4. Initial site occupation at low temperature

When the temperature is too low to activate the surface migration of the adsorbed molecules, the site occupation of the adsorbed CO may reflect the elementary adsorption dynamics rather than the thermodynamic equilibrium.

Gaseous CO was pulse-dosed on Ni(100) at 20 K and the IRAS spectra were observed. Below 20 K, no observable change in the spectra was detected within ~600 s after the adsorption at t = 0 s, indicating that the surface migration of adsorbed CO on Ni(100) at 20 K is negligibly small. Two bands are observed at ~2019-2029 cm^{-1} and at ~1863-1875 cm^{-1}, which are assigned to CO stretching modes of terminal CO and bridged CO species, respectively. Since the ratio of the dynamic dipole moment of terminal CO to that of bridged CO is ~1.1 [6, 8], the fractional coverage of each site is estimated from the integrated absorbance of each CO stretching band. Figure 2 shows the ratio between the fractional coverage of bridged CO and that of terminal CO on Ni(100) at 20 K as a function of total coverage. At a very low coverage (θ_{total} = 0.003 ML) the ratio is located at ~2.8. With increasing θ_{total} the ratio rapidly decreases to 1.0 at θ_{total} = 0.05 ML, and it gradually decreases to about 0.8 above θ_{total} = 0.1 ML. Note that the observed occupation ratio is completely different from the equilibrated state, which is shown in Fig. 1.

There are several factors which affect the elementary adsorption dynamics, such as (1) the potential energy surface (PES) of adsorption system, (2) the initial conditions

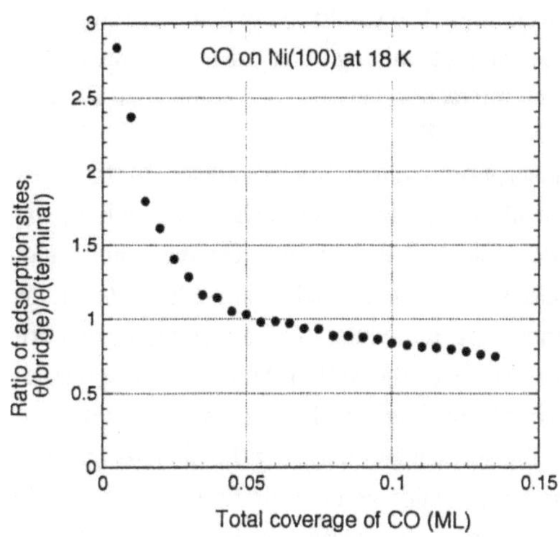

Fig. 2
Occupation ratio of adsorption sites, *bridged CO to terminal CO* on Ni(100) at 20 K, as a function of total coverage.

of incident gas molecules (kinetic energy, internal energy, incident angle, etc.), (3) the energy dissipation processes, (4) the surface temperature. In the case of gaseous CO molecules (~300 K) approaching the Ni(100) surface at low temperature, the potential energy surface and the energy dissipation processes are important, because the adsorption energy (>1 eV) is much larger than the thermal kinetic energies and also the substrate heat, which may induce the migration of adsorbed CO, is negligibly small.

One may expect that a mobile molecule just before the accommodation at a certain site prefers the site with higher density-of-states (DOS) including hindered motions and phonons, because energy could be dissipated more effectively, resulting in being accommodated in the potential well. Although the multi-dimensional PES should be considered essentially [15], it seems that the terminal sites are occupied more than the bridge sites, judging from the shape of PES. However, in the case of Ni(100), the probability of terminal CO is less than that of bridged CO below θ_{total} = ~0.07 ML [Fig. 2]. These results suggest that the above simple low-dimensional DOS picture does not explain the phenomena. In the case of CO on Pt(111), the terminal sites are more favorably occupied at low coverages [10, 14, 16], on the contrary, where the binding-energy for the terminal site is larger than the bridge site. As far as the above two cases tell us, at the zero-coverage limit, CO molecule seems to predominantly occupy the site with larger binding-energy.

In order to study the mixing of preadsorbed (accommodated) CO (0.015 ML) and post-dosed CO (0.015 ML) on Ni(100), the isotope experiments were performed at a low coverage (θ_{total} = 0.03 ML) at 20 K (Fig. 3). First, 0.015 ML $^{12}C^{16}O$ (or $^{12}C^{18}O$) was adsorbed on Ni(100). Then, this preadsorbed surface was exposed to 0.015 ML $^{12}C^{18}O$ (or $^{12}C^{16}O$, respectively). The results show that these two final spectra resemble each other and almost complete mixing about adsorption sites takes place, indicating that there is substantial interaction between accommodated CO and incoming CO molecules on Ni(100) even at 20 K.

This is an indication of "indirect adsorption" even at low coverages. In the direct adsorption mechanism, the elementary adsorption dynamics is controlled only by the interaction between the incoming molecule and the surface. On the other hand, "indirect mechanisms" involve the interactions between adsorbate-adsorbate and/or between adsorbed molecules and incoming molecules. Once the molecules are accommodated on a low temperature surface, the self migration is suppressed. However, it is possible that transient "mobile" molecules interact with the molecules already accommodated on the surface and inducing the site-change. The interaction probability depends on the cross-section and the life time of such mobile state, i.e., the traveling length. As a result of the interaction (or collision on the surface), lateral translational energy of transient mobile CO could be dissipated effectively, and thus it is finally accommodated in close proximity to the adsorbed CO on the surface. In this situation, the site-conversion of the preadsorbed CO may

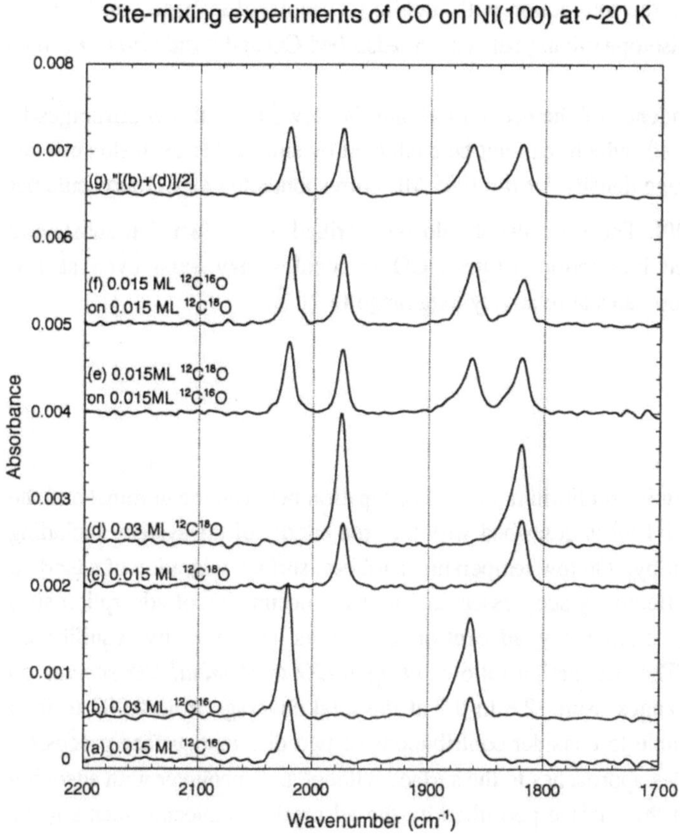

Fig. 3

IRAS spectra of adsorbed CO on Ni(100) at ca. 20K. Absorption bands at ~2020 cm^{-1} and ~1860 cm^{-1} are due to the $^{12}C^{16}O$ stretching vibration of terminal CO and bridged CO, respectively (a and b). These two bands are observed at ~1980 cm^{-1} and ~1820 cm^{-1} for $^{12}C^{18}O$ (c and d). It is clear that the intensity ratio between the terminal and bridge CO is different when the coverages increases from 0.015 ML to 0.03 ML (a, b and c, d). Spectrum (e) represent 0.015 ML $^{12}C^{18}O$ adsorbed on 0.015 ML $^{12}C^{16}O$ preadsorbed surface and (f) represent the other way round. Artificially convoluted spectrum of (b) and (d) is shown in (g). See the similarity in spectra (e), (f) and (g).

also take place by the interaction with mobile CO molecules. Based on "indirect mechanisms", the isotope mixing between preadsorbed CO and postdosed CO can be explained.

Coverage dependence of the occupation ratio is very strong at low coverages (< 0.05 ML) on Ni(100), which can not be explained by simple site-exclusion effects. Note that the average density for $\theta = 0.05$ ML corresponds to one CO molecule per ~10 $Å^2$ on Ni(100). These results should be ascribed to the fact that substantial substrate mediated interaction between CO molecules may exist even at low coverages, which operates at relatively long range (> 10 Å).

5 Summary

At the thermodynamic equilibrium, the site occupation between the terminal and the bridged sites on Ni(100) is described with the free energy of the system, including the vibrational entropy. On low temperature surfaces, surface migration of adsorbed CO molecules is effectively suppressed, and thus the occupation of adsorption sites is controlled by elementary adsorption dynamics and not by equilibrium thermodynamics. The occupation ratio of *bridged CO to terminal CO species* on Ni(100) at 20 K ranges from ~2.8 to 0.7 at the total coverage from 0.003 to 0.15 ML. It seems suitable to consider contributions of two different surface processes. When CO molecules approaches to the surface without any encounter with adsorbed molecule nor with the surface perturbed by the adsorption, molecule seems to be trapped preferentially to the site with larger binding-energy. On the other hand, when molecule happens to encounter the preadsorbed molecule or sites perturbed by the preadsorbed molecule, interaction between the adsorbed molecules may induce the stabilized preadsorbed molecules to hop to a different site.

References

[1] J. T. Yates, Jr., Surf. Sci., **299/300** (1994)731.

[2] G. A. Somorjai, "Introduction of Surface Chemistry and Catalysis", (John Wiley& Sons, New York, 1994).

[3] G. D. Billing, Chem. Phys., 86 (1984)349.

[4] D. J. Doren and J. C. Tully, J. Chem. Phys. **94** (1991)8428.

[5] E. Schweizer, B. N. J. Persson, M. Tüshaus, D. Hoge and A. M. Bradshaw, Surf. Sci., **213** (1989) 49.

[6] J. Yoshinobu, N. Takagi and M. Kawai, Chem. Phys. Lett., **211** (1993)48.

[7] J. Yoshinobu, N. Takagi and M. Kawai, Phys. Rev., **B49** (1994) 16670.

[8] A. Grossman, W. Erley and H. Ibach, Phys. Rev. Lett., **71** (1993) 2078.

[9] A. Grossman, W. Erley and H. Ibach, Appl. Phys. A **57** (1993) 499.

[10] J. Yoshinobu and M. Kawai, Surf. Sci. (1996) in press.

[11] J. Yoshinobu and M. Kawai, J. Electron Specrosc. Related Phenomena, **64/65** (1993) 207.

[12] G. Hähner, J. P. Toennies and C. Wöll, Appl. Phys. A **51** (1990) 208.

[13] J. E. Reutt-Robby, D. J. Doren, Y. J. Chabal and S. B. Christman, J. Chem. Phys., **93** (1990) 9113.

[14] J. V. Nekrylova and I. Harrison, J. Chem. Phys., **101** (1994) 1730.

[15] J. C. Tully, Surf. Sci., **299/300** (1994) 667.

[16] J. V. Nekrylova, C. French, A. N. Artsyukhovich, V. A. Ukraintsev, and I. Harrison, Surf. Sci. Lett., **295** (1993) L987.

Part II

Beam-Surface Scattering and Charge and Energy Transfer

The Interaction of O_2 with Ag(111) Probed by Beam Experiments

A.W. Kleyn

FOM Institute for Atomic and Molecular Physics, Kruislaan 407, 1098 SJ Amsterdam, The Netherlands, fax X-31-20-6684106, e-mail kleyn@amolf.nl

Ion beam and molecular beam studies involving the interaction of (hyper-)thermal O_2 with Ag(111) are reported. Several different scattering paths are observed: direct ionisation, physisorption followed by desorption, direct-inelastic scattering from both the physisorption and the chemisorption potential, transient trapping-desorption, molecular chemisorption and dissociative chemisorption. The underlying dynamics of all of these processes is reviewed. The results are discussed in terms of simple potential energy surfaces.

Introduction

The interaction between O_2 molecules and silver surfaces has been studied for a long period in the history of surface science. These studies are important for a number of reasons. Firstly, the system is an example of precursor mediated dissociative chemisorption, in which more than one precursor can be involved. The O_2-Ag systems are examples of so-called triple well systems. As such, the system can serve as a prototype of oxidation reactions [1]. In addition, the epoxidation reaction between ethene and oxygen is carried out at silver catalysts [2]. Apparently, a unique state of weakly bound oxygen exists exclusively on silver surfaces, which is very hard to characterise.
Four different states of oxygen on Ag(111) have been identified, as recently summarised [1]. Most weakly bound is physisorbed O_2 which desorbs at a surface temperature $T_s \leq 50K$. The corresponding trapping coefficient for thermal O_2 is of the order of unity [3, 4]. Molecular chemisorbed O_2 desorbs at $T_s \approx 200K$ [5, 6]. Its vibrational frequency is very much lower than that of gas phase O_2 and O_2^-, which has been interpreted as due to significant electron transfer to the molecule [7, 8]. Here we will refer to this state by O_2^-. The sticking coefficient for molecular chemisorption is on the order of 10^{-6} [5]. Dissociative chemisorption can take place at the surface leading to recombinative desorption at $T_s \approx 600K$ and in subsurface sites, desorbing at $T_s \geq 700K$ [1, 9]. The dissociative sticking coefficient is on the order of 10^{-7} [5, 10].
A one dimensional representation of the O_2-Ag(111) interaction derived by Campbell [5] is shown in figure 1. The physisorption (O_2), molecular chemisorption (O_2^-), and dissociative (O + O) states are clearly indicated. The diagram indicates that molecular chemisorption is an activated process. There seems to be no large energetic barrier between molecular and atomic chemisorption. The precise nature of the reaction coordinate in figure 1 is not known.

Fast beam experiments

Over a period of more than ten years we have tried to unravel some of the characteristics of the potential energy surface of the oxygen-silver system using beam studies. At first we were amazed by the fact that the molecularly chemisorbed species was referred to as a negative ion ($O_2^{\partial-}$), where $0<\partial<2$. From gas phase harpooning reactions of the type $K + Br_2 \rightarrow KBr + Br$ it was known that at sufficiently high energies the process

Springer Series in Solid-State Sciences, Volume 121
Elementary Processes in Excitations and Reactions on Solid Surfaces
Eds.: A. Okiji, H. Kasai and K. Makoshi

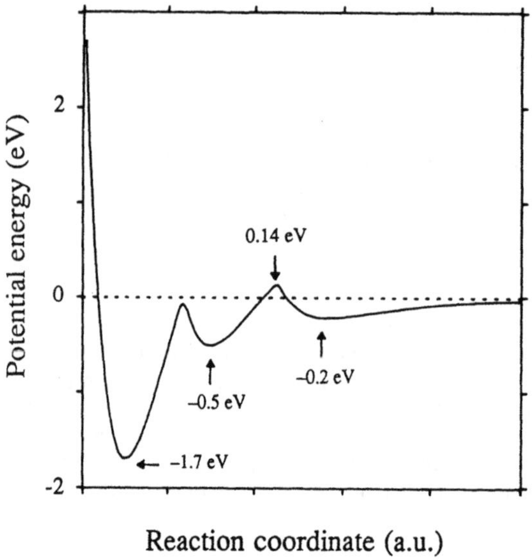

Figure 1. Schematic potential energy diagram for the interaction between O_2 and Ag(111). The three states (see text) into which O_2 can adsorb at the surfaces are depicted as a function of a reaction coordinate. Adapted by Spruit et al. [4] from Campbell [5].

$K + Br_2 \rightarrow K^+ + Br_2^-$ was also possible and that in fact the electron transfer occurring in this process is the first step of the chemical reaction [11-13]. By analogy we decided to search for production of O_2^- in collisions of fast O_2 and Ag(111). Because the negative ion formation process is endothermic (as in the gas phase) O_2 and O_2^+ beams with energies in the 10-100 eV range were used. Facile production of O_2^- was observed [14-18]. Because the interaction of a molecule with the surface is a more complex process than the interaction of an alkali atom with an electronegative molecule, the striking similarity between gas phase experiments and surface scattering experiments does not have to implicate that the ion production mechanisms are similar in both cases. Nevertheless, the similarity does suggest that the first step in the molecular chemisorption of O_2 is the transfer of an electron to this molecule.

Direct-inelastic scattering

The fast ion beam experiments are insensitive to any subtle barriers to electron transfer and molecular chemisorption because of the large excess translational energy. Therefore, we attempted to study adsorption by a scattering experiment of hyperthermal O_2. If the barrier to molecular chemisorption in figure 1 can be overcome by translational energy, a dramatic decrease of scattered O_2 should be observed upon an increase of the beam energy above 0.2 eV. In an early calculation of a potential energy surface for O_2-Ag(110) Lin and Garrison found that the barrier was indeed in the entrance valley of the potential energy surface and translational activation should stimulate molecular chemisorption[19]. Hayden and Godfrey have observed a dramatic decrease in the number of scattered CO molecules from Cu(110), when the beam energy exceeded the barrier for molecular chemisorption of CO [20, 21]. Namiki and co-workers have made similar observations for other systems [22-24]. However, Spruit et al. could not ob-

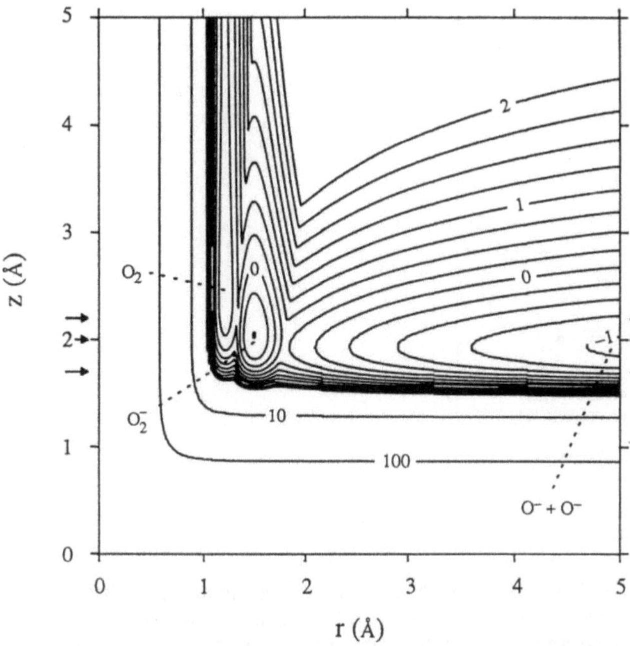

Figure 2. Topographical plot of the lowest adiabatic potential energy surface for the O_2-Ag(111) system, as a function of the intramolecular separation in the O_2 molecule (r), and the molecule-surface separation (z). Equipotential lines are drawn from -1 eV up to 2 eV with steps of 0.25 eV. Additional equipotential lines are drawn for 10 eV and 100 eV. From Spruit [27, 28].

serve any major change when the beam energy of O_2 was increased substantially above 0.2 eV [25, 26]. In a direct study of dissociative chemisorption Spruit and Kleyn only observed sticking at the 10^{-5} level at energies around 1 eV [10].

Based upon these observations Spruit has suggested connecting the diagram in figure 1 to a possible potential energy surface, which is depicted in figure 2 [27, 28]. Here only the molecule-surface separation and the internuclear distance in the molecule are parameters. The molecular chemisorption (O_2^-), and dissociative (O + O) states are simply accessed from the physisorption (O_2) state by stretching the molecular bond in a vibrationally activated process. Thus, according to this potential surface the barrier to O_2^- formation is lowest along the vibrational degree of freedom. This suggests a dramatic increase of sticking when increasing the vibrational energy content of the beam, and could account for the isotope effect found in dissociative chemisorption [9, 29]. A large effect of vibrational excitation has not been found more recently by Raukema et al. [30].

Spruit et al. did observe a change in the width of the angular distribution for direct-inelastic scattering with increasing beam energy [25, 26]. Their studies have recently been extended by Raukema et al. [31]. They studied the yield of several atoms and molecules as a function of the final scattering angle θ_f for several angles of incidence θ_i and incidence energies E_i. The scattering angles are measured from the surface normal. The dependence of the width $\Delta\theta_f$ (FWHM) of the angular distribution is plotted as a function of E_i in figure 3. A slight decrease of $\Delta\theta_f$ with E_i is seen for N_2 and Ar, which only interact through a physisorption potential. Very recently, this behaviour of

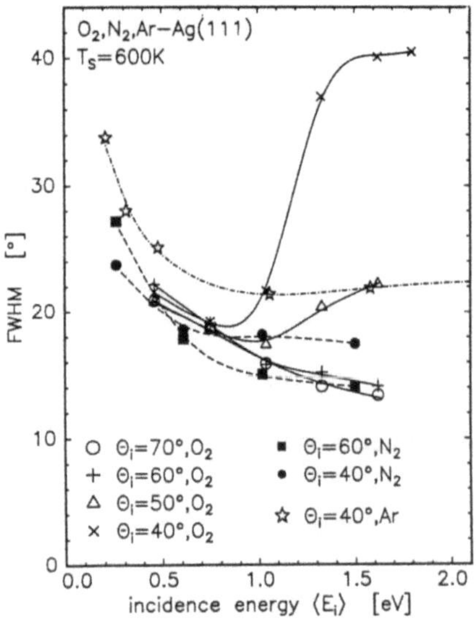

Figure 3. Angular width (Full Width at Half Maximum) of the angular flux distributions of O_2, N_2, and Ar scattering from Ag(111) as a function of the incidence energy. Lines drawn through the data points serve to guide the eye only. From Raukema et al. [31].

Ar has been reproduced in classical molecular dynamics calculations by Lahaye et al. [32, 33]. The observed decrease in $\Delta\theta_f$ with E_i is due to the decreasing influence of the potential well and of the thermal surface motion on the scattering pattern. Without this thermal roughening the surface appears to be effectively very flat, which was also apparent from a computation of the Ar-Ag interaction potential [34]. The dependence of $\Delta\theta_f$ on E_i closely follows the prediction of cube models [35, 36].

At low E_i the angular distributions for O_2 scattering are very similar to those of N_2 [31]. However, for E_i>0.7 eV clear differences become apparent. The peak intensity decreases, the angular distribution shows an asymmetric broadening and the final energy exhibits a different dependence on θ_f. For E_i>1.5 eV there are no further changes; the width is stabilised at a large value such as is also observed for NO scattering from Pt(111), a system dominated by a large chemisorption well [37, 38]. Therefore, Raukema et al. have attributed the very sudden change in $\Delta\theta_f$ to the fact that part of the flux no longer follows the N_2-like physisorption potential. Lahaye et al. have theoretically studied $\Delta\theta_f$ for Ar scattering from Ag(111) in the energy range 0.1-100 eV [32, 33]. Such a drastic change in $\Delta\theta_f$ as a function of E_i is not seen for a single repulsive potential. This observation supports the interpretation that the sudden increase followed by a stabilisation of $\Delta\theta_f$ as a function of E_i is due to a change in the relevant interaction potential: the change in $\Delta\theta_f$ is attributed to the presence of a second repulsive wall, connected to the chemisorption potential. Whether the second repulsive wall is connected to the O_2^- state or to the O + O state cannot be determined from this scattering experiment. These observations are not consistent with the simple potential of figure 2, because there any transition from O_2 to O_2^- induced by translational energy seems very unlikely.

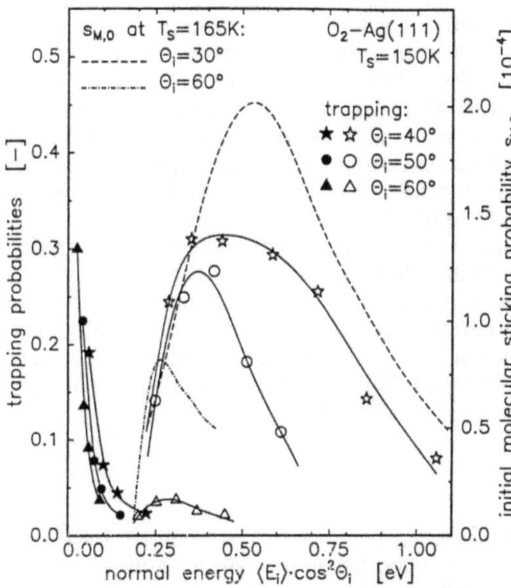

Figure 4. Normal energy dependence of the probabilities for physisorption trapping (closed symbols) and transient trapping-desorption (open symbols) for three different values of θ_i. The full lines through the data points serve to guide the eye only. The molecular chemisorption probability S_M is shown by the dashed and the dash-dotted line. Note the different Y-axes. From Raukema and Kleyn [3].

Adsorption-desorption

In the studies of direct-inelastic scattering, discussed above, attention was also paid to the energy exchange between molecule and surface. This can be done using a pulsed molecular beam and time-of-flight (TOF) techniques [39]. Such experiments have been carried out recently by Raukema et al. [3, 40]. These authors focused their attention on experiments at several energies, for T_s=150 K, θ_i rather large and θ_f small. For E_i=0.1 eV, the TOF spectrum is dominated by a single broad distribution, which can be fitted very well with a Maxwell-Boltzmann distribution, corresponding to a temperature slightly below the surface temperature as has been observed previously for this system [4]. Clearly, physisorption and complete equilibration followed by desorption is the only important process at these given conditions. When E_i is increased to 0.2 eV, the signal intensity corresponding to physisorption-desorption decreases and is no longer a pure Maxwell-Boltzmann distribution . The TOF distribution can be fitted by two distributions, one of which represents physisorption trapping-desorption. From such experiments the relative intensity of the trapping-desorption signal can be measured as a function of E_i. The result is given in figure 4 as a function of the 'normal energy' $E_n=E_i*\cos^2(\theta_i)$. Scaling of results with E_n indicates that parallel momentum is conserved in the interaction. The characteristic monotonic decreasing intensity distributions familiar for physisorption trapping-desorption are seen.

At higher energies it is observed that other processes arise and the TOF spectra show two narrow peaks, a slow one and a fast one. More detailed analysis suggests that the fast peak corresponds to direct-inelastic scattering of an O_2 molecule with essentially a single Ag atom, and that in the collision considerable rotational excitation occurs [41, 42]. The slow TOF peak exhibits a roughly E_i independent final energy $E_f \approx 0.14eV$, which suggests that these molecules result from activated desorption, that exhibits no measurable surface residence time (> 10^{-6} s.). Activated desorption is known to result in very peaked angular distributions [43, 44]. As expected, the angular distribution is strongly peaked to the surface normal, and is described very well by $I(\Theta_f,\phi_f) \propto \cos^{15}(\Theta_f)$. Additional information concerning the activated desorption process can be obtained from the beam energy dependence of its intensity. This is shown in figure 4 as a function of E_n. A threshold of 0.2eV is observed for all θ_i. This strongly supports the inference that the slow peak is due to an activated adsorption-desorption process. In figure 4 it can also be seen that there is a strong θ_i dependence on the magnitude of the adsorption process. Parallel momentum decreases the adsorption probability. Summarising: a fast activated adsorption-desorption process is observed. The O_2 molecules are transiently trapped in a metastable state at the surface. We refer to this process by transient trapping-desorption (TTD). More information on the nature of the state follows from the study of molecular chemisorption.

Molecular chemisorption

To get further information about which (metastable) state is responsible for TTD we studied the energy dependence of the sticking coefficient for molecular chemisorption S_M in the O_2^- state, which desorbs at 200K [30]. It is also reproduced in figure 4. It shows a strong similarity to the transient trapping-desorption probability. This suggests that transient trapping-desorption and molecular chemisorption are mediated by the same state. Molecular chemisorption is also an activated process with a threshold energy of about 0.2 eV. It is very hard to molecularly adsorb O_2 at the Ag(111) surface. The Genoa group did not observe any molecular adsorption using EELS [45]. We found a maximum in the molecular sticking coefficient with energy of about $2x10^{-4}$, as is shown in figure 4. There is no clear explanation for this discrepancy. We feel it might be related to the step density of the crystal, as will be discussed later. The structure of the overlayer formed is unknown. Our study indicates that the layer is dispersed and that adsorption takes place at the terraces [30]. The He reflectivity of the layer is remarkably high. Transient trapping-desorption is about three orders of magnitude more probable than molecular chemisorption.

Dissociative chemisorption

In contrast to the (110) face, dissociative chemisorption on the (111) face is very improbable at all experimental conditions. At thermal energies values between 10^{-7} and 10^{-5} are seen [5, 9, 29]. In the energy dependence of the dissociative sticking coefficient at zero coverage S_D for Ag(111) two channels for dissociative chemisorption are seen [30]. At low E_i, S_D is much lower than the molecular chemisorption probability S_M. In addition the T_s dependence of S_D is suggestive of a precursor mediated process. Therefore, it is assumed that for E_i<1 eV molecular chemisorption precedes dissociative chemisorption. For E_i> 1 eV the sticking mechanism is different. Increasing T_s leads to enhanced sticking, as does an increase in the energy, and $S_D \gg S_M$. Here a tunnelling or surface motion mediated direct sticking seems to be prominent [9, 29, 46-51]. Formation of an excited molecule-surface complex that accesses the barrier to dissociation statistically, may also occur [52].

Interaction dynamics

Perhaps the most remarkable and new observation for the interaction between O_2 and Ag(111) is the occurrence of transient trapping-desorption. Its understanding may shed

some light on the question why the chemisorption probabilities on this face are so low. The state involved in transient trapping-desorption seems to be closely related to the molecular chemisorbed state. This inference is based on the very similar energy and angular dependence of the transient trapping probability t_{TTD} and S_M. Clearly transient trapping in this state does in most cases not lead to molecular chemisorption. Several explanations for the nature of this state can be proposed[28]. We feel that most likely the O_2^- like state is metastable with respect to desorption. It exists as a local binding minimum in the potential. Recent calculations of the bonding of O_2^- to Ag(110) suggest that the O_2^- binds to Ag-atoms in the same plane. Those are abundant on Ag(110) but absent at the Ag(111) face [53]. Therefore the O_2^- state might be a resonance on Ag(111) and decays after transient trapping via desorption. Molecular chemisorption, stabilisation of O_2^- at the surface may only occur at special sites, such as at steps or in the presence of ad atoms. The stabilisation of the O_2^- by ad-atoms or surface imperfections could explain the difference between the results of the Genoa group for this system and our results. This could be due to a larger step density for the Amsterdam crystal or a better decoration of the steps in the Genoa crystal.
Because the O_2^- well is rather shallow (depth less than 0.5 eV [5, 30]), (transient) trapping in this well will only occur for a limited energy range. This is clearly observed for transient trapping-desorption and for molecular chemisorption in figure 4 [3, 30, 40]. Above 1 eV trapping in the O_2^- well becomes less probable. Because dissociative chemisorption is also improbable above 1 eV ($S_D < 5 \cdot 10^{-3}$) most of the incident flux has to be scattered directly. According to figure 3 most of the flux is now scattered along the chemisorption potential. However, if dissociation involves defects or restructuring of the lattice, it is very unlikely that dissociation can be described using a two dimensional potential energy surface. Therefore, we believe that the corrugated potential probed in hyperthermal O_2 scattering is related to the O_2^- state at the surface and not by the O + O potential. It should be noted that at even higher energies ($E_n > 5$ eV) direct formation of free O_2^- ions has been observed [14, 16, 18]. This suggests that O_2^- formation (harpooning [12, 54]) at the surface occurs readily at energies above 0.2 eV. Formation of free O^- occurs above E_n=10 eV; according to classical trajectory calculations the dissociation is due to impulsive energy transfer in the collision of an O_2^- and the surface and is not due to spontaneous dissociation at some repulsive $O^- + O^-$ potential energy surface [14, 55]. All of this suggests that for the Ag(111) surface a potential energy surface such as displayed in figure 2 cannot describe the interaction. Instead, a potential energy surface such as shown in figure 5 is more likely, where the dissociation channel is absent in this 2 dimensional representation because it needs more degrees of freedom to be described properly.

Potential energy surfaces

A number of potential energy surfaces have been calculated for the O_2-Ag interaction [19, 27, 53, 56-58]. In all of these surfaces the O_2^- state is considerably stretched with respect to the O_2 state, as shown in figure 2. The barrier between the O_2 and O_2^- states is in most surfaces rather low. The experimental results for both Ag(110) and Ag(111) indicate that access to the O_2^- well is activated by about 0.2 eV. Since translational energy easily helps the molecules overcome the barrier and because desorption from the O_2^- (and O + O) state leads to excess translational energy, the location of this barrier is rather early in the potential energy surface. Such a barrier is absent in the potential energy surface in figure 2. The barrier is indicated in figure 5, where a sketch of a potential energy surface is shown. Note that the binding energy of the O_2^- state in this surface is very small, making it metastable. Because at low energy dissociation of the O_2^- species is improbable and the STM studies point to mass transport at the surface, the barrier between O_2^- and O + O is high and not shown in this figure. This is also found in the calculations by Nakatsuji and Nakai [56, 57] for Ag(111).

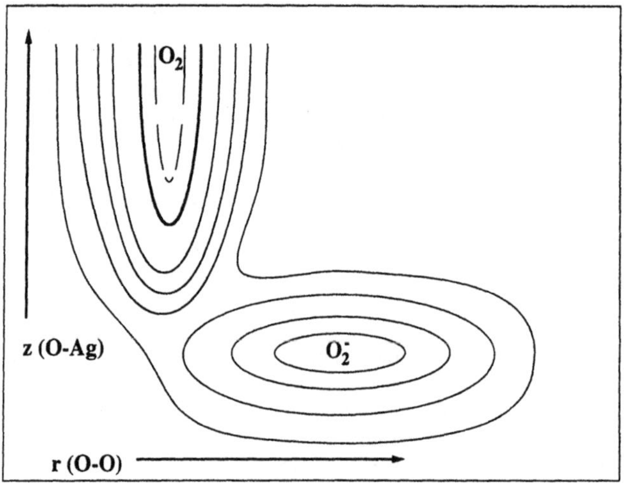

Figure 5. Potential energy surface describing the O_2-Ag(111) interaction as a function of the oxygen surface distance and the intramolecular separation. Dashed lines show negative energies. The thick line shows the zero of energy around the physisorption well. The full lines represent the repulsive part of the potential and are taken to be 0.1 eV apart, to reproduce the barrier height for transient trapping-desorption and molecular chemisorption. Note that the O_2^- part of the potential is not bound with respect of the zero of energy, far from the surface. From Kleyn et al. [28].

Conclusions

A variety of processes can occur in the interaction of O_2 molecules and Ag(111). At first scattering from and trapping in the physisorption potential can occur. Secondly, scattering from the chemisorption (O_2^-) potential occurs, together with transient trapping-desorption. The chemisorption potential well is very shallow. From being transiently trapped the molecule can be captured in the molecular chemisorption well; presumably surface imperfections are necessary to stabilise the molecular adsorbate in this case. From the molecular chemisorption well the molecule can proceed to dissociation. In this step ad atoms may be involved on Ag(111). Finally, there is a small probability for direct dissociative chemisorption of O_2 at Ag(111).

Acknowledgements

This work is part of the research program of FOM and is supported financially by NWO. The author gratefully acknowledges all his co-workers, who have studied the O_2 silver interaction, and whose work forms the basis of this report.

References:

1. F. Besenbacher, and J.K. Nørskov, Prog. Surf. Sci. **44** (1993) 1.
2. R.A. VanSanten, and H.P.C.E. Kuipers, Adv. Catal. **35** (1987) 265.
3. A. Raukema, and A.W. Kleyn, Phys. Rev. Lett. **74** (1995) 4333.

4. M.E.M. Spruit, E.W. Kuipers, F.H. Geuzebroek, and A.W. Kleyn, Surf. Sci. **215** (1989) 421.
5. C.T. Campbell, Surf. Sci. **157** (1985) 43.
6. K.C. Prince, G. Paolucci, and A.M. Bradshaw, Surf. Sci. **175** (1986) 101.
7. C. Backx, C.P.M. de Groot, and P. Biloen, Surf. Sci. **104** (1981) 300.
8. A.F. Carley, P.R. Davies, M.W. Roberts, and K.K. Thomas, Surf. Sci. **238** (1990) L467.
9. P.H.F. Reijnen, A. Raukema, U. VanSlooten, and A.W. Kleyn, Surf. Sci. **253** (1991) 24.
10. M.E.M. Spruit, and A.W. Kleyn, Chem. Phys. Lett. **159** (1989) 342.
11. A.W. Kleyn, J. Los, and E.A. Gislason, Phys. Rep. **90** (1982) 1.
12. A.W. Kleyn. in Invited papers of the XVI International Conference on the Physics of Electronic and Atomic Collisions, New York, 1989,. 1990, Eds: A. Dalgarno, *et al.*, New York: American Institute of Physics Conference Proceedings, 451.
13. D.R. Herschbach, Adv. Chem. Phys. **10** (1966) 319.
14. P.H.F. Reijnen, P.J. VanDenHoek, A.W. Kleyn, U. Imke, and K.J. Snowdon, Surf. Sci. **221** (1989) 427.
15. P. Haochang, P.H.F. Reijnen, T.C.M. Horn, and A.W. Kleyn, Rad. Eff. Def. Sol. **109** (1989) 41.
16. P. Haochang, T.C.M. Horn, and A.W. Kleyn, Phys. Rev. Lett. **57** (1986) 3035.
17. P. Haochang, T.C.M. Horn, and A.W. Kleyn, J. Electron Spectrosc. Relat. Phenom. **45** (1987) 361.
18. P.H.F. Reijnen, U. VanSlooten, and A.W. Kleyn, J. Chem. Phys. **94** (1991) 695.
19. J.-H. Lin, and B.J. Garrison, J. Chem. Phys. **80** (1984) 2904.
20. B.E. Hayden, and D.C. Godfrey, Surf. Sci. **232** (1990) 24.
21. B.E. Hayden, and D.C. Godfrey, J. Electron Spectrosc. Relat. Phenom. **45** (1987) 351.
22. F.H. Geuzebroek, Y. Babasaki, M. Tanaka, T. Nakamura, and A. Namiki, Surf. Sci. **297** (1993) 141.
23. A. Namiki, S. Suzuki, H. Kato, Y. Babasaki, M. Tanaka, T. Nakamura, and T. Suzaki, J. Chem. Phys. **97** (1992) 3781.
24. A. Namiki, H. Hayashi, Y. Ukai, and F.H. Geuzebroek, J. Chem. Phys. **102** (1995) 9689.
25. M.E.M. Spruit, P.J. VanDenHoek, E.W. Kuipers, F.H. Geuzebroek, and A.W. Kleyn, Phys. Rev. B **39** (1989) 3915.
26. M.E.M. Spruit, P.J. VanDenHoek, E.W. Kuipers, F.H. Geuzebroek, and A.W. Kleyn, Surf. Sci. **214** (1989) 591.
27. M.E.M. Spruit, in *Thesis: Probing Gas-surface interactions with a molecular beam.* 1988, University of Amsterdam: Amsterdam. p. 83.
28. A.W. Kleyn, D.A. Butler, and A. Raukema, Surf. Sci. (accepted)
29. P.H.F. Reijnen, A. Raukema, U. VanSlooten, and A.W. Kleyn, J. Chem. Phys. **94** (1991) 2368.
30. A. Raukema, D.A. Butler, F.M.A. Box, and A.W. Kleyn, Surf. Sci. **347** (1996) 151.
31. A. Raukema, R.J. Dirksen, and A.W. Kleyn, J. Chem. Phys. **103** (1995) 6217.
32. R.J.W.E. Lahaye, A.W. Kleyn, S. Stolte, and S. Holloway, Surf. Sci. **338** (1995) 169.
33. R.J.W.E. Lahaye, S. Stolte, S. Holloway, and A.W. Kleyn, Surf. Sci. (accepted)
34. E.J.J. Kirchner, A.W. Kleyn, and E.J. Baerends, J. Chem. Phys. **101** (1994) 9155.

35. R.M. Logan, in *Solid state surface science,* M. Green, Editor. 1973, Marcel Dekker: New York. p. 1.
36. F.O. Goodman, and H.Y. Wachman, *Dynamics of Gas-Surface Scattering.* 1976, New York: Academic press.
37. A.E. Wiskerke, and A.W. Kleyn, J. Phys. Condens. Matter **7** (1995) 5195.
38. R.J.W.E. Lahaye, S. Stolte, S. Holloway, and A.W. Kleyn, J. Chem. Phys. (accepted)
39. D.J. Auerbach, in *Atomic and Molecular Beams Methods II,* G. Scoles, *et al.*, Editors. 1988, Oxford University Press: Oxford. p. 444.
40. A. Raukema, and A.W. Kleyn, J. Chem. Phys. (to be published)
41. G.O. Sitz, A.C. Kummel, and R.N. Zare, J. Chem. Phys. **89** (1988) 2558.
42. A.W. Kleyn, Surf. Rev. Lett. **1** (1994) 157.
43. W. VanWilligen, Phys. Lett. **28A** (1968) 80.
44. G. Comsa, and R. David, Surf. Sci. Rep. **5** (1985) 145.
45. F. Buatier de Mongeot, U. Valbusa, and M. Rocca, Surf. Sci. **339** (1995) 291.
46. J. Harris, J. Simon, A.C. Luntz, C.B. Mullins, and C.T. Rettner, Phys. Rev. Lett. **67** (1991) 652.
47. J. Harris, and A.C. Luntz, Surf. Sci. **287** (1993) 56.
48. A.C. Luntz, and J. Harris, Surf. Sci. **258** (1991) 397.
49. R. Kosloff, and O. Citri, Faraday Discuss. **96** (1993) 175.
50. E. Por, G. Haase, O. Citri, R. Kosloff, and M. Asscher, Chem. Phys. Lett. **186** (1991) 553.
51. O. Citri, and R. Kosloff, Surf. Sci. (in press)
52. V.A. Ukraintsev, and I. Harrison, J. Chem. Phys. **101** (1994) 1564.
53. P.A. Gravil, J.A. White, and D.M. Bird, Surf. Sci. (accepted)
54. J.W. Gadzuk, Comments Atom. Mol. Phys. **16** (1985) 219.
55. P.J. VanDenHoek, and A.W. Kleyn, J. Chem. Phys. **91** (1989) 4318.
56. H. Nakatsuji, and H. Nakai, Chem. Phys. Lett. **174** (1990) 283.
57. H. Nakatsuji, and H. Nakai, J. Chem. Phys. **98** (1993) 2423.
58. P.J. VanDenHoek, and E.J. Baerends, Surf. Sci. **221** (1989) L791.

On the Molecular Orientation Dependence of Dynamical Processes on Solid Surfaces: Dissociative Adsorption and Scattering

H. Kasai, A. Okiji, and W.A. Diño

Department of Applied Physics, Osaka University,
Suita, Osaka 565, Japan

Abstract. Recently molecular orientation dependence has been observed in several dynamical processes on solid surfaces. In the first half of this paper, the dissociative adsorption of a hydrogen molecule interacting with a metal surface is cited as an example of a dynamical process on solid surface which shows strong molecular orientation dependence in the range of translational energy comparable to the height of the activation barrier around 1 eV. On the basis of the results of numerical calculation, it is shown how the orientation dependence results in what is called steering effects. In the latter half, the dissociative scattering of a hydrogen molecule interacting with a metal surface is used as a representative dynamical process which shows strong molecular orientation dependence in the range of translational energy around a few hundred eV. On the basis of the results of numerical calculation, it is shown how the orientation dependence manifests itself.

1. Introduction

When a diatomic molecule is made to react with a metal surface, it goes through various dynamical processes. Depending on its initial state preparation and the kind of metal surface it is made to react with, the impinging molecule might end up sitting adsorbed on the surface, either as a whole molecule or dissociated into constituent atoms. It is also possible that instead of being adsorbed, it is scattered back either as a whole molecule or, again, dissociated into constituent atoms. And during the time that these processes are taking place, there are energy exchanges between the various degrees of freedom involved. Of these different phenomena, plus some more that might have been left out, we would like to pay attention to the molecular orientation dependence of dynamical processes on metal surfaces, citing as examples the dissociative adsorption and the dissociative scattering of a hydrogen molecule when it reacts with metal surfaces.

The effective reaction potential between the impinging molecule and the surface shows molecular orientation dependence that gives rise to what is now generally called steering effect (cf., e.g., [1]). In section 2, we concentrate on the orientation

Springer Series in Solid-State Sciences, Volume 121
Elementary Processes in Excitations and Reactions on Solid Surfaces
Eds.: A. Okiji, H. Kasai and K. Makoshi

dependence of the dissociative adsorption of D_2 on Cu and show how this steering effect influences the behaviour of the adsorption/sticking probability. In section 3, we concentrate on the orientation dependence of the dissociative scattering of H_2 from metal surfaces, and show that the orientation dependence of the dissociative scattering probability is determined by what characteristics the reaction partners possess, i.e., it depends on, e.g., the wave vectors of the substrate electrons directly involved in the dissociative scattering process. These electrons are, in turn, determined by what may be called surface-projected bond-length, which depends on the orientation of the impinging molecule relative to the metal surface. Finally, we provide a summary in section 4.

2. Dissociative Adsorption

A good starting point for investigating the interaction of a molecule with a surface theoretically is provided by the introduction of the Born-Oppenheimer approximation, on the assumption that the electrons are capable of following the nuclear motion adiabatically. Then, the electron system provides some potential energy (hyper-) surface (PES) for the molecular motion. Thus, it is standard theoretical practice to formulate the problem of molecules interacting in some environment in terms of the motion of some representative particle or quasi-particle over a PES. The problem of studying the interaction of a molecule with a surface then reduces to (1) obtaining the relevant PES and (2) solving the corresponding equations of motion. From the above reasoning, it is obvious that before one can achieve a complete understanding of any dynamical process, it is vital that a good PES is at hand [2].

To pinpoint how the molecular orientation affects the dissociation process, one can consider the model system shown in Fig.1, where a diatomic molecule (e.g., H_2 or D_2) with center-of-mass (CM) distance Z above the metal surface, internuclear distance or bond-length r, and molecular orientation θ, is impinging on the surface perpendicularly. For the interaction potential between the impinging diatomic molecule and the metal surface, a typical PES is shown in Fig.2 [1], which corresponds to the case when the orientation of a hydrogen molecule is artificially kept parallel or perpendicular to the surface throughout the reaction.

It can be seen that the features of these PESs differ greatly when the orientation of the molecule with respect to the surface is varied from parallel (Fig.2(a)) to perpendicular (Fig.2(b)). We concentrate first on the parallel-oriented molecule in Fig.2(a). Initially, at distances considerably far from the surface, the interatomic distance of the hydrogen molecule is equivalent to its equilibrium interatomic distance in the gas phase (1.4 a.u.) and may, if vibrationally excited, vibrate along r. As the distance of the molecule to the surface decreases, the interatomic distance increases relative to its equilibrium interatomic distance. It is customary to take advantage of

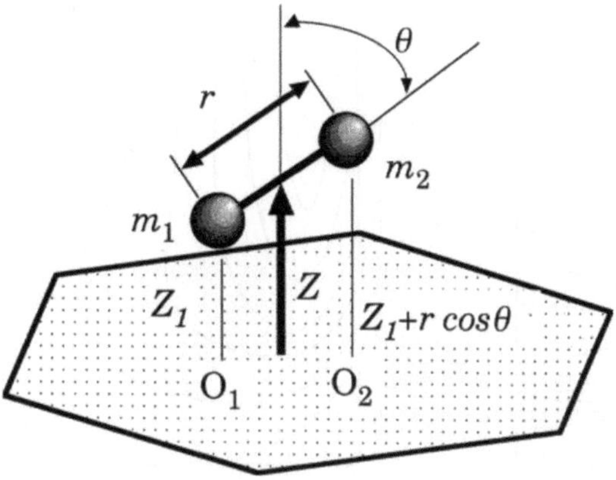

Fig.1 The model system showing a diatomic molecule (molecular orientation with respect to the surface normal given by θ, surface to center-of-mass (CM) distance Z, and bond-length r) approaching a flat surface perpendicularly. m_1 and m_2 correspond to the masses of the constituent atoms of the diatomic molecule, located at distances Z_1 and $Z_1 + r\cos\theta$ from the surface, respectively. O_1 and O_2 are the surface projections of the two constituent atoms.

the reaction path (dashed line in Fig.2(a), cf., eg., [1]) and imagine that as the hydrogen molecule approaches the surface, there is a corresponding quasi-particle moving along the reaction path. In connection with this, it can be seen that in the transition from a whole molecule to adsorbed, dissociated atoms, the quasi-particle will encounter a so-called dissociation/activation barrier, with a maximum height of approximately 1eV. Thus, with enough energy, the quasi-particle may overcome this activation barrier and, as a result, the molecule may end up as two dissociated atoms sitting on the surface, which, if vibrationally excited, vibrate along Z. However, when the molecular orientation is kept perpendicular to the surface throughout the reaction process (Fig.2(b)), the potential barrier increases continuously with increasing r, and it will be hard, if not impossible, for the molecule to dissociate, as the quasi-particle cannot overcome the activation barrier in this case.

If we restrict the diatomic molecule in Fig.1 to its vibrational ground state throughout the dissociation process, then the corresponding Hamiltonian for this system, after transformation using the reaction path coordinate S [1], is given by

$$H(S,\theta) = -\frac{\hbar^2}{2M}\frac{\partial^2}{\partial S^2} + \frac{\hbar^2}{2I(S)}L^2 + U(S,\theta)\,, \tag{2.1}$$

where M is the total mass of the molecule and L is the angular momentum operator.

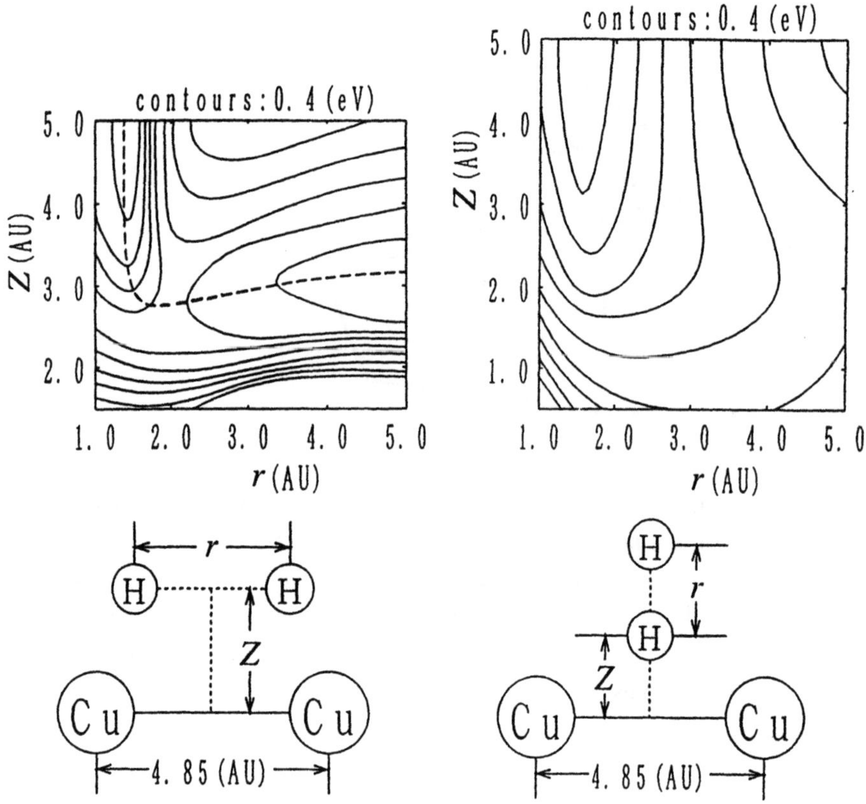

Fig.2 Results of ab-initio potential energy calculation (GAUSSIAN88) for H_2 /Cu(100) showing (a) parallel, and (b) perpendicular orientations of the H_2 molecule relative to the copper surface. Contour spacing is 0.4 eV and barrier height in (a) is approximately 1.6 eV. Dashed line in (a) corresponds to the reaction path. Adapted from [1].

The moment of inertia I is given by $I(S)=\mu r^2(S)$, where μ is the reduced mass of the molecule. And the orientation dependent PES is given by

$$U(S,\theta)=\frac{E_a}{\cosh^2(\alpha S)}\left(1-\beta\cos^2\theta\right)+U_1\cos^2\theta\frac{1+\tanh(\alpha S)}{2}. \tag{2.2}$$

Here, we got (2.2) by combining the features of the PESs shown in Fig.2 (a) and (b), and assuming that the activation barrier increases continuously and smoothly as the orientation of the hydrogen molecule changes from parallel to perpendicular.

This anisotropy of the effective PES gives rise to a mechanism called steering effect, which pertains to the ability of the anisotropic PES to reorient the molecule from an initially unfavorable orientation to a favorable one, and vice-versa. Thus, in certain situations, this steering effect can promote dissociation via reorientation of the impinging molecule towards a favorable orientation (parallel). Under certain conditions, it can also hinder dissociation. Two different ways in which steering can reduce or hinder dissociation are as follows-

1) when the amount of time that the molecule spends in a favorable orientation is short, or

2) when the molecule approaches the surface with an initially unfavorable orientation and reorients to a favorable one, but in the process losing some translational energy needed for overcoming the barrier.

As mentioned earlier, during the various dynamical processes on surfaces, energy exchanges between the different internal degrees of freedom take place. These energy exchanges result in another mechanism which can be called rotational to translational (R-T) energy transfer effect. This effect pertains to the effective transfer of rotational energy to translational energy to help the quasi-particle overcome the activation barrier to dissociation. This effect arises because of the occurrence of bond-length extension. Recall that during the dissociation process, the interatomic distance or the bond-length of H_2 stretches beyond its equilibrium interatomic distance in the gas phase. This increases the moment of inertia of the dissociating hydrogen molecule and decreases the rotational constant and in effect the rotational energy, E_{ROT}. For fixed kinetic energy, $E_{TRANS} + E_{ROT}$, of the molecule, the energy difference resulting from a decrease in E_{ROT} can then be transferred to the translational degree of freedom, to E_{TRANS}, and an increase in E_{TRANS} can aid the quasi-particle in overcoming the activation barrier. As a result, the R-T energy transfer effect promotes dissociation.

To get a better feel of what these two mechanisms really are, and how they influence the adsorption/sticking probability, the results of numerical calculation for the dissociation/sticking probability as a function of the incident translational energy of the impinging diatomic molecule (D_2) in the present model system mentioned above (see (2.1) and (2.2), and Fig.1) are shown in Fig.3. It can be seen that without PES anisotropy (Fig.3(a)), the effect of energy transfer is to increase the adsorption probability as the initial rotational state (energy) of the impinging molecule is increased. But, upon introducing PES anisotropy (Fig.3(b)), steering effects are observed and a marked decrease in the dissociation probability can be seen, when the initial rotational energy of the impinging molecule is increased, as compared to the results shown in Fig.3(a). In fact, it is observed experimentally that there is a non-monotonous dependence of the adsorption probability on the initial rotational state of D_2 [3]. There is an initial decrease in the adsorption probability with a slight increase in the initial rotational state j of the impinging molecule, and finally an increase in the adsorption probability with further increases in j. With proper choice of initial conditions, one can see in Fig.4 the expected initial decrease of the adsorption probability of D_2, and its eventual increase as the initial rotational state j of the impinging molecule is increased.

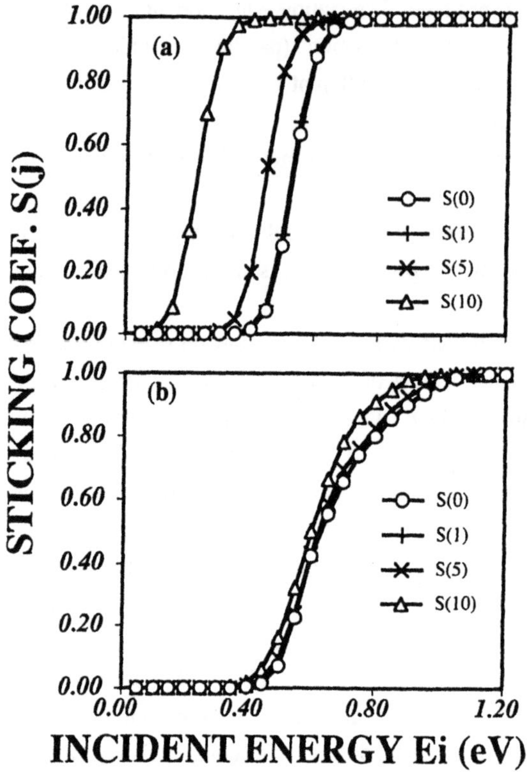

Fig.3 *j*-dependent adsorption probability for D_2 in the vibrational ground state (a) without PES orientational anisotropy (β=0.0, U_1=0.0 eV), and (b) with PES orientational anisotropy (β=0.25, U_1=1.0 eV). Width of the potential barrier (α=4.5 Å^{-1}) is the same for the two cases. Activation barrier E_a=0.536 eV. Adapted from [1].

To conclude this first half, it can be said that, as a result of the orientation dependence of the PES for H_2 (D_2) interacting with metal surfaces and the energy exchange between the rotational and the translational degree of freedom, two factors or mechanisms dominate the dissociative adsorption process when the rotational degree of freedom is taken into account, viz., steering and R-T energy transfer. The first mechanism, steering, is a direct manifestation of the orientation dependence of the PES for the impinging molecule and results in a reorientation of the molecule. Depending on the initial conditions it promotes or hinders adsorption. The second mechanism, R-T energy transfer, results from bond-length extension and always promotes adsorption. As a result of the interplay between these two mechanisms, there is an observed initial decrease in the sticking probability of hydrogen molecules with a slight increase in the initial rotational state *j*, and then a final increase when *j* is further increased.

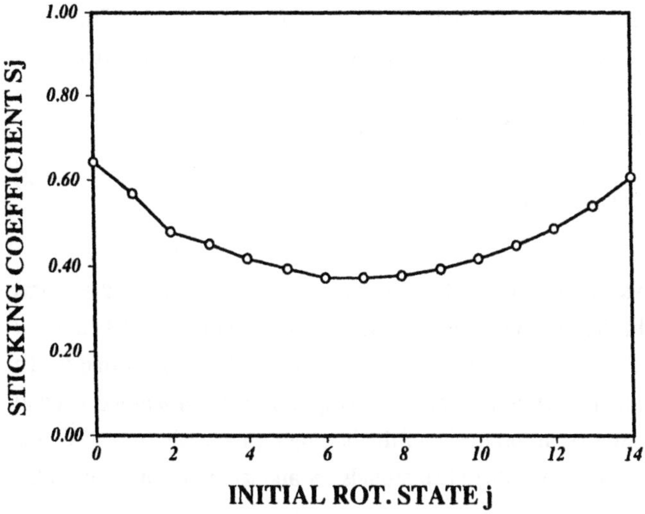

Fig.4 Sticking coefficient S_j as a function of the initial rotational state j of the impinging diatomic molecule for E_{TRANS}=0.6 eV. (α=1.5 Å^{-1}, β=0.25, E_a=0.536 eV, U_1=1.0 eV). Adapted from [1].

3. Dissociative Scattering

Now we come to the latter half where the dissociative scattering of H_2 from a metal surface is considered in the framework of the system shown in Fig.1. In the range of the translational energy involved (a few hundred eV) in the experiments [4,5], the velocity of the molecular CM motion is comparable to the Fermi velocity as an order of magnitude, and much faster than molecular rotation. Then, the coupling between the molecular CM motion and degrees of freedom of metal electron system should be taken into account explicitly [2]. Furthermore, it is assumed in the following calculation that the PES shape (see Fig.2) loses its effect on the CM motion and thus the surface fills its role as a hard wall to the CM motion, and that the orientation of the molecule is fixed throughout the scattering process.

The Hamiltonian for this model system can be expressed as a sum of the Hamiltonians corresponding to the CM motion, H_{CM}, and to that of the electron system, $H_{E(lectrons)}$, i.e., $H=H_{CM}+H_{E(lectrons)}$, where

$$H_{CM} = -\frac{\hbar^2}{2M}\frac{\partial^2}{\partial Z^2}. \tag{3.1}$$

The corresponding Hamiltonian for the electron system is further separated into that describing the electrons of the impinging molecule, $H_{M(olecule)}$, and that describing the electrons of the metal substrate, $H_{S(ubstrate)}$. On top of that, the molecule-surface interaction term, $H_{M(olecule)\text{-}S(substrate)}$, is added, i.e., $H_E = H_M + H_S + H_{M\text{-}S}$. Then, the Hamiltonian for the electrons of the impinging molecule, H_M, is expressed as

$$H_M = \sum_\sigma \varepsilon_a(r)(n_{1\sigma} + n_{2\sigma}) + t(r)\sum_\sigma \left(C_{1\sigma}^+ C_{2\sigma} + h.c.\right) + U(r)\left(n_{1\uparrow}n_{1\downarrow} + n_{2\uparrow}n_{2\downarrow}\right) + J(r)\sum_\sigma \sum_{\sigma'} n_{1\sigma}n_{2\sigma'}, \tag{3.2}$$

where ε_a gives the corresponding energy of the electron in the orbital localized at atoms 1 and 2, t is the hybridization energy, U is the intra-atomic Coulomb interaction and J is the inter-atomic Coulomb interaction. $C_{1\sigma}(C^+_{1\sigma})$ and $C_{2\sigma}(C^+_{2\sigma})$ correspond to the single electron annihilation (creation) operators with spin σ in the orbital localized at atoms 1 and 2, i.e., ϕ_1 and ϕ_2, respectively, of the impinging molecule ($n = C^+C$, the number operator). It is noted that these two orbitals, ϕ_1 and ϕ_2 are modified from the 1s orbitals of two hydrogen atoms, ψ_1^{1s} and ψ_2^{1s}, such that they are orthogonal to each other (cf., [6]), i.e.,

$$\phi_1 = \lambda\psi_1^{1s} + \zeta\psi_2^{1s}, \quad \phi_2 = \zeta\psi_1^{1s} + \lambda\psi_2^{1s}, \tag{3.3}$$

$$\lambda = \frac{1}{2}\left(\frac{1}{\sqrt{1+Q}} + \frac{1}{\sqrt{1-Q}}\right) \text{ and } \zeta = \frac{1}{2}\left(\frac{1}{\sqrt{1+Q}} - \frac{1}{\sqrt{1-Q}}\right). \tag{3.4}$$

Here, Q is the overlap integral given by $Q = \langle \psi_1^{1s} | \psi_2^{1s} \rangle$. For the electron system of the metal substrate, the kinetic energy of the conduction electrons is taken into account, i.e., $H_S = \sum_{k\sigma} \varepsilon_k C_{k\sigma}^+ C_{k\sigma}$, where $C_{k\sigma}\left(C_{k\sigma}^+\right)$ corresponds to the single electron annihilation (creation) operator in the energy band of the metal with wave vector k and spin σ. Then, the Hamiltonian $H_{M\text{-}S}$ is expressed as

$$H_{M-S} = \sum_\sigma \left[V_1(Z,\theta)C_{1\sigma}^+ C_{O_1\sigma} + V_2(Z,\theta)C_{2\sigma}^+ C_{O_2\sigma} + h.c.\right], \tag{3.5}$$

where

$$C_{O_1\sigma} = \frac{1}{\sqrt{N}}\sum_k C_{k\sigma} \exp\left(ik\frac{r}{2}\sin\theta\right), \text{and } C_{O_2\sigma} = \frac{1}{\sqrt{N}}\sum_k C_{k\sigma} \exp\left(-ik\frac{r}{2}\sin\theta\right) \tag{3.6}$$

The Hamiltonian, $H_{M\text{-}S}$, above describes the hybridization between an electron in the orbitals of the molecule and an electron in the Wannier orbitals (cf., (3.6)) located at surface sites which are closest to atoms 1 and 2 - O_1 and O_2 (cf., Fig.1), respectively. The hybridization potentials are assumed to have exponential forms

$$V_1(Z,\theta)=\exp(-\gamma|Z|),\text{ and } V_2(Z,\theta)=V_1(Z,\theta)\exp(-\gamma r\sin\theta). \tag{3.7}$$

Here, the positive Z-region corresponds to the hybridization potential seen when the impinging molecule approaches the surface and the negative Z-region as that seen when the resulting product of the reaction scatters back. To get more insight, the interaction Hamiltonian, H_{M-S}, is re-expressed in terms of the bonding and antibonding states of the impinging molecule, i.e.,

$$H_{M-S}=\frac{1}{\sqrt{N}}\sum_{k,\sigma}\left[V_{ak}(Z,\theta)C^{+}_{a\sigma}C_{k\sigma}+V_{bk}(Z,\theta)C^{+}_{b\sigma}C_{k\sigma}+h.c.\right], \tag{3.8}$$

where

$$\left.\begin{matrix}V_{ak}(Z,\theta)\\ V_{bk}(Z,\theta)\end{matrix}\right\}=\frac{V_O u(Z)}{\sqrt{2}}\begin{bmatrix}[1\pm\exp(-\gamma r\sin\theta)]\cos\left(k\frac{r}{2}\cos\theta\right)\\ -i[1\mp\exp(-\gamma r\sin\theta)]\sin\left(k\frac{r}{2}\cos\theta\right)\end{bmatrix}, \tag{3.9}$$

$$C_{a\sigma}=\frac{1}{\sqrt{2}}(C_{1\sigma}-C_{2\sigma}),\text{ and } C_{b\sigma}=\frac{1}{\sqrt{2}}(C_{1\sigma}+C_{2\sigma}). \tag{3.10}$$

Now one can see clearly from the form of the interaction Hamiltonian (3.8) and (3.9) that electronic transitions depend on the orientation of the impinging molecule, θ, its bond-length, r, and the wave vector of the electrons in the metal substrate, k.

The simple model described above is then used to investigate the dissociation of a H_2 scattering from a metal surface into 2 neutral H atoms. It is assumed that dissociation is triggered/induced by a two-step process involving the excitation of an electron from the singlet bonding state, $^1\Sigma_g$, of the impinging H_2 to the triplet antibonding state, $^3\Sigma_u$, during the scattering process [6]. Initially, (1) one of the electrons in the filled bonding orbital of the impinging molecule, $^1\Sigma_g$, is excited into a state above the surface Fermi level, after which (2) an electron from the energy band of the metal substrate is excited into the antibonding state of the molecule (or (2) and then (1)), with the impinging molecule finally ending up in the triplet state, $^3\Sigma_u$. The corresponding wave function and total energy when the electron system of the impinging molecule is in the state $^1\Sigma_g$ or $^3\Sigma_u$ are given by

$$|\Phi_i\rangle=C^{+}_{b\uparrow}C^{+}_{b\downarrow}|F\rangle|\Psi_i\rangle, \tag{3.11}$$

$$E_{TOT,i}=\frac{\hbar^2K_i^2}{2M}+2\sum_{|k|\le k_F}\varepsilon_k+2(\varepsilon_a+t)+\frac{U+J}{2}, \tag{3.12}$$

$$|\Phi_f\rangle=\frac{1}{\sqrt{3}}\begin{bmatrix}C^{+}_{a\downarrow}C^{+}_{b\downarrow}C_{k'\downarrow}C^{+}_{k\uparrow}-C^{+}_{a\uparrow}C^{+}_{b\uparrow}C_{k'\uparrow}C^{+}_{k\downarrow}\\ +\frac{1}{2}\left(C^{+}_{a\uparrow}C^{+}_{b\downarrow}+C^{+}_{a\downarrow}C^{+}_{b\uparrow}\right)\left(C_{k'\uparrow}C^{+}_{k\uparrow}-C_{k'\downarrow}C^{+}_{k\downarrow}\right)\end{bmatrix}|F\rangle|\Psi_f\rangle, \tag{3.13}$$

and

$$E_{TOT,f} = \frac{\hbar^2 K_f^2}{2M} + 2 \sum_{|k''| \le k_F} \varepsilon_{k''} + \varepsilon_k - \varepsilon_{k'} + 2\varepsilon_a + J , \tag{3.14}$$

respectively. The ket F corresponds to the ground state of the electron system in the substrate, the ket Ψ corresponds to the state of the CM motion, and k_F corresponds to the Fermi wave number.

Up to 4th order in the perturbation term $H_{M\text{-}S}$, the expression for the probability that the electrons of the impinging molecule, initially in the state ${}^1\Sigma_g$, end up in the state ${}^3\Sigma_u$ is $\omega_{if}=F_1F_2$,. Here, F_1 is an orientation independent term given by

$$F_1 = \left| I\left(\varepsilon_{CM,i}, \varepsilon_{CM,f}; B_+ - \frac{U+J}{2} \right) + I\left(\varepsilon_{CM,f}, \varepsilon_{CM,i}; B_- \right) \right|^2 \frac{\rho(\varepsilon_F)\rho(\varepsilon_F - A)A}{K_i K_f \left[1 - \exp(\beta_s A)\right]}, \tag{3.15}$$

and F_2 an orientation dependent term given by

$$F_2 = \left(f_+^2 g_{k_1}^+ + f_-^2 g_{k_1}^- \right)\left(f_-^2 g_{k_2}^+ + f_+^2 g_{k2}^- \right), \tag{3.16}$$

where

$$I\left(\varepsilon_{CM,f}, \varepsilon_{CM,i}; \varepsilon_{CM,\ell} \right) = \sum_j \frac{\langle \Psi_f | u(z) | \Psi_j \rangle \langle \Psi_j | u(z) | \Psi_i \rangle}{\varepsilon_{CM,i} - \varepsilon_{CM,j} + \varepsilon_{CM,\ell} + i0^+}, \tag{3.17}$$

$$f_\pm(\theta) = \frac{1 \pm \exp(-\gamma r \cos\theta)}{2}, \quad g_k^\pm(\theta) = 1 \pm \frac{\sin(kr \sin\theta)}{kr \sin\theta}, \tag{3.18}$$

$$A = \varepsilon_{CM,i} - \varepsilon_{CM,f} + 2t + (U - J)/2 , \ \varepsilon_{CM,\ell} = \hbar^2 K_\ell^2 / (2M), \quad B_\pm = \pm(\varepsilon + t + U + J + \varepsilon_F)$$

and $\beta_S=(k_B T_S)^{-1}$. The orientation independent term, as shown above, depends only on ρ, the density of states of the substrate electrons, $\varepsilon_{CM\text{-}i}$ and $\varepsilon_{CM\text{-}f}$, the initial and final energy of the impinging molecule, (K_i and K_f, the wave-vectors), and T_S, the surface temperature. On the other hand, the orientation dependent term also depends on the surface-projected bond-length of the impinging molecule, which determines or chooses which substrate electron wave vectors are involved in the transition process.

In Fig.5, the corresponding orientation independent and orientation dependent parts of the dissociation probability are shown, as functions of the final translational energy of the impinging molecules. It can be seen here that the orientation independent part of the dissociation probability peaks at around 175 eV of the final translational energy (cf., curve F_1 in Fig.5). And from the orientation dependent part, a parallel (90°) orientation preference for dissociation can be seen at around 175 eV (cf., curves F_2 in Fig.5). Shown in the upper region of Fig.6 are the corresponding total dissociation probabilities as a function of the final energy of the CM motion after scattering from the metal surface, for the case where the final products are neutral atoms. In the

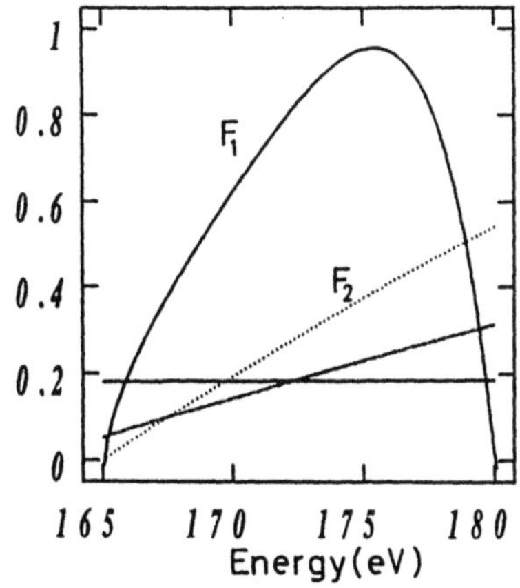

Fig.5 Dissociation probability as a function of the final energy of the CM motion. The curve F_1 shows the orientation (θ) independent part of the transition probability ω_{if}, and, the curves F_2 show the corresponding orientation (θ) dependent part (dots: θ=90°, dash-dot: θ=70°, solid: θ=10°). In the calculation, the following parameter values are used, i.e., ε_a=−15 eV, t=−15 eV, U=16 eV, J=8 eV (these values are roughly related to those of the molecule at equilibrium interatomic distance), and $\varepsilon_{CM\text{-}i}$=200 eV, for the impinging molecule, and ε_F=−5 eV, D=20 eV, T_S=500 K for the metal substrate. The vacuum level is chosen as the origin of energy. A plane wave function was chosen to represent the state of the CM motion, i.e., $\Psi_f(Z) \propto \exp(-iK_fZ)$, and $u(Z)=\exp(-\gamma|Z|)$, γ=1.3 Å^{-1}.

calculation, the impinging neutral molecule is considered to have an initial orientation of 90°, 70° and 10° with respect to the surface normal. One can see here a strong orientation dependence of the total dissociation probability (peak at around 175 eV), with the parallel orientation (90°) preferred. It can be also seen here that the orientation independent part of the dissociation probability determines where, along the final energy of the CM motion of the impinging molecule, a dissociation peak can be observed (cf., curve F_1 in Fig.5 and curves in the upper region of Fig.6). On the other hand, the orientation dependent part of the dissociation probability determines which initial orientation yields the highest dissociation peak (cf., curves F_2 in Fig.5 and the upper region of Fig.6).

Upon changing the characteristics of the substrate, by changing the position of the bottom of the substrate energy band $-D$ from −20 eV to −10 eV relative to the vacuum level (i.e., the value of the filled-band width, ε_F+D, is changed from 15 eV to 5 eV), the behaviour of the orientation independent and the orientation dependent parts of the dissociation probability is affected considerably. We observe a remarkable decrease in yield, a shift of the dissociation peak towards higher final translational energies in the orientation independent part, and a near perpendicular (10°) orientation preference for

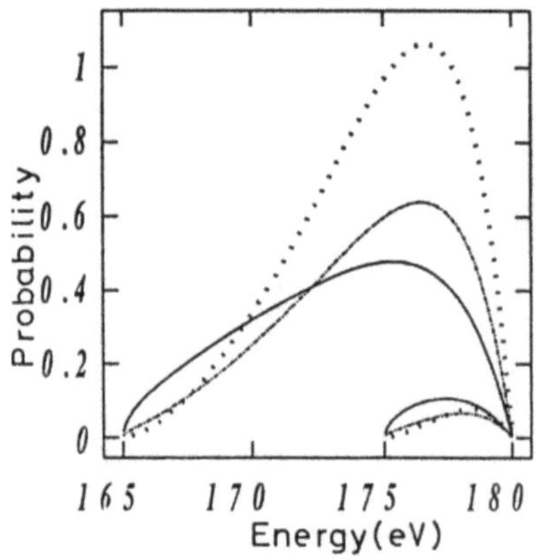

Fig.6 Comparison between the dissociation probability curves as a function of the final energy of the CM motion of the molecule scattered from (upper region) a metal surface with D=20eV, and (lower right-hand region) another metal surface with D=10 eV. The following parameter values are used in the calculation, (ε_a=−15 eV, t=−15 eV, U=16 eV, J=8 eV, $\varepsilon_{CM\text{-}i}$=200 eV, ε_F=−5 eV, T_S=500 K). (dots: θ=90°, dash-dot: θ=70°, solid: θ=10°)

dissociation in the orientation dependent part. As a result, besides a marked decrease in yield, two additional things happen to the total dissociation probability (cf., upper and lower right-hand region of Fig.6). First, there is a general shift of the dissociation peaks towards higher final translational energy region. And second, there is a change in orientation preference for dissociation, from parallel orientation (90°, when D = 20eV) to near perpendicular (10°, when D = 10 eV).

The same calculation was done for the case when the final products are ions, i.e., H^+ and H^-. In this case, it is assumed that the process is triggered/induced by a two-step process involving the excitation of an electron from the singlet bonding state, $^1\Sigma_g$, of the impinging H_2 to a singlet antibonding state, $^1\Sigma_u$, during the scattering process. Except for a difference in the order of magnitude of the dissociation yield, the same strong orientation dependence of the dissociation probability is observed as that shown in the upper region of Fig.6. When the characteristics of the substrate is changed (in the manner mentioned above), the same change is observed in orientation preference as that shown in Fig.6.

To conclude this latter half, it can be said that there are two factors that may help us determine which orientation is preferred by a particular reaction process. The first is the orientation of the impinging molecule. As one can see from the interaction term, whether or not the electrons take a transition to an antibonding state of the impinging molecule (corresponding to a dissociation into individual constituent particles) or to a bonding state is determined by whether the surface-projected bond-length increases or decreases the matrix element V_{ak} or V_{bk}. (cf., (3.9)). The bigger V_{ak} and V_{bk} are, the easier dissociation becomes. The second is the filled-band width (the position of the bottom of the substrate energy band). One can see here that a deeper band (D big) means a bigger k, and like the bond-length, this also affects how V_{ak} and V_{bk} behave for a particular orientation.

4. Summary

In the first half, the dissociative adsorption of hydrogen molecules on metal surfaces is cited as an example of the manifestation of molecular orientation dependence in dynamical processes. In the dissociative adsorption of hydrogen molecules on copper surfaces, a parallel orientation is preferred to a perpendicular orientation. This orientation dependence results in a mechanism called steering effect. This pertains to the ability of the effective anisotropic PES to reorient the hydrogen molecule from an unfavorable orientation (perpendicular) towards a favorable one (parallel), and vice-versa. As a result of the interplay between steering effect and energy transfer effect (from rotation to translation, R-T energy transfer), the dissociation probability of hydrogen on copper is a non-monotonous function of the initial rotational state of the impinging hydrogen molecule. In the second half, the dissociative scattering of hydrogen molecules from metal surfaces is used as a representative dynamical process occurring on surfaces to study how the orientational dependence manifests itself. It can be said that the molecular orientation dependence of a reaction is determined by what characteristics the reaction partners possess [7], i.e., it depends on the bond-length of the impinging molecule and the wave vectors of the substrate electrons involved in the reaction process. Depending on the characteristics and initial state of the reaction partners involved, a perpendicular orientation or parallel orientation will be preferred in the dissociative scattering of hydrogen molecules from metal surfaces (cf., e.g., [4,5]).

References

[1] W.A. Diño, H. Kasai, A. Okiji: J. Phys. Soc. Jpn. **64**, 2478 (1995).

[2] A. Okiji, H. Kasai, and H. Nakanishi, in this issue

[3] H.A. Michelsen, C.T.Rettner, D.J. Auerbach: Phys. Rev. Lett. **69**, 2678 (1992).

[4] U. van Slooten, D.R. Andersson, A.W. Kleyn, E.A. Gislason: Surf. Sci. **274**,1 (1992).

[5] A. Nesbitt, R. Harder, A. Golichowski, G. Herrmann, K.J. Snowdon: Chem. Phys. **179**, 215 (1994).

[6] J.C. Slater: Quantum Theory of Molecules and Solids, (McGraw-Hill, New York, 1962) Vol. 1.

[7] A. Okiji, H. Kasai, A. Kato: J. Phys. Soc. Jpn. **51**, 3291 (1982).

Dynamics and Energy Transfer in Scattering of Xe, O_2 and Cl_2 from Graphite Surface

K. Shobatake[1,2], *K. Ito*[1], *H. Yoshikawa*[2], *T. Ogi*[1], *H. Ariga*[1], *H. Ohashi*[2], *and T. Fujimoto*[1,3]

[1]School of Engineering, Nagoya University, Chikusa-ku, Nagoya 464-01, Japan
[2]Institute for Molecular Science, Myodaiji, Okazaki 444, Japan
[3]Present address: School of Science and Technology, Meijo University, Tenpaku-ku, Nagoya 468, Japan

Abstract. The energy transfer processes in scattering of Xe, O_2, and Cl_2 from graphite (Gr) surface are studied by measuring the velocity distributions using a time-of-flight (TOF) technique. Angular and velocity distributions of scattered atoms and molecules were measured as a function of translational energy and angle of the incident beam as well as substrate temperature. Inelasticity and the degree of the energy transfer are discussed in light of the interaction potential and dynamics of the atoms and molecules involved.

1. Introduction

Scattering experiment is known to provide information on the interaction potential energy hypersurfaces between the molecules in question [1]. Despite a lot of efforts spent for the molecule-solid surface systems studied the interaction potential surfaces have been determined for only a few systems, especially for heavy atom [2-5]. On the other hand dynamics of surface reactions has attracted a great attention of the chemical physicists and molecular dynamicists from the viewpoints of clarifying surface reaction dynamics. Eley-Rideal reactions proceeding via direct collision between the gas phase molecules or atoms with surface adsorbed atoms have been found [6-7]. To understand the dynamics information on the interaction potential is indispensable.

In the course of our scattering study of a chlorine molecule with Si(111) surface bimodal distributions have been observed in the time-of-flight spectra of nonreactively scattered Cl_2 and the observations have been interpreted to be due to deeper interaction potential well between Cl_2 and chlorine-covered Si surface than van der Waals interaction would give [8]. Furthermore in case of molecule-surface scattering inelastic processes can occur via rotational excitation of a projectile molecule as well as vibrational excitation of the surface photons. Therefore it is not always obvious if high inelasticity, when observed, is due to deep potential well or to efficient energy transfer from translational mode to rotational mode or vibrational motions of the surface atoms [8-9]. In order to clarify this point we have studied the scattering of Xe, O_2, and Cl_2 from graphite surface which systems are expected not to have a deep

Springer Series in Solid-State Sciences, Volume 121
Elementary Processes in Excitations and Reactions on Solid Surfaces
Eds.: A. Okiji, H. Kasai and K. Makoshi

chemical well. In the present study the energy transfer processes are studied by measuring the velocity distributions using a time-of-flight (TOF) technique. O_2 and Cl_2 molecules have been chosen since they are not easily detected with lasers to obtain state-selected distribution of the scattered molecules. Angular and velocity distributions of scattered atoms and molecules were measured as a function of translational energy and angle of the incident beam as well as substrate temperature. Inelasticity and the degree of the energy transfer are discussed considering the interaction potential and dynamics of the atoms and molecules involved.

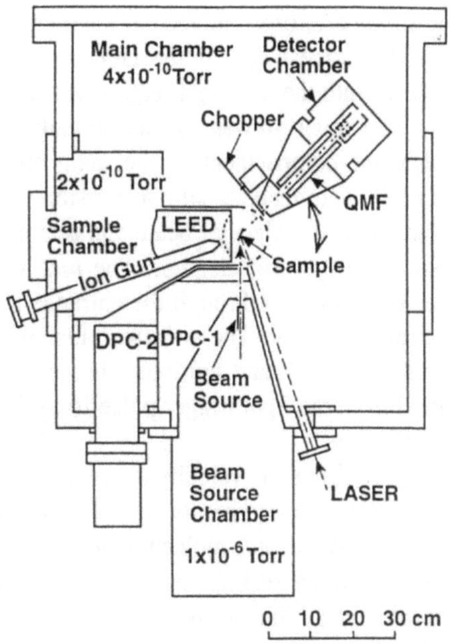

Figure 1. Schematic top view of the molecular beam-surface scattering apparatus. QMF, DPF-1, and DPC-2 denote a quadrupole mass filter, the first and second differential pumping chamber.

2. Experimental

Figure 1 illustrates the schematic top view of the molecular beam apparatus constructed to study molecule-surface scattering processes and photoexcited surface reactions under ultrahigh vacuum conditions. The details of the apparatus are described elsewhere [10] and in this paper only the relevant points will be briefly mentioned. This apparatus consists of a) a main chamber, b) a sample chamber, c) a triply differentially pumped molecular beam source, d) a triply differentially pumped rotatory detector chamber with a quadrupole mass spectrometer equipped with an electron bombardment ionizer.

In the *sample chamber* a home-made reverse view type LEED for characterizing the sample surface is installed. The substrate surface is kept vertical and the incident angle of a beam is varied by rotating the sample around the z-axis. The molecules scattered in the principal plane go out of the sample chamber through a radial slit with a slit height of 5 mm. The pressure differential of a factor of about 20 can be maintained, and thus the pressure of the sample chamber can be kept at 2.0×10^{-10} Torr even when the main chamber is at 3.0×10^{-9} Torr.

A velocity-selected beam was produced by a supersonic expansion of a gas mixture using a triply differentially pumped molecular beam source. The nozzle source chamber is pumped by a 5000 *l*/s diffusion pump (Varian Model VHS-10) using Fomblin oil as diffusion pump oil (Model YH-Vac 25/9). Above the diffusion pump a liquid N_2 trap is installed to lower the vapor pressure of the diffusion pump oil in the nozzle source chamber. A Pt aperture of 2 mm ϕ with a nozzle in 30 μm in diameter was used.

The design of the *rotatory mass spectrometer detector* is based on that of Yuan Lee's type of crossed molecular beam apparatus [11]. The ionizer chamber however is at the farthest end of the detector chamber and then the ions formed there is extracted into the reverse direction of the molecular beam through a quadrupole mass filter, which design was adopted to make the neutral flight length as long as possible despite the compactness of the main chamber. The beam source and the detector chambers were made as slim as possible so that the angle subtended by the directions of the incident molecular beam and scattered beam to be detected spans a wide angle range from 0 to 145 degrees. The ions which passed through the quadrupole mass filter were focused through the entrance lens into the cathode of and detected by an electron multiplier. The negative pulse was amplified, discriminated, and converted to the TTL level, and fed into a home-made multichannel scaler (MCS) which was started from the light trigger pulse from the beam chopper and operated at a constant time interval.

To determine the velocity distributions of scattered species two modes of time-of-flight (TOF) measurements were employed. First in a pulsed incident beam mode a continuous supersonic molecular beam was modulated in the first differential pumping chamber using a rotating chopper disc with symmetrically spaced four raidial slits (1.0 mm wide at 80 mm from the center). The pulsed molecular beam goes through the second differential pumping chamber and then enter the sample chamber. The distance from the chopper to the sample center is 123 mm and that from the sample center to the ionizer is 353 mm.

In a scattered beam modulation mode a beam chopper is mounted on the rotatory detector chamber so that the molecules scattered or desorbed from the sample surface are modulated before entering the detector chamber. In this case a pseudorandom disc with 255 elements mounted on a hysteresis synchronous motor [12] was employed. The distance from the chopper to the ionizer was 283 mm.

The crystalline graphite sample which was used as a substrate is high-quality, highly oriented graphite (Commercial name: Super Graphite, MB grade with a mosaic spread of 0.6°) purchased from Matsushita Research Institute Tokyo, Inc. The topmost layer of the graphite crystal surface (11 mm wide x 15 mm high) was pealed off with a scotch tape in the ambient atmosphere before inserting it into the sample chamber. Before scattering experiments the sample was cleaned by resistive heating to 500 K.

3. Results and Discussion

3.1. Xe-graphite scattering

Figure 2 illustrates the TOF spectra of the scattered Xe atoms measured using the pulsed incident beam mode observed at four scattering angles, θ_f = 0, 50, 60, and 80° for the translational energy and angle of the incident beam, E_i = 0.45 eV and θ_i =

60°, and for the substrate temperature, T_s = 298 K. A well depth of D = 162 ± 4 meV has been suggested by Vidali et al. [13] for the Xe-graphite system. Therefore an incident energy of 0.45 eV is about 2.8 times as large as the well depth. From the TOF spectrum observed at θ_f = 50° one notes that there are three velocity components. Therefore all of the TOF spectra for scattered Xe atoms have been analyzed, assuming that each spectrum is composed of a linear sum of three velocity components, two shifted Maxwell-Boltzmann (MB) distributions (Eq. 1) and a MB distribution (Eq. 2). Each of the shifted-MB distributions is assumed to have a form:

$$f_i(v) = v^3 \exp[-\frac{M}{2RT_i}(v-\alpha_i)^2] \quad (1)$$

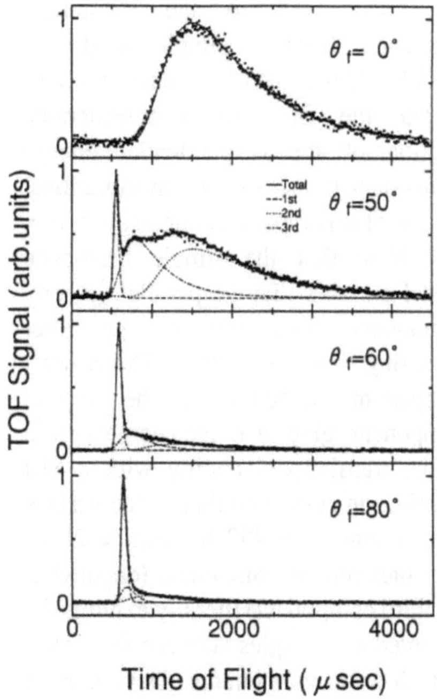

Figure 2. TOF spectra of Xe atoms scattered from Gr surface at T_s = 298 K for E_i = 0.45 eV and θ_i= 60°. Dots: observed, lines: results of the deconvolution analysis.

where α_i, T_i (i = 1, 2), and M are the drift velocity, the temperature representing the velocity spread of the distribution, and the mass of the scattered species, respectively. The slowest component (defined here as the second component) is assumed to obey a MB form:

$$f(v) = v^3 \exp[-\frac{Mv^2}{2RT_B}] \quad (2)$$

where T_B and M are the translational temperature which is expected to be close to the surface temperature, T_s, and the mass of the scattered species, respectively. The velocity distribution observed for scattering angle θ_f = 0° was found to comprise of a MB distribution with T_B very close to the surface temperature T_s. At larger scattering angles, however, one finds that even the TOF spectra measured can be fitted well only by calling for the three velocity components mentioned above.

The fastest component (defined here as the first component) exhibits a rather sharp distribution and must be due to direct inelastic scattering whereas the slowest component may be interpreted as the atoms scattered via trapping-desorption as explained by Hurst et al. for bimodal distributions of Xe-Pt(111) scattering [14]. From our experiment three components instead of two have been found.

Figure 3 illustrates the flux-averaged translational energies for all three components vs. scattering angle, θ_f. The angular distribution for the first component fits fairly well to the theoretical prediction from a simple hard-cube model [16] (denoted by a dash-dot line) despite the fact that a considerable fraction of the perpendicular energy component is lost to surface vibrational motion at superspecular angles, $\theta_f > \theta_i = 35°$. Note that the simple hard-cube model assumes that the parallel velocity component does not change after scattering from the surface. The slowest component (denoted as the second component) gives average energies close to the translational energy with a MB distribution corresponding to the surface temperature, $T_s = 298$ K, $\langle E_s \rangle = 2k\,T_s$. The intermediate component (denoted as the third component) literally exhibits the intermediate energies between the fastest and slowest components. Deviation of the observed energy from the simple hard-cube model for the fastest component is seen at subspecular angles, $\theta_f < \theta_i = 35°$.

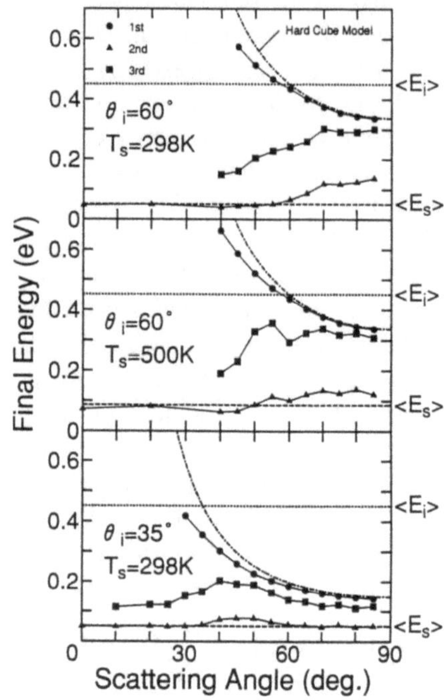

Figure 3. Flux-averaged final energy of scattered Xe vs. θ_f for each velocity component. The dash-dot curve: theory assuming a hard-cube model. $\langle E_i \rangle$: incidence energy, $\langle E_s \rangle = 2k\,T_s$.

Figure 3 also shows the angular distribution of the average energies under different experimental conditions, T_s and θ_i. The general trends are similar for all the scattering processes performed at E_i = 0.45 eV, except that the energy loss is larger for lower incident angle, $\theta_i = 35°$ than 60°. Scattering experiments were also performed for E_i = 0.17 eV and 70 meV in the pulsed incident beam mode. It has been found that the trend is similar for E_i = 0.17 and 0.07 eV, except that 1) the agreement is better between the hard-cube model and the experiment for the first component and 2) the presence of the third component was not conspicuous for the scattering at E_i = 70 meV for which the TOF spectra were analyzed assuming that there are only two components, the shifted-MB and MB distributions.

Figure 4 illustrates the polar plot of the flux intensity for all the experiments carried out. For the lowest collision energy E_i = 70 meV the peak angles are smaller than the incident angles while those for E_i = 0.17 and 0.45 eV are larger than the specular angles, which indicates that a large fraction of the incidence translational energy E_i is lost to vibrational motions of the surface atoms. In the atom-surface scattering the effective mass of the surface cube varies with collision energy. At low incidence energies the incident atom feels the force due to rather smooth surface and the surface atoms can follow the slow movement of the incident atom. As the incidence energy is increased, the incident atom start to feel the repulsive part of the interaction potential and the surface corrugation. In the high energy limit the incident atom feels only the repulsive force due to individual surface atoms. Therefore effective mass varies with incidence energy. Under the assumption of the hard cube model one can estimate the effective mass of the surface cube, M_s, from the two conditions, the conservation of the total momentum in the perpendicular direction and the total energy before and after the collision. The effective mass relative to the mass of the incident atom, m_g, can be estimated as a function of incident angle, θ_i, and scattering angle, θ_f. The effective mass for a certain scattering process may be reasonably defined as that for the peak scattering angle, θ_f^P:

$$\frac{M_s}{m_s} = \frac{\tan\theta_f^P + \tan\theta_i}{\tan\theta_f^P - \tan\theta_i}. \quad (3)$$

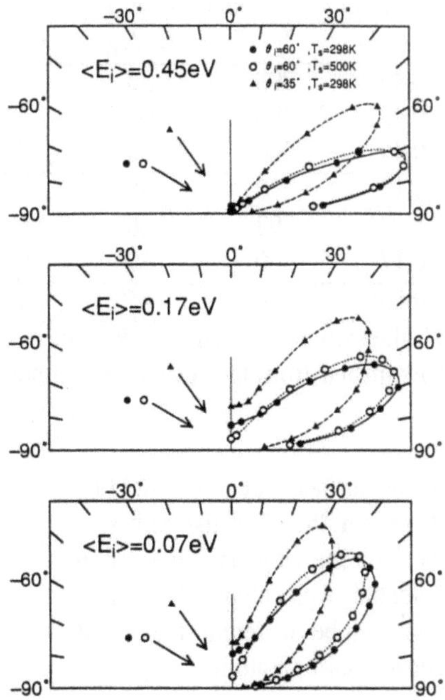

Figure 4. Polar plot of the scattered Xe intensity for all incidence energies studied $<E_i>$. ▲: θ_i=35°, T_s = 298K, ●: θ_i=60°, T_s = 298K, and ○: θ_i=60°, T_s = 500K.

Evaluated ratios of M_s/m_s using Eq. (3) are summarized in Table 1 along with the results for the Xe-Pt(111) scattering experiments by Rettner et al. [17]. A typical effective mass for E_i = 0.45 eV and θ_i = 35° is estimated as M_s = 2.9 Xe which is about 32 carbon atoms.

At this point it would be worthwhile to speculate the origin of the third component observed in the Xe-graphite scattering at E_i = 0.45 eV. Two components have been observed in the TOF spectra by other investigators, especially when the incidence energy is small compared with the well depth of the atom-surface interaction potential and surface temperature is low. Our experiment was performed for a long flight length from the surface to the ionizer region, L = 353 mm and more accurate TOF spectra have been obtained. However we know no cases in which three velocity components have been observed in the TOF spectrum except the cases where phonon transitions have been observed in He or H_2 scattering from surfaces [18]. In the scattering at E_i = 0.170 eV the third component is not so conspicuous in TOF spectra as those for E_i = 0.45 eV. It may be because at fairly low collision energies the Xe-Gr pair starts to feel the surface corrugation, although the surface is rather smooth. Further the graphite surface is not as close-packed, say as close-packed

Table 1. Effective mass of the surface cube relative to Xe mass, M_s/m_s, calculated from Eq. (3).

E_i (eV)	T_s (K)	θ_i(deg.)	θ_f^P(deg.)	M_s/m_s
0.45	298	35	55	2.9
0.45	298	60	72	3.6
0.45	500	60	73	3.3
0.17	298	35	72	4.8
0.17	298	60	68	5.7
0.17	500	60	64	11.9
0.50[a]	200[a]	45[a]	58[b]	4.3[b]

[a] Taken from the results by Rettner et al. [Ref. 17].
[b] Estimated from the results by Rettner et al.

metal surface such as Ni(111); there is a "hole" at the center of the benzene ring, the interaction potential well along the center line can be different from that above a carbon atom.

3.2. O_2-graphite scattering

The potential well of O_2-Gr pair has been estimated as 102 meV [13]. The TOF measurements were performed in the scattered beam modulation mode. Figure 5 shows the TOF spectra for E_i = 0.29 eV at $\theta_i = 60°$ and T_s = 300 K. The data have been analyzed assuming each spectrum consists of two components, the shifted-MB and pure MB formulae. The incident energy E_i = 0.29 eV is about 2.8 of the well depth

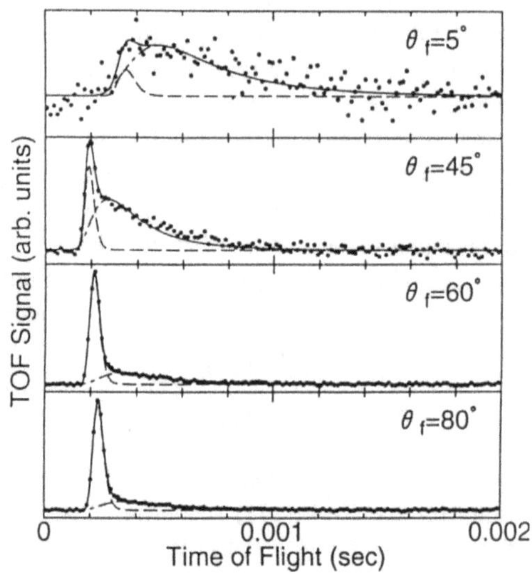

Figure 5. TOF spectra of O_2 scattered from Gr at T_s = 300 K for E_i = 0.29 eV and θ_i= 60°. Dots: experiment, and lines: results of the deconvolution analysis.

which is close to that for Xe-graphite scattering carried out for E_i = 0.45 eV, θ_i = 60° and T_s = 298 K, as already shown in Figure 2. The angular distribution observed is peaked at 68°, where-as that in a Xe-graphite case is peaked at 72° as shown in Table 1. The TOF spectra at 60 and 80° in Figure 5 look very similar to those in Figure 2. Therefore in O_2-graphite scattering it seems that the second components (which exhibit distributions for a translational temperature close to the surface temperature, T_s) observed in Figure 2 for Xe-Gr system are missing, and only the intermediate velocity components appear.

Figure 6 shows the angular distributions of the flux-averaged translational energies for the O_2 scattered. The experimental conditions are the same as those shown in Figure 5. The intensities at the subspecular angles are so low that reliability of the values are low. Solid circles denote the average final energies for the fast component, open circles for the slow components and the open squares denote those for both components. The dotted curve represents the simple hard-cube model calculations. It is rather surprising that the angular distribution of the final translational energy for the faster component fits fairly well to that for the simple hard-cube model. The observed distribution tends to be flatter than the hard-cube model predicts. Although not shown, the angular distribution for the scattering at θ_i = 35° also fits fairly well to the simple hard-cube model, which means that the translational energy for the first component becomes about 40% of the incidence energy at θ_f = 85° and is very efficiently transferred to rotational motion of O_2 molecule as well as phonon modes. Although only the translational energy distributions have been measured, anticorrelation may have occurred, that is, as rotational mode is more excited the final translational energy is decreased, as was observed for NO scattered from Ag(111) [19].

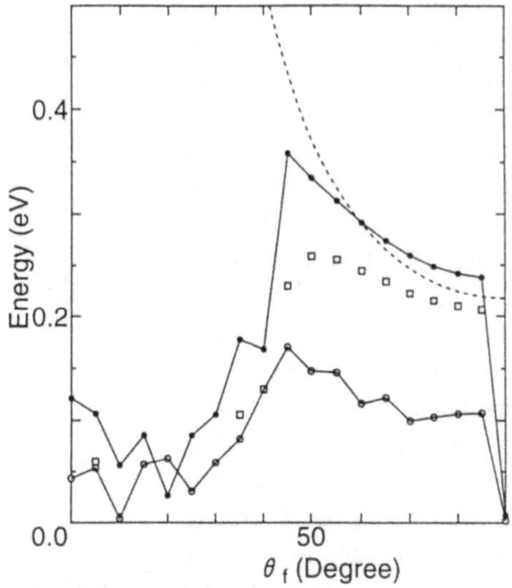

Figure 6. Angular distribution of the final energy of O_2 scattered from Gr at T_s = 300 K for E_i = 0.29 eV and θ_i= 60°. For details see the text.

3.3. Cl_2-graphite scattering

The Cl_2-graphite scattering experiments have been performed since the bond distance for Cl_2 molecule is long (1.99 Å) and the anisotropy in the interaction potential is expected to be large compared with O_2 scattering. Figure 7 summarizes the averaged final energy of Cl_2 scattered from graphite surface at T_s = 500 K for E_i = 0.571 eV and θ_i = 35°. The experiment was also carried out in a scattered beam modulation mode. The solid circles and open squares are for flux-averaged final energies of the fast component and the slow component, respectively. The dotted curve represents calculated values assuming a hard-cube model. It is obviously seen that the fit of the observed final values for the fast component to those of theory is rather excellent at larger scattering angles. The fit is rather poor at smaller angles, especially for subspecular angles. It is because a large fraction of the incidence energy is lost upon scattering. The agreement between theory and experiment is good for superspecular angles, $\theta_f > \theta_i$. Although angular distribution is not shown here, it would be suffice to mention that the scattered Cl_2 flux is peaked at 55°. The potential well for Cl_2-Gr pair has not been obtained. It would be reasonable to assume $D \sim 0.17$ eV. $E_i = 0.571$ eV is about three times as large as the well depth, D. Therefore one may be able to compare the scattering data for Cl_2-Gr with those for O_2-Gr and Xe-Gr described above. The TOF profiles are similar for the three systems except that for the Xe-Gr system one finds the slow components which can be represented by the translational temperature, T_B close to T_s.

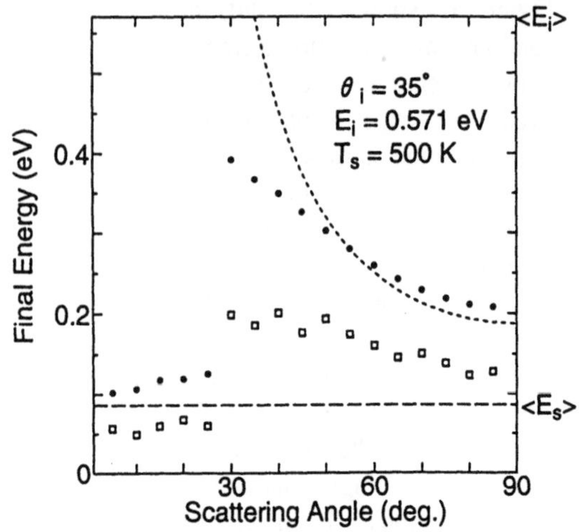

Figure 7. Angular distribution of the final energy of Cl_2 scattered from Gr at $T_s = 500$ K for $E_i = 0.571$ eV and $\theta_i = 35°$. For details see the text.

This trend is in remarkable contrast to the scattering of Cl_2 from chlorine-covered Si(111) for $\theta_i = 35°$, $E_i = 0.56$ eV, and $T_s = 420$ K, which exhibits almost a flat average energy vs. scattering angle and furthermore a fraction of the trapping-desorption component is dominant [8-9]. Therefore we conclude that the interaction between Cl_2 and Cl-covered Si(111) surface is of chemical nature rather than van der Waals nature.

4. Concluding Remarks

The scattering processes of Xe, O_2, and Cl_2 from graphite surface have been studied by measuring the velocity distribution of the scattered species using a TOF technique. The velocity distributions of scattered Xe atoms for $E_i = 0.45$ eV exhibit three velocity components, whereas those for the O_2 and Cl_2 show only two components. The TOF spectra are similar for three projectiles at superspecular angles except that, 1) in Xe scattering, we observe a slow component which is represented by T_B close to substrate temperature, T_s, in addition to two components, the directly scattered component and another component which is directly scattered but with lower translational energy than the fastest one, and 2) at subspecular angles deviation from the hard-cube model is largest for Cl_2 scattering than Xe and O_2, which means the rotational excitation is higher for Cl_2 than O_2. The average translational energy of the directly scattered component agrees fairly well with the hard-cube model.

5. Acknowledgements

The authors thank Messrs. T. Oguri, H. Deguchi, and K. Yoshimine for their experimental assistance in obtaining scattering data. Enlightening discussions with Prof. Y. Matsumoto and Mr. N. Yamanishi of the Univ. of Tokyo are greatly acknowledged. This work is partially funded by the Grant-in-Aid from the Ministry of Education, Science, and Culture (No. 07404033).

References

1. U. Buck, Advan. Chem. Phys. **30**, 313 (1975).
2. J. A. Barker and D. J. Auerbach, Surf. Sci. Rep. **4**, 1 (1985).
3. G. P. Brivio and T. B. Grimley, Surf. Sci. Rep. **17**, 1 (1993).
4. a) D. S. Bethune, J. A. Barker, and C. T. Rettner, J. Chem. Phys. **92**, 6847 (1990); b) J. A. Barker and C. T. Rettner, J. Chem. Phys. **97**, 5844 (1992); and see the references therein.
5. J. Harris, "Mechanical Energy Transfer in Particle-Surface Collision, in Dynamics of Gas-Surface Interactions," eds. C. T. Rettner and M. N. R. Ashfold, (Royal Soc. Chemistry, 1992), pp. 1.
6. E. W. Kuipers, A. Vardi, A. Danon, and A. Amirav, Phys. Rev. Lett. **66**, 116 (1991).
7. C. T. Rettner, J. Chem. Phys. **101**, 1529 (1994) and references therein.
8. H. Yoshikawa, "Dynamics of Surface Reactions of Si(111) with Chlorine Molecules" Ph. D. Dissertation, The Graduate University for Advanced Studies, September, 1994.
9. H. Yoshikawa and K. Shobatake (unpublished results).
10. H. Yoshikawa, H. Ohashi, K. Tabayashi[a], M. Suzui, T. Horigome, K. Hayakawa[b], S. Kato, K. Shobatake, "Modification of a Molecular Beam-Surface Scattering Apparatus with a Rotatable Mass Spectrometer Detector" (to be published).
11. Y. T. Lee, J. D. McDonald, P. R. LeBreton, and D. R. Herschbach, Rev. Sci. Instrum., **40**, 1402 (1969).
12. V. L. Hirschy and J. P. Aldridge, Rev. Sci. Instrum. **42**, 381 (1971).
13. G. Vidali, G. Ihm, H-Y. Kim, M. W. Cole, Surf. Sci. Rep. **12**, 133 (1991)
14. J. E. Hurst, C. A. Becker, J. P. Cowin, K. C. Janda, and L. Wharton, Phys. Rev. Lett. **43**, 1175 (1979).
15. C. T. Rettner, E. K. Schweizer, and C. B. Mullins, J. Chem. Phys. **90**, 3800 (1989).
16. a) R. M. Logan and R. E. Stickney, J. Chem. Phys. **44**, 195 (1960); b) E. K. Grimmelmann, J. C. Tully, and M. J. Cardillo, J. Chem. Phys. **70**, 1039 (1980)
17. C. T. Rettner, J. A. Barker, and D. S. Bethune, Phys. Rev. Lett. **67**, 2183 (1991).
18. J. P. Toennies, "Phonon Interactions in Atom Scattering from Surfaces", in Dynamics of Gas-Surface Interaction, eds. G. Benedek and U. Valbusa, (Springer-Verlag, Berlin, 1982), pp. 208.
19. J. Kimman, C. T. Rettner, D. J. Auerbach, J. A. Barker, and J. C. Tully, Phys. Rev. Lett. **57**, 2053 (1986)

Collision-Induced Desorption of Xe Adsorbed on Pt(111) by Ar Atoms

D. Kulginov[1], *M. Persson*[1], *C.T. Rettner*[2], *and D.S. Bethune*[2]

[1]Department of Applied Physics, Chalmers University of Technology and University of Göteborg, S-412 96 Göteborg, Sweden
[2]IBM Almaden Research Center, 650 Harry Road, San Jose, USA

Abstract. We present a short description of our recent studies of collision-induced desorption of Xe adsorbed on Pt(111) by hyperthermal Ar atoms. We combine molecular beam experiments with molecular dynamics calculations in an effort to quantify fully the dynamics of this model system.

1. Introduction

Most of our knowledge of the dynamics of gas–surface reactions has been obtained by studying carefully chosen elementary systems. The results of such studies can provide great insight into the mechanisms of important catalytic cycles. If we are to bridge the "pressure gap" between (low pressure) studies and (high-pressure) catalysis, however, we must be sure to consider all relevant processes. In many "real" systems, it seems likely that collision-induced chemistry can play an important role. Any comprehensive model of heterogeneous catalysis must therefore include reactions induced by the bombardment of impinging species. The simplest of these processes is collision-induced desorption (CID), where an incident species transfers sufficient energy to an adsorbate to cause it to leave the surface. Despite the importance of such phenomena, there have been relatively few relevant experimental or theoretical studies [1–6]. The first experimental observation of CID was reported by Beckerle *et al.* [2], who scattered Ar beams from CH_4 adsorbed on Ni(111). More recently, Szulczewski and Levis [3] observed the CID of NH_3 from Pt(111). Both groups analyzed the results in terms of simple hard cube and sphere models [2–4]. Beckerle *et al.* were able to rationalize dependence of the CID probability on incidence energy and angle. Szulczewski and Levis argued that it is possible to determine the binding energy from the energy of the desorption threshold. Zeiri *et al.* [1] had pointed out earlier that analysis of the states of the scattered and desorbed species should provide information on the binding properties of adsorbate. Zeiri and coworkers [1, 5, 6] have carried out detailed dynamical studies of the CID process using a classical molecular dynamics. Although instructive, the value of this work is limited by uncertainties in the form of the chosen (Lennard-Jones) interaction potentials.

In this short paper, we describe our own recent studies of CID dynamics, concerning the CID of Xe from Pt(111) induced by collisions of hyperthemal

Springer Series in Solid-State Sciences, Volume 121
Elementary Processes in Excitations and Reactions on Solid Surfaces
Eds.: A. Okiji, H. Kasai and K. Makoshi

Ar. We have combined molecular beam experiments with detailed molecular dynamics calculations in an effort to quantify fully the dynamics of this model system. The calculations employ accurate semi-empirical potential energy functions for both Xe/Pt(111) and Ar/Pt(111) interactions. Here we present the first results of this study, a comparison of measured and calculated desorption yields for monolayer Xe coverage.

2. Experiments

The apparatus and methods for studying gas–surface scattering have been described in detail previously [7]. A supersonic beam of Ar seeded in He was directed at the Pt(111) sample held on a manipulator that gave fine control of the incidence angle, θ_i, and surface temperature. The translational energy, E_i, was varied by changing the seeding ratio and by varying the tungsten source temperature up to 2300 K. The surface was cleaned and cooled to 90 K before exposing it to a saturation dose of Xe from an auxiliary beam. Desorption yields were determined by the initial rate of decline in the Xe coverage relative to the flux of each incident Ar beam. Xe coverages were measured using temperature programmed desorption. Measured desorption yields, N_d, for different incidence energies and angles are shown in Fig. 1. The data exhibit several notable features, such as deviation from scaling with "normal energy", $E_i \cos^2 \theta_i$; an energy threshold (0.5 eV) of about twice the binding energy of Xe in the overlayer; and a desorption yield of greater than unity for high E_i. This behaviour will be discussed in detail in Sec. 4.

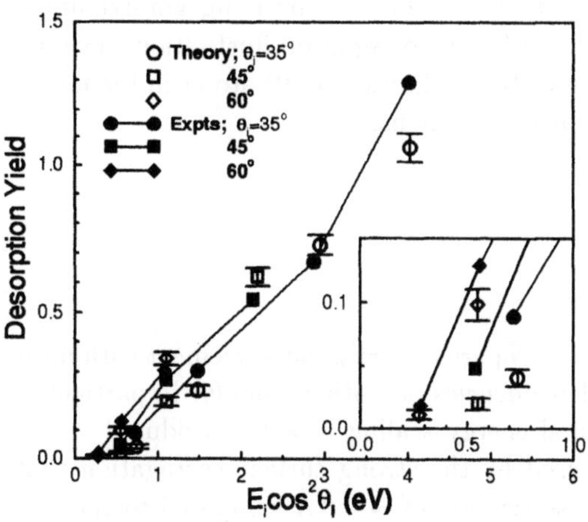

Fig. 1. Comparison between measured and calculated desorption yields for different incidence energies, E_i and angles, θ_i.

3. Theory

One of the main motivations for choosing the Ar–Xe/Pt(111) system was that the relevant gas–surface potentials are very well characterized with respect to interactions and structure. In the case of the Xe–Pt(111) interactions, we have employed the potential energy function developed by Barker and Rettner [8] that is consistent with a larger range of dynamical and equilibrium data than for any other gas-atom–surface system. This function combines the pair-wise interaction that consists of an exponential repulsive branch and a van der Waals branch with a non-pair-wise flat repulsive term that mimics the interaction with the delocalized conduction electrons of Pt(111). The potential energy function that we employed for the Ar–Pt(111) interaction has a similar but simpler form where the interaction represented by the non-pair-wise term is instead modeled by making the the potential in the repulsive branch noncentral. This function was recently developed by Kulginov *et al.* [7] and is also consistent with scattering and trapping data over a large range of incidence energies and angles that covers the range of interest in this work. The Ar–Xe and Xe–Xe interactions were modeled by well-characterized gas-phase pair potentials developed by Aziz and Slaman [9] and Aziz and van Dalen [10]. Finally, the interactions between the Pt ion-cores were described by a simple nearest neighboring central harmonic force model, which gives a good representation of the high frequency part of the bulk phonon spectrum [11].

Our classical molecular dynamics approach follows closely that by Arumainayagam *et al.* [12]. The Pt(111) surface and the $(\sqrt{3}\times\sqrt{3})R30°$ overlayer of Xe atoms were represented by a slab with periodic boundary conditions in the lateral directions. The slab consisted of 4 layers of Pt atoms with 192 atoms in each layer and an overlayer of 64 Xe atoms. The temperature of the slab and the energy dissipation into the bulk was modeled by Langevin dynamics; random and friction forces, as related by the dissipation-fluctuation theorem, were applied to the atoms in the third layer whereas the Pt atoms in the fourth layer were clamped at their equilibrium positions.

4. Results and discussion

As shown in Fig. 1, the calculated N_d are in very good agreement with measured values with respect to the dependencies on both E_i and θ_i. In particular, the observed deviation from normal energy scaling is well-reproduced in the calculation. This deviation is caused by the strong surface corrugation that is introduced by the large Xe–Xe separation of ~ 4.8 Å compared to the Ar–Xe dimer distance of ~ 4.1 Å. The calculations also reproduce the desorption threshold. We find that the value of this threshold is determined by a delicate competition between the energy transfers from the Ar atom to an adsorbed Xe atom and from this Xe atom to the substrate. In fact, this competition was found to be very sensitive to the surface temperature. The calculated yields are in good overall agreement with measurements. The mean yield does indeed exceed unity as E_i exceeds 4 eV. We find that at $E_i = 5.8$ eV and $\theta_i = 35°$, where $N_d > 1$, there are collision events that induce desorption of 4 atoms and at $E_i = 2.2$ eV there are events where two Xe atoms desorb. The possibility to desorb a large number of Xe atoms in a single collision was found to be an effect of the adsorbate–adsorbate interactions. In summary, we find good overall agreement between calculated and measured results for this model CID system. More detailed results of this study, including the final-angle-resolved Xe flux and velocity distributions at different Xe coverages will be presented in the near future.

Acknowledgements

The work by MP has been supported by the Swedish Natural Science Research Council (NFR). DK acknowledges support from the Chalmers fund for research collaboration with Eastern Europe and the Swedish Royal Academy of Sciences (KVA). CTR and DSB thank Daniel Auerbach and John Barker for many useful discussions.

References

1. Y. Zeiri, J. J Low, and W. A. Goddard III, J. Chem. Phys. **84**, 2408 (1986).
2. J. D. Beckerle, A. D. Johnson, and S. T. Ceyer, Phys. Rev. Lett. **62**, 685 (1989); J. Chem. Phys. **93**, 4047 (1990).
3. G. Szulczewski and R. J. Levis, J. Chem. Phys. **101**, 11070 (1994).
4. G. Szulczewski and R. J. Levis, J. Chem. Phys. **98**, 5974 (1993).
5. Y. Zeiri, Surf. Sci. **231**, 404 (1990).
6. Y. Zeiri and R. R. Lucchese, Surf. Sci. **264**, 197 (1992).
7. J. A. Barker and C. T. Rettner, J. Chem. Phys. **97**, 5844 (1992).
8. D. Kulginov, M. Persson, C. T. Rettner, and D. S. Bethune, J. Phys. Chem. (in press).
9. R. A. Aziz and M. J. Slaman, Mol. Phys. **57**, 825 (1986).
10. R. A. Aziz and A. van Dalen, J. Chem. Phys. **78**, 2402 (1983).
11. J. E. Black, F. C. Shanes, and R. F. Wallis, Surf. Sci. **133**, 199 (1983).
12. C. R. Arumainayagam, R. J. Madix, M. C. McMaster, V. M. Suzawa, and J. C. Tully, Surf. Sci. **226**, 180 (1990).

Spin-Polarized Atomic Beam as a Probe of Spin Correlation at the Topmost Layer of Surfaces

E. Torikai and H. Hori

Faculty of Engineering, Yamanashi University,
Takeda 4-3-11, Kofu, Yamanashi, 400 Japan

Abstract. A spin-polarized and energy-tunable cesium atomic beam in the ground state is under development to study the dynamical atom-surface interactions attempting the application to probing the spin correlation at the topmost layer of surfaces and adsorbates. The kinetic energy of the beam is either thermal or tunable between 0.1 and 10 keV, where the spin as well as the velocity parallel to a surface play important roles on resonant charge exchange. By means of laser optical pumping, almost complete spin-polarization has been attained for the thermal energy atomic beam. We also discuss the basic concept of spin sensitivity in the resonant charge exchange process speculatively.

1. Introduction

Dynamical space-time correlation of spins at the topmost layer of a solid surface is a key parameter to understand characteristic surface phenomena such as reaction, catalysis, friction and conduction. However, very little information is available so far on the behavior of surface spins except for ferromagnetism on transition metal surfaces. Freeman et al. predicted significant enhancement of the magnetic moment at the topmost layer of 3d and 4f transition metals by ab initio band calculations[1]. Alvarado et al. observed the ferromagnetic transition and critical behavior of Ni(100) at a temperature equivalent to the bulk T_c by spin-polarized low energy electron diffraction(SPLEED)[2]. Electron capture spectroscopy (ECS) developed by Rao et al., on the other hand, revealed the existence of short range correlation up to $2T_c$[3]. Magnetic properties of Fe and Fe/W, Fe/Ag and Fe/GaAs surfaces measured by SPLEED, ECS, spin-polarized angle-resolved photoemission (SPARES) and conversion electron spectroscopy (CEMS) also show discrepancies with each other[4]. These results strongly suggest the existence of the characteristic spin fluctuation at the topmost layer of the transition metal surface.

In order to study the dynamical space-time correlation of spins on metal surface, we are developing a spin-polarized cesium (Cs) atomic beam with an energy range between 0.03 eV(thermal) and 10 keV by using a laser optical pumping method. In this paper, we report on the characteristic properties of the spin-polarized atomic beam as well as the present status of development.

Springer Series in Solid-State Sciences, Volume 121
Elementary Processes in Excitations and Reactions on Solid Surfaces
Eds.: A. Okiji, H. Kasai and K. Makoshi

2. Principles of Optical Pumping of Cesium Atomic Beam[5,6]

Figure 1 shows the schematic diagram of the spin-polarization of the cesium atomic beam. First, a single mode diode laser selectively excites atoms in the ground state $|6\ ^2S_{1/2}, F=3>$ to the excited state $|6^2P_{3/2}, F=4>$, and the subsequent spontaneous decay populates the $|6\ ^2S_{1/2}, F=4>$ ground hyperfine state (hyperfine pumping) as shown in Fig. 1(a). Consecutively the atomic spin orientation $|m_F>$ is selected to the highest quantum number $m_F=4$ by using a circularly polarized σ^+ laser which drives the cyclic transition between the ground state $|6\ ^2S_{1/2},F=4>$ and the excited state $|6^2P_{3/2}, F=5>$. An average quantum number $< m_F>$ increases by one in two pumping cycles (Zeeman pumping) as shown in Fig.1(b). The final spin state of the atomic beam has, in principle, the perfect spin-polarization of $|6\ ^2S_{1/2}, F=4, m_F=4>$ which corresponds to the pure state of $|S=1/2, I=7/2, m_s=+1/2, m_I=+7/2>$.

Spin direction is reversed by using the σ^- polarized light. Other spin states can also be realized flexibly by changing the laser polarization for Zeeman pumping, e.g. the linearly polarized σ light pumps the atomic beam to a specific alignment with a multipole moment.

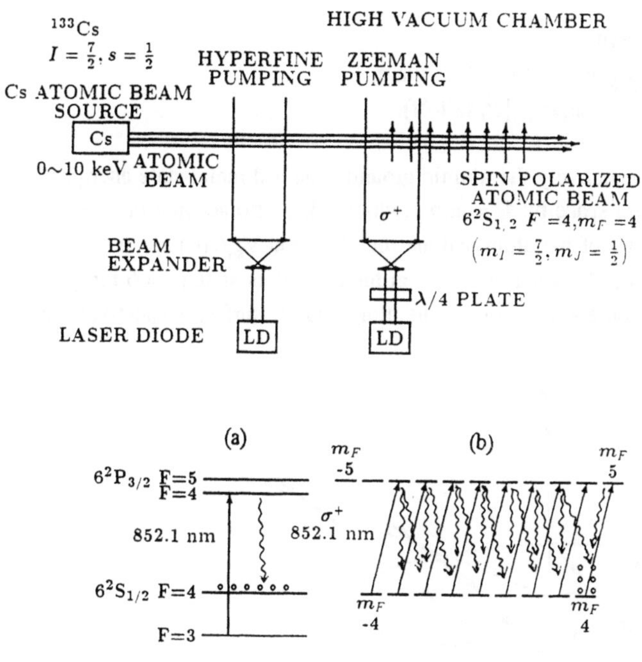

Fig. 1 Schematic diagram of the spin-polarization of the Cs atomic beam (top). The energy levels and transitions used for (a) the hyperfine pumping and (b) Zeeman pumping (bottom).

3. Basic Concept

In atom-surface scattering, the interactions between projectile and surface atoms are strongly influenced by their spin states as well as their relative kinetic energies.

When a spin-polarized alkali atom is scattered from a magnetic surface at grazing angles of incidence and reflection as shown in Fig. 2, we expect that resonant charge exchange between a moving atom and a surface takes place selectively by their spin states as discussed in the following sections. The selectivity of the electron transfer affects various surface phenomena such as the chemisorption probability, the spin and charge states of the reflected particle and the Auger electron emission from the surface. By observing the change in these parameters as a function of the spin and the kinetic energy of the impinging atom, we can gain information about the spin correlation of the electron spins at the topmost layer of the surface.

3.1 Spin Selectivity in Resonant Electron Exchange

As a simple model of the spin and kinetic energy dependent alkali atom-surface interaction, we modify the Newns Hamiltonian in second quantization[7] as follows,

$$\begin{aligned} H(z(t)) = & \sum_{\sigma}\varepsilon_0 n_\sigma + U n_\uparrow n_\downarrow \\ & + \sum_{k,\sigma}\varepsilon_{k+Q} n_{k\sigma} + \sum_{k,\sigma}\left(V_{k\sigma}(z(t)) C_{k\sigma}^{+} C_{a\sigma} + \text{h.c.} \right) \\ & - \{ 1 - n_{a\uparrow} - n_{a\downarrow} + 2 n_{a\uparrow} n_{a\downarrow} \} e^2/4z(t-\delta t), \end{aligned}$$

where ε_0 and ε_{k+Q} are respectively the atomic ground state and metal state energies shifted in the momentum space by $\mathbf{Q} = m_e \mathbf{v}_{//}$,which takes into account the relative velocity of the rest frames of the atom and the metal. U and $V_{k\sigma}(z(t))$ are the Coulomb repulsive energy between electrons in the atomic ns state (n = 6 for Cs) and the transfer interaction energy between the atom and the surface respectively. n_σ

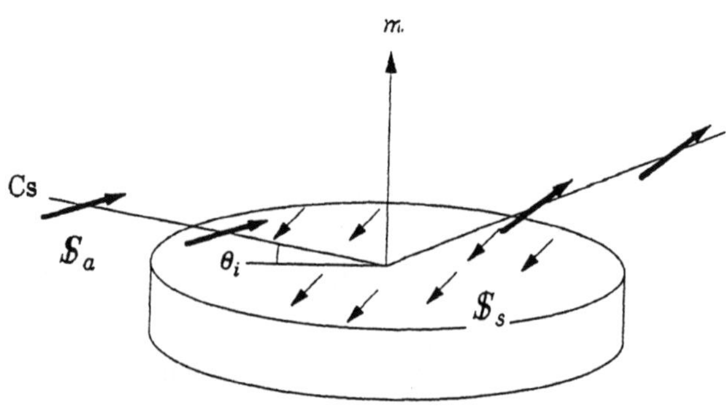

Fig. 2 Sketch of the atom-surface grazing angle scattering in spin ordering system. $\mathbf{S_a}$ and $\mathbf{S_s}$ indicate spin directions of an atom and a surface, respectively.

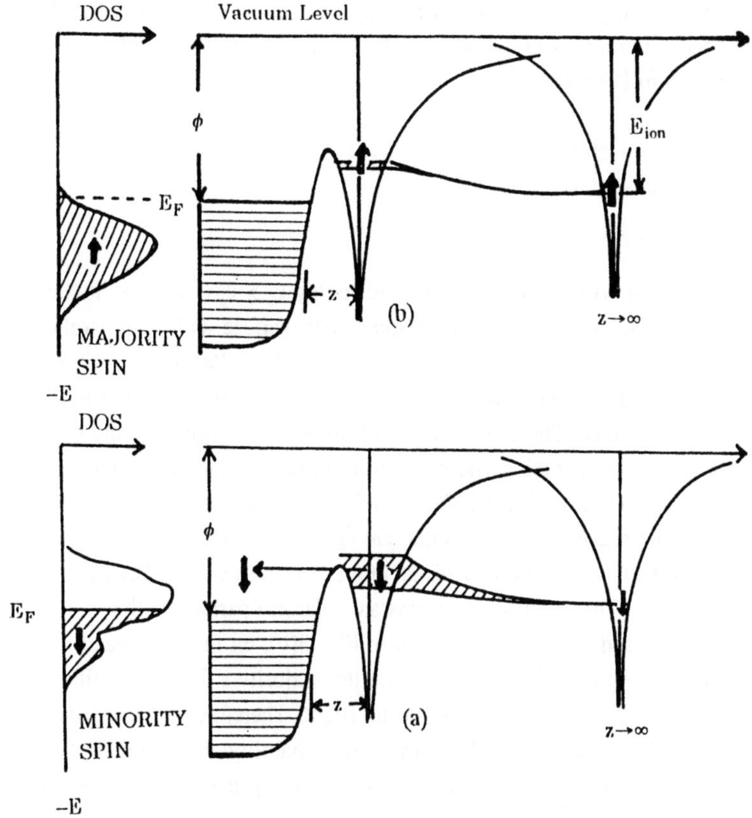

Fig. 3 Schematic diagram of the resonant electron exchange processes that can occur when $\varepsilon_0 > \varepsilon_F$. Spin direction of atoms is along (a) the minority spin and (b) majority spin.

and $n_{k\sigma}$ are the electron occupation numbers in the atom and in the surface band states |k>, respectively. The occupation numbers are written as $n_\sigma = C_{a\sigma}^+ C_{a\sigma}$ and $n_{k\sigma} = C_{k\sigma}^+ C_{k\sigma}$, by using the creation and annihilation operators. σ denotes the spin direction of an electron either up spin ($\sigma = \uparrow$) or down spin ($\sigma = \downarrow$). The last term of the Hamiltonian shows the image potential which has a delay time of δt if the velocity of the atom is in the order of the inverse surface plasmon frequency or above.

Since there is no rigorous theory to deal with this subject, we first consider the resonant charge transfer between a thermal or hyperthermal atom and a ferromagnetic metal surface in the quasi-static limit. In a transition metal surface carrying a large magnetic moment, the work function (ϕ) decreases in a ferromagnetic phase; a ferromagnetic Fe(001) surface yields $\phi = 4.29$ eV, whereas $\phi = 4.86$ eV in a paramagnetic film [8]. Furthermore, the spin resolved density of states are different for minority and majority electrons as schematically shown in the left hand side of Fig. 3 (a) and (b) respectively[9]. In the quasi-static limit, the model Hamiltonian is simplified by the Hartree approximation, taking the ensemble average of the occupation numbers $n_\uparrow$ and $n_\downarrow$. The atomic levels for up and down spins are obtained as follows[10],

$$\varepsilon_{a\uparrow} = \varepsilon_0 + U_{eff}\langle n_\downarrow \rangle + e^2/4z,$$
$$\varepsilon_{a\downarrow} = \varepsilon_0 + U_{eff}\langle n_\uparrow \rangle + e^2/4z,$$

Here $U_{eff} = U - e^2/4z$.

When an alkali atom with ground state atomic energy $\varepsilon_0 > \varepsilon_F$ (where ε_F is the Fermi energy of the metal) approaches the ferromagnetic surface with spin parallel to the minority spins ($\langle n_\uparrow \rangle=0$ and $\langle n_\downarrow \rangle=1$) as shown in Fig. 3(a), the s electron tunnels into an unfilled level above the Fermi surface. The resulting positive ion continues towards the surface. The energy levels of up and down spins $\varepsilon_{a\uparrow}$ and $\varepsilon_{a\downarrow}$ become closer as they approach the surface, resulting in the formation of a covalent chemical bond due to electron exchange. The chemisorption probability and the final charge state of the scattered particle, thus, are functions of the temperature, work function of the surface and velocity of the atom as in case of unpolarized atom-surface scattering. When impinging atoms are polarized parallel to the majority spin ($\langle n_\uparrow \rangle=1$ and $\langle n_\downarrow \rangle=0$) as shown in Fig. 3(b), on the other hand, electron tunneling probability is very low because the majority spin level is filled up. The value of $\varepsilon_{a\uparrow} - \varepsilon_{a\downarrow}$ remains finite during the flight. Therefore we expect that the depolarization of the reflected particle, ionization probability and chemisorption probability are much less than those in the former case, independent on the velocity of atoms.

If $\varepsilon_0 < \varepsilon_F$, the resonant ionization occurs only in the closest region to the surface where the image potential gives rise to a sufficient level shift. However, an electron capture scheme in an s-orbital is plausible. The electron affinities of alkali-series atoms were estimated to be 0.47 eV for Cs, 0.62 eV for Li and 0.75 eV for H in singlet ns^2 states[11]. So far, there is no report on negative Cs ion formation in the atom-metal scattering process. In scattering from W(110) surface, where ϕ =1.45 eV, the Li^- fraction in the scattered particles was more than 20 %[12] while the H^- fraction attained 40 %[13]. In spin-polarized Cs atom scattering we expect the maximum negative ion formation, if one exists, to occur when the spin direction is parallel to the majority spin. The minimum would occur for the reversed spin direction, because the negative ion has only one bound state as proved by R. N. Hill for H^- [14]. Thus, the electron capture process is also spin sensitive.

The spin sensitivity in atom-surface scattering was found in an Auger process of spin-polarized metastable $He(2^3S)$[15]. This effect attracts emerging interests in application for surface magnetism, namely the spin-polarized metastable atom deexcitation spectroscopy (SPMDS). As an origin of the spin sensitivity, either the Auger deexcitation (AD) process at the ferromagnetic Co/W(110) surface[16] or the Auger neutralization process subsequently following the resonance ionization process (RI + AN) at ferromagnetic Ni(110), Fe(110) and paramagnetic Cu(110) surfaces [17-19] were suggested. Together with studies on Auger process in SPMDS, our study on the spin selectivity in resonant process will complementary reveal the nature of spin-dependent atom-surface interactions.

3.2 Role of Kinetic Energy

The kinetic energy of the impinging atom plays a major role in determining whether the impinging particle will scatter from the surface or remain trapped in a potential well at the surface. The role of the energy for atoms with velocities normal to the surface was discussed in the review article by S. R. Kasi et al. for ion-surface collision processes [20]. Generally speaking, the sticking probability of alkali atoms is close to unity below the thermal kinetic energy due to chemisorption as well as above 1 MeV because of implantation. The sticking is expected to go through a minimum in the intermediate region between one and several hundred electron volts. Above 30 eV, however, the physical sputtering process becomes active[21].

The atomic velocity parallel to a surface, on the other hand, causes a translational shift of the Fermi sphere by $\mathbf{Q} = m_e \mathbf{v}_{//}$ relative to the electron momentum bound to atomic orbitals in reciprocal space, hence influences the number of metal electrons which are in resonance with the atomic level. J. Los et al. discussed the effect of the parallel velocity in the resonant charge exchange process quantitatively[22]. In the case of K^+ ion - Al(111) surface scattering, where the ionization potential of the atom is 0.1 eV higher than the work function of the surface, the neutral fraction increases drastically with increasing parallel energy up to a maximum at around 30 KeV[23]. A similar effect is seen in the negative ion formation process: the Li^- fraction in scattered particles from a cesiated W(100) surface (ϕ= 1.45 eV) at grazing angles depends significantly both on the parallel energy between 1 and 3 keV and on the normal velocity in the range from 3 to 20 km/s [24]. The kinetic energy of the impinging atom is, therefore, useful to control the resonant charge exchange interaction, hence the spin sensitivity of the process. The effective correlation time and length between an atom and a surface can be also controlled widely by changing the kinetic energy.

Finally, in the further discussions, many-body effects in the atom-surface system should be taken into account. This has been studied extensively with respect to the surface Kondo effect. We would like to call for theoretical studies on the spin and the kinetic energy dependencies of the atom-surface interactions with a ferromagnet, an antiferromagnet and a paramagnet of a metal, a semiconductor and an insulator.

4. Specification of the Spin-polarized Atomic Beam under Development

Based on the speculative concept in the above section, we are developing a spin-polarized and energy-tunable atomic beam with the following specifications. The details of the experimental apparatus are described elsewhere[25].

a. atomic beam

species: cesium ($^{55}Cs_{133}$)

lithium (Li) will also be available in the future.

kinetic energy range: 0.1 - 10 keV or thermal energy(30 meV)

beam characteristics: 1 mm in diameter and 1 mrad in divergence angle (nominal value)

spin direction: transverse or longitudinal polarization

method of spin-polarization: optical pumping

source: surface ionization using a porous hot tungsten disk 5mm in diameter

focusing: acceleration and collimation by using a set of electrostatic lenses

neutralization: resonant neutralization in a Cs vapor cell 20 cm in length

collimation of the neutral beam: two diamond collimators with an aperture 1 mm in diameter set at 1.2 m in distance

b. detection systems

spin-polarization of the incident beam: the laser absorption method using circularly polarized photons

spin detection of the reflected particles: optical ionization spectroscopy

charge state of the reflected particles: position sensitive multi-channel plate

c. environmental condition

back pressure: $2x10^{-9}$ Pa

magnetic field: residual magnetic field along the path of the atomic beam after polarization < 0.1 mG by using 3 pairs of orthogonally-aligned and feedback-controlled double-Helmholtz coils and a double layered magnetic shield of μ-metal installed in the vacuum chamber

d. laser systems for optical pumping and detection

laser : 3 AlGaAs diode lasers, 852 nm in wavelength

frequency locking: electronic servo control using frequency reference of Doppler free saturated absorption spectrum[6] for hyperfine pumping.

novel self-locking system based on frequency selective optical feedback similar to Faraday filter for the Zeeman pumping[26].

5. Spin-polarization Measurement of Thermal Energy Beam

Spin-polarization experiments were carried out for a thermal kinetic energy beam using the experimental set up shown in Fig. 4. The lower part of the sketch is a

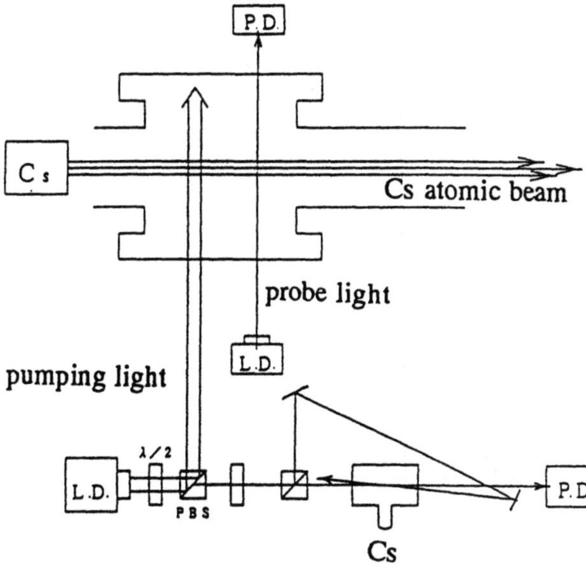

Fig. 4 A schematic view of the experimental arrangement used for measurement of the spin-polarization and hyperfine pumping rate of the Cs atomic beam with thermal kinetic energy.

diagram of the novel self-locking system developed for Zeeman pumping based on frequency selective optical feedback. The optical feedback technique provides simultaneous spectral narrowing and locking at the same time. The laser frequency was within 1 MHz of the resonance frequency ($3.5x10^8$ MHz) for the sweep ranging to 300 MHz. It was found to be adequate for producing spin-polarized Cs atomic beam with a highly controlled reproducibility during operation.

Figure 5 shows the absorption spectra in derivative form without pumping (Fig. 5a) and with hyperfine pumping (Fig. 5b). Frequency of the probing laser was swept in the range of resonance from the ground state F=4 to the excited states F=3, 4 and 5. It is seen that the absorption intensities increase proportionally for the three resonance frequencies by the hyperfine pumping. Hyperfine pumping rate is estimated by the following equation,

$$\text{hyperfine pumping rate} = 9\Delta I_p / 16\Delta I,$$

where ΔI_p and ΔI are absorption intensities with and without pumping laser, respectively. The average absorption rate in Fig. 5a and Fig. 5b is 16:9 which means the hyperfine pumping rate is 100 %.

The spin-polarization was measured by means of light absorption for a σ^+ - and σ^-- switched probe beam as shown in Fig. 6. Frequency range swept corresponds to those in Fig. 5. Clearly, the peaks which occur for other than F=5 absorption

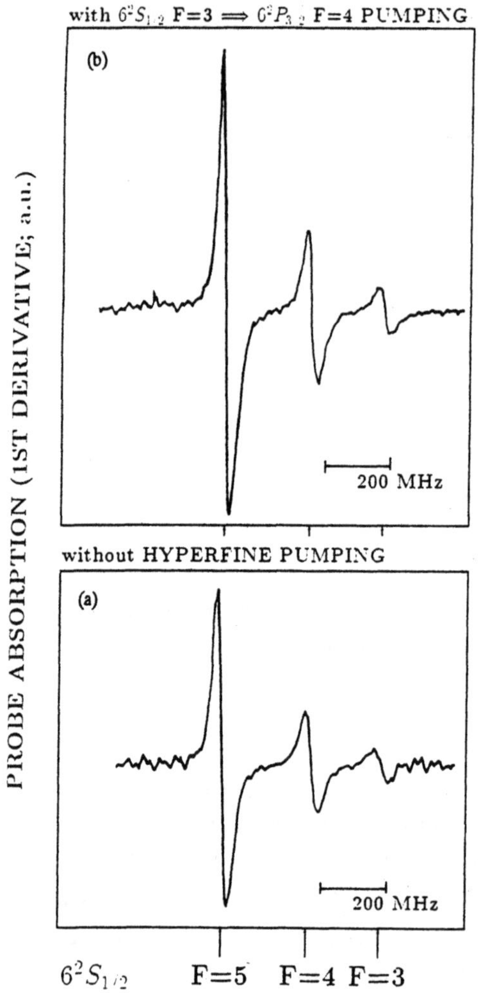

Fig. 5 Light absorption spectra of the Cs atomic beam with thermal kinetic energy. Main peaks correspond to the transition from the ground state $|6\ ^2S_{1/2}, F=4>$ to the excited states $|6\ ^2P_{3/2}, F=3, 4$ and $5>$ (a) without pumping light and (b) with hyperfine pumping light.

nearly vanish for the σ^+ probe light (b) while the peak at F=5 absorption decreases drastically for the σ^- probe light (a). One can estimate the expectation value of the spin-polarization with use of the following equation derived by the calculation of the transition probability

$$<m_F>/F = 5(\Delta I_+ - \Delta I_-)/12\Delta I,$$

where ΔI_+ and ΔI_- respectively represent the observed light absorption rates probed

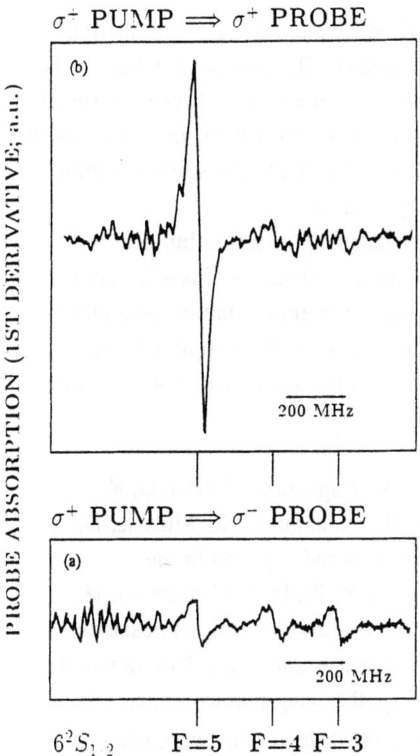

Fig. 6 The light absorption spectra of the Cs atomic beam with thermal kinetic energy for (a) σ^+ and (b) σ^- switched probe light. The Cs atomic beam was polarized by σ^+ light. The frequency range is the same with that in Fig. 5.

by σ^+ and σ^- polarized lasers. The results shown in Fig. 6 gives a polarization rate of more than 98 % which is good enough for surface spin measurements. For an accelerated atomic beam, we need an expanded high intensity laser beam to increase the time and cross section of the interaction. The spin-polarization rate of the 10 keV beam is estimated to attain about 90 % of maximum by use of a pumping laser with 150 mW/cm² power and a 20 cm diameter. A polarimator which can be used efficiently with an accelerated beam is under development.

6. Summary

In this report, we have introduced the spin-polarized ground state atom which is under development as a probe for spin correlation at the topmost layer of a magnetic surface. Spin sensitivity in the resonant electron exchange process are discussed speculatively. Effects of spin sensitivity are expected to be observed in the spin and charge states of reflecting atoms from a surface and in chemisorption probability.

At the present stage, the transversely polarized Cs atomic beam with thermal kinetic energy is ready for use in experimental studies. The spin-polarization of the beam is above 98 %. An accelerated beam with an energy range between 0.1 and 10 keV is under development . As a starting experiment to study the spin sensitivity of the resonant charge exchange, we are preparing experimental setups for scattering from single crystalline Ni (100),(110) and(111) surfaces.

This method is applicable not only to the conducting materials but also to insulators. The spin-polarized atomic beam is attractive because of nearly perfect polarization, flexibility of orientation as well as alignment of atomic spins along preferable directions, expected sensitivity to the spin state of the topmost layer of the surface and plausibility of observing multipole interactions with surface state due to the total spin of probe larger than unity.

The authors are grateful to professors K.Koura, K. Nagamine, Y. Murata, K. Hayakawa, E. Hayashi, T. Nagata, T. Sakurai, S. Suganomata, K. Mukasa, Dr. T. Briere and Mr. M. Inoue for the valuable discussions and supports in the development. We also thank to misters H. Hirose, Y. Ikedo, T. Hashimoto, H. Iwata, S. Mitsuya, H. Tanaka, K. Kobayashi and A. Kato for their enthusiastic collaboration in developing the spin-polarized cesium beam. The collaboration of Dr. M. Tokita of Sumitomo coals Co. Ltd. in synthesizing porous tungsten by Spark Plasma Sintering(SPS) method is also acknowledged. This work was supported by the precursory research for embryonic science and technology (PRESTO) project of the research development corporation of Japan, the grant-in-aid of the Ministry of Education, Science, Sports and Culture and Saneyoshi scholarship foundation.

References

[1] A. J. Freeman, C. L. Fu, M. Wiener and S. Ohnishi, Hyperfine Int. **33**, 53 (1987).
[2] S. Alvarado et al., Phys. Rev. Lett. **47**, 51 (1981).
[3] C. Rau and S. Eichner, Phys. Rev. Lett. **47**, 939 (1981).
[4] J. Mathon, Rep. Prog. Phys. **51**, 1 (1988) and references therein.
[5] C. Wieman and T. W. Hansh, Phys. Rev. Lett. **36**, 1170 (1976).
[6] H. Hori, Y. Kitayama, M. Kitano, T. Yabuzaki and T. Ogawa, IEEE J. Quantum Electron. **QE-19**, 169 (1983).
[7] D. M. Newns, Phys. Rev. **178**, 1123 (1969).
[8] M. Weinert and A. J. Freeman, J. Mag. Magn. Mattls. **38**, 23 (1983).
[9] S. Ohnishi, A. J. Freeman and M. Weinert, Phys. Rev. **B29**, 6741 (1983).
[10] M. Tsukada, in: Theories at Surface II: absorption and dynamical process (Maruzen, Tokyo, 1995) pp. 44-49, in Japanese.
[11] H. Hotop and W. C. Lineberger, J. Phys. Chem. Ref. Data **14**, 731 (1985).

[12] J. J. C. Geerings, R. Rodnik, J. Los and J. P. Gauyacq, Surf. Sci. **186**, 15 (1987).
[13] J. N. M. van Wunnik, J. J. C. Geerings and J. Los, surf Sci. **131**, 1 (1983).
[14] R. N. Hill, Phys. Rev. Lett. **38**, 643 (1977).
[15] M. Onelion, M. W. Hart, F. B. Dunning and G. K. Walters, Phys. Rev. Lett. **52**, 380 (1984).
[16] M. Getzlaff, D. Egert, H. Steidl, G. Baum and W. Raith, Z. Phys. **D30**, 245 (1994)
[17] D. R. Penn and P. Apell, Phys. Rev. **B41**, 3303 (1990).
[18] M. S. Hammond, F. B. Dunning and G. K. Walters, Phys. Rev. **B45**, 3674 (1992-I)
[19] D. M. Oro, Q. Lin, P. A. Soletsky, X. Zhang, F. B. Dunning and G. K. Walters, Phys. Rev. **B46**, 9893 (1992-I).
[20] S. R. Kashi, H. Kang, C. S. Sass and J. W. Rabalais, Surf. Sci. Rept. **10**, 1 (1989).
[21] J. Roth, in : Sputtering by Particle Bombardment II, Ed. R. Bchrisch, Vol. **52** of Topics in Applied Physics (Springer, New York, 1983).
[22] J. Los and J. J. C Geerings, Physics Reports, **190**, 133 (1990).
[23] H. Winter and R. Zimmy, in: Coherence in Atomic Collision Physics. eds. H. J. Beyer, K. Blum and R. Hippler (Plenum, New York, 1988) p.283.
[24] J. J. C. Geerings, R. Rodnik, J. Los and J. P. Gauyacq, Surf. Sci. **181**, L177 (1987).
[25] E. Torikai: to be published in Ad. Colloid and Interface Sci.
[26] M. Kozuma, M. Kourogi and M. Ohtsu, Appl. Phys. Lett. **61**, 1895 (1992) and T. Kiguchi, H. Hori, T. Inoue, Y. Inoue and T. Sakurai, to be published.

The Effects of Electronic Excitations on the Behaviour of Particles Scattered by Metal Surfaces

A. Okiji, H. Kasai, and H. Nakanishi

Department of Applied Physics, Osaka University, Suita, Osaka 565, Japan

Abstract. The effects of electronic excitations on the motion of particles (atoms or molecules) scattered by metal surfaces are investigated in the framework of a model consisting of the Newns-Anderson Hamiltonian for the electron system and the translational energy of the moving particle. Through the results of perturbational calculation, it is clearly shown that the direct connection between the state change of the particle motion and that of the electron system suppresses the divergent behaviour, due to the Fermi surface effects, in the final translational energy distribution of moving particles with the charge-state change of particles, in so far as the particle velocity is finite. In summary, a brief comparison is made between the various approaches to investigations of the time dependent behaviour of particles scattered by metal surfaces at the present stage of our knowledge.

1. Introduction

From a theoretical point of view, the time-dependent behaviour of the electron system should be of first consideration in the case where particles (atoms or molecules) are approaching toward or moving away from the surface, in order to describe the dynamical processes at the metal surface properly. We took three approaches for this purpose. The first is an application of the Bethe ansatz solution obtained by one of the present authors[1] for the Anderson model[2,3]. The second is a potential energy surface (PES) picture approach[4,5]. And the third is an ordinary time evolution method[6-10].

The main purpose of investigations done with the use of the first approach is to examine the many-body effects induced by an intra-atomic Coulomb interaction. In ref.[3], the temperature dependence of the friction coefficient of moving particles near metal surfaces was investigated on the basis of the local Fermi-liquid theory, at low temperatures in particular, presupposing a quasistatic motion of the particle. It was shown that the friction coefficient strongly depends on the temperature as well as on the magnetic field. It is desirable to clarify more carefully these many-body effects in consideration of the interplay between the state change of the particle motion and that of the electron system during the dynamical processes.

In the PES picture approach, the potential energy for a moving molecule is calculated for each coordinate of that molecule, that is to say, the state of the electron system, which is usually the ground state, is given as a function of the molecular

Springer Series in Solid-State Sciences, Volume 121
Elementary Processes in Excitations and Reactions on Solid Surfaces
Eds.: A. Okiji, H. Kasai and K. Makoshi

coordinates. The couplings between various degrees of freedom of the molecule are introduced simultaneously through the PES in the dynamical processes. For example, in ref.[4,5], it is shown that the coupling between the rotational and the translational degrees of freedom through the PES for H_2 (D_2) molecule interacting with Cu surface gives rise to two effects which dominate the dissociative adsorption process, viz. the steering effect and the rotational-translational energy transfer effect. As a result of the interplay between these two effects, there is an initial decrease in the dissociative adsorption probability with a slight increase in the initial rotational energy, E_{ROT}, and then a final increase when E_{ROT} is further increased. This approach is adequate to investigate the dynamical processes of molecules at metal surfaces, when the particle motion is slow, in particular. Many people are using this approach and are giving successful explanations about the experimental findings.

In the time evolution method, a simple model Hamiltonian is usually introduced. Then, the classical trajectory approximation is employed for the molecular center of mass (CM) motion. As a result, the CM coordinate dependence of the Hamiltonian for the electron system is replaced by a time-dependence. In this method, the molecular internal degrees of freedom can be coupled directly with the electron system. For example, the deexcitation energy of thermally excited electron-hole pairs at metal surfaces can be transferred from the electron system to the molecular vibration. This is one of the possible mechanisms working for the vibrational excitation of NO scattered from Ag(111) [11,12].

In order to understand how this mechanism operates, numerical calculations of the vibrational excitation probability of NO scattering from Ag(111) were done with the use of (I) rate equation method[6,11] and (II) iterative method for Dyson equations[7]. For relatively low velocity region of NO, both results obtained by (I) and by (II) agree well with each other, and reproduce the experimental findings. As the velocity of NO is further increased, however, the discrepancy between the results obtained by (I) and by (II) becomes remarkable, and only the results obtained by (II) can reproduce the experimental findings. The origin of the discrepancy can be explained as follows: As the velocity of NO is increased, an appreciable uncertainty of the energy should arise in the intermediate stage of the scattering process[10]. As a result, the number increases of electron-hole pairs contributing to the vibrational excitation. This effect is taken into account in (II) but not in (I), because (I) is based on the assumption that energy conservation within the molecular vibration and the electron system is satisfied even in the intermediate stage of the scattering process (see ref.[13]). It can be said that one should treat the time-dependent behaviour of the electron system properly, taking into account that the electron states involved in the intermediate stage of the scattering process increase in number as the particle velocity is increased.

In this paper, the effects of electronic excitations on the motion of particles (atoms or molecules) scattered by metal surfaces are investigated in the framework of a model consisting of the Newns-Anderson Hamiltonian for the electron system and the translational energy of particles[14]. Perturbational calculation for the final translational energy distribution of moving particles is done up to the fourth order with respect to the hybridization matrix element between conduction electron states in metals and an electron state on moving particles. It is shown that the direct connection between the state change of the particle motion and that of the electron system gives rise to an increase of the number of electron states involved in the intermediate stage of the scattering process. As will be shown in section 3, the divergent behaviour due to the Fermi surface effects is suppressed in the final translational energy distribution of particles scattered by the metal surfaces with the charge-state change of particles, in so far as the particle velocity is finite. By analogy with the case of the X-ray absorption by metals[15] where the divergence of the X-ray adsorption intensity caused by the charge excitation is weakened by the exchange interaction, it is considered that the divergence does not come out in the translational energy distribution of scattered particles, even if the exchange interaction is taken into account. Furthermore, it is expected that the Fermi surface effect is suppressed in the final translational energy distribution of moving particles even in the case of no charge-state change of scattered particles, in so far as the particle velocity is finite.

2. Model

The following model Hamiltonian is introduced in order to describe the energy transfer process as well as the charge transfer process in the scattering of particles by metal surfaces[14].

$$H_{\text{total}}(Z) = H_{\text{particle}}(Z) + H_{\text{electron}}(Z). \tag{1}$$

The first term is the translational energy of a moving particle:

$$H_{\text{particle}}(Z) = -\frac{1}{2M}\frac{\partial^2}{\partial Z^2}, \tag{2}$$

where, Z and M are the coordinate and mass of that moving particle. Here, one dimensional motion normal to the surface is assumed for the particle motion. Henceforth a unit of $\hbar = 1$ is used. $H_{\text{electron}}(Z)$ is the energy of the electron system:

$$H_{\text{electron}}(Z) = H_0 + H'(Z), \tag{3}$$

$$H_0 = \sum_{k,\sigma} \varepsilon_k n_{k\sigma} + \sum_{\sigma} \varepsilon_a n_{a\sigma} + U n_{a\uparrow} n_{a\downarrow}, \tag{4}$$

$$H'(Z) = \sum_{k,\sigma}\left[V_{ak}(Z) c^{+}_{a\sigma} c_{k\sigma} + h.c.\right], \tag{5}$$

where $c^{+}_{k\sigma}$, ($c^{+}_{a\sigma}$) is the electron creation operator for the state $|k\rangle$ with spin σ and energy ε_k of the conduction electron in the metal substrate (for the localized orbital $|a\rangle$ on the moving particle with spin σ and energy ε_a), $n = c^{+}c$, and V_{ak} is the matrix element of the hybridization between $|a\rangle$ and $|k\rangle$, and $U n_{a\uparrow} n_{a\downarrow}$ represents an intra-atomic Coulomb interaction for two electrons on the moving particle.

In the following calculation, the sum $H_{\text{particle}}+H_0$ is chosen as the unperturbed Hamiltonian. The eigenfunction, Φ, and the energy eigenvalue, E_{total}, for the unperturbed Hamiltonian obey the equation,

$$\left(H_{\text{particle}} + H_0\right)\Phi(\mathbf{x}, Z) = E_{\text{total}}\,\Phi(\mathbf{x}, Z), \tag{6}$$

where $\mathbf{x}$ represents the coordinates of electrons in the present system. It is assumed that Φ can be decomposed into Ψ for the moving particle and ϕ for the electron system,

$$\Phi(\mathbf{x},Z) = \Psi(Z)\cdot\phi(\mathbf{x};Z), \tag{7}$$

where Ψ and ϕ are eigenfunctions of H_{particle} and H_0,

$$H_{\text{particle}}\Psi(Z) = E_{\text{particle}}\,\Psi(Z), \tag{8}$$

$$H_0\,\phi(\mathbf{x};Z) = E_{\text{electron}}\,\phi(\mathbf{x};Z) \tag{9}$$

and

$$E_{\text{total}} = E_{\text{paricle}} + E_{\text{electron}}\,. \tag{10}$$

Here, ϕ involves Z only parametrically.

We consider the case where the moving particle is initially a positive ion and is finally a neutral atom after the scattering by the metal surface, and the electron system of the metal substrate is in its ground state initially. Wave functions for the total system in the initial state and the final state are written as $\Phi_{\text{i}} = \Psi_{K_{\text{i}}}(Z)\cdot\phi_{\text{i}}(\mathbf{x};Z)$ and $\Phi_{\text{f}} = \Psi_{K_{\text{f}}}(Z)\cdot\phi_{\text{f}}(\mathbf{x};Z)$. Here, $\Psi_{K_{\text{i}}}$ ($\Psi_{K_{\text{f}}}$) is a plane wave function, $\Psi_K(Z) = \exp[iK\cdot Z]$, for the moving particle with wave vector K=K_{i} (K_{f}) in the initial (final) state, and ϕ_{i} (ϕ_{f}) is a wave function for the electron system in the initial (final) state.

The transition probability from the initial state to the final state per unit time is given by

$$\omega_{\text{i,f}} = 2\pi\delta(E_{\text{total,f}} - E_{\text{total,i}})\left|\left\langle\Phi_{\text{f}}\middle|H'\middle|\tilde{\Phi}_{\text{i}}\right\rangle_{\text{N}}\right|^2, \tag{11}$$

where $\tilde{\Phi}_{\text{i}}$ is a wave function in the scattering state, written as

$$\left|\tilde{\Phi}_{\text{i}}\right\rangle = \left(1 + G_{\text{total}}H'\right)\left|\Phi_{\text{i}}\right\rangle, \tag{12}$$

and normalized as

$$\left|\tilde{\Phi}_{\text{i}}\right\rangle_{\text{N}} = \left(\left\langle\tilde{\Phi}_{\text{i}}\middle|\tilde{\Phi}_{\text{i}}\right\rangle\right)^{-1/2}\left|\tilde{\Phi}_{\text{i}}\right\rangle. \tag{13}$$

Here, G_{total} is the Green function for the total system,

$$G_{\text{total}} = \left(E_{\text{total,i}} - H_{\text{total}} + i0^+\right)^{-1}, \tag{14}$$

and $E_{\text{total,i}}$ ($E_{\text{total,f}}$) is the initial (final) total energy.

Expanding G_{total} in terms of the Green function, G_0, for the unperturbed system as

$$G_{\text{total}} = G_0 + G_0H'G_0 + G_0H'G_0H'G_0 + \cdots, \tag{15}$$

where

$$G_0 = \left(E_{\text{total,i}} - (H_{\text{particle}} + H_0) + i0^+\right)^{-1}, \tag{16}$$

one obtains the following expression for the transition matrix element from the initial state to the final state,

$$\left\langle\Phi_{\text{f}}\middle|H'\middle|\tilde{\Phi}_{\text{i}}\right\rangle = \left\langle\Phi_{\text{f}}\middle|H' + H'G_0H' + H'G_0H'G_0H' + H'G_0H'G_0H'G_0H' + \cdots\middle|\Phi_{\text{i}}\right\rangle. \tag{17}$$

In the present scattering case, perturbation terms of the odd-order give rise to the charge-state change of the moving particle from the positive state to the neutral state and make contributions to the transition probability amplitude. In the case of the Kondo problem, it is noted that perturbation terms of the odd-order with respect to H' do not make any contribution to the scattering amplitude of conduction electrons by an impurity in metals described by the Anderson model.

3. Translational Energy Distribution of Particles Scattered by Metal Surfaces

Here, we present the results for the case where we neglect an intra-atomic Coulomb interaction term. Integrating over the electron coordinates in (17), the transition matrix element obtained up to the third order with respect to H' can be written by,

$$\left\langle \Phi_{\mathrm{f}} \middle| H' \middle| \tilde{\Phi}_{\mathrm{i}} \right\rangle \approx \left\langle \Psi_{\mathrm{f}} \middle| u \middle| \Psi_{\mathrm{i}} \right\rangle \cdot v_{ak}^{0} + \sum_{l} \sum_{l'} \left\langle \Psi_{f} \middle| u \middle| \Psi_{l'} \right\rangle \cdot \left\langle \Psi_{l'} \middle| u \middle| \Psi_{l} \right\rangle \cdot \left\langle \Psi_{l} \middle| u \middle| \Psi_{i} \right\rangle$$

$$\times \sum_{k'} v_{ak'}^{0} \frac{1 - f(\varepsilon_{k'})}{E_{\mathrm{particle,i}} - E_{\mathrm{particle},l'} + \varepsilon_{k} - \varepsilon_{k'}} v_{k'a}^{0} \frac{1}{E_{\mathrm{particle,i}} - E_{\mathrm{particle},l} + \varepsilon_{k} - \varepsilon_{a}} v_{ak}^{0}, \quad (18)$$

where $f(\varepsilon)$ is the Fermi distribution function. It is assumed that the k and Z dependence of $V_{ak}(Z)$ can be separable, $V_{ak}(Z) = v_{ak}^{0} \cdot u(Z)$. Since the conduction electron state $|k'\rangle$ in the metal substrate is occupied at the intermediate stage of the electronic transitions, it should be empty initially. This is the reason why a factor $1 - f(\varepsilon_{k'})$ appears in the second term of (18).

It is assumed for simplicity that the resonance level width of the localized orbital on the moving particle, $\Delta(\varepsilon) = \pi \sum_{k} \left| v_{ak}^{0} \right|^{2} \delta(\varepsilon - \varepsilon_{k})$, is given by

$$\Delta(\varepsilon) = (\pi \left| v_{ak}^{0} \right|^{2} / \Gamma) \cdot \theta((\Gamma/2) - |\varepsilon|), \quad (19)$$

where Γ is the conduction band width of the metal substrate (flat band), and θ is a step function. Inserting (18) into (11), one obtains the translational energy distribution of moving particles in the final state at the absolute zero of temperature[14],

$$F(\delta E) = \frac{M(M+m)\left| v_{ak}^{0} \right|^{2}}{\Gamma K_{i} \cdot K_{f}} \cdot \left| \hat{u}(K_{\mathrm{i}} - K_{\mathrm{f}}) - \frac{\left| v_{ak}^{0} \right|^{2}}{4\pi^{2}\Gamma} \cdot \int_{-\infty}^{\infty} dK \int_{-\infty}^{\infty} dK' \, \hat{u}(K_{\mathrm{f}} - K') \, \hat{u}(K' - K) \, \hat{u}(K - K_{\mathrm{i}}) \right.$$

$$\left. \times \frac{1}{(K_{\mathrm{f}}^{2} - K^{2}) / 2(M+m)} \ln \left| 1 - \frac{\Gamma}{2\left[K_{\mathrm{f}}^{2} / 2(M+m) - K'^{2} / (2M) + \varepsilon_{a} \right]} \right| + \cdots \right|^{2}, \quad (20)$$

for $-\varepsilon_{a} - \Gamma/2 < \delta E < -\varepsilon_{a}$, and $F(\delta E) = 0$, otherwise, where $\delta E = K_{\mathrm{f}}^{2} / 2(M+m) - K_{\mathrm{i}}^{2} / (2M)$ and $\hat{u}(q)$ is the Fourier transform of $u(Z)$. In fig.1, the numerical results are shown for $F(\delta E)$ obtained up to the fourth order with respect to v_{ak}^{0} (solid line). Although there is a logarithmic term in (20), it does not show the divergent behaviour in $F(\delta E)$ for the case of the finite particle velocities, because the translational energy difference of moving particles between the initial state and the intermediate states appears in the energy denominator of the second term of (18) and the integration over these intermediate states of the particle motion suppresses the divergent behaviour in (20). It can be said that the suppression of the divergent behaviour results from the effects of the interplay between the state change of the particle motion and that of the electron system during the scattering process, that is, no anomalous Fermi surface effect appears in this case.

In the slow motion limit of the moving particle, however, the anomalous Fermi surface effect might appear. In this case, for example, one can assume that H_{particle}

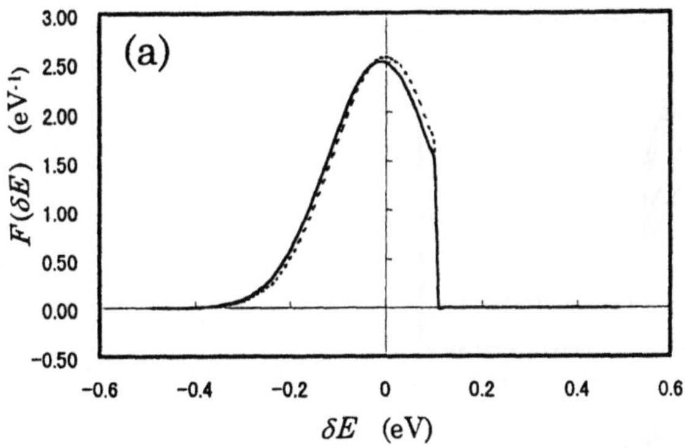

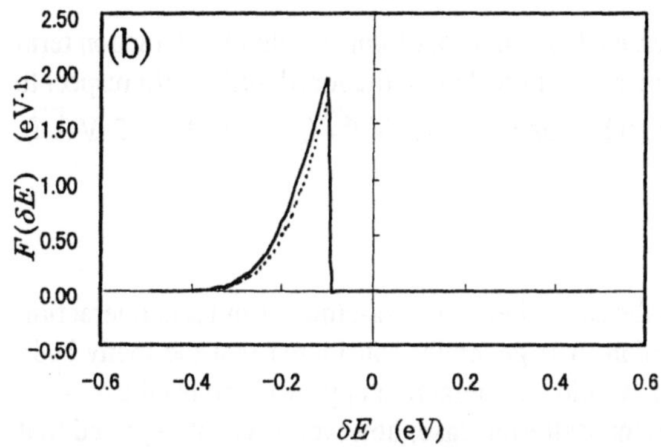

Fig. 1. The translational energy distribution, $F(\delta E)$, of moving particles after the scattering process, where positive ions are neutralized, by the metal surfaces at absolute zero of temperature in the case where we neglect an intra-atomic Coulomb interaction term. The numerical results obtained up to the fourth order (the second order) is indicated by solid line (dashed line). It is assumed that mass of the moving particle corresponds to that of a lithium atom and that $u(Z)$ is Gaussian, $u(Z) = \exp(-\gamma^2 \cdot Z^2)$ ($\hat{u}(q) = (\sqrt{\pi}/\gamma) \cdot \exp[-(q/2\gamma)^2]$). Used parameters are $E_{\text{particle,i}}$=100eV, $v_{ak}{}^0$= -0.1eV and γ= 0.32 Å^{-2}. The energy of the localized orbital on the moving particle measured from the Fermi energy, $\varepsilon_a - \varepsilon_F$, is −0.1eV (a) and +0.1eV (b).

commutes with H' approximately, which means that the translational energy of moving particles does not change in the intermediate stage of the scattering process. One obtains the following expression for the transition matrix element,

$$\left\langle \Phi_f \middle| H' \middle| \tilde{\Phi}_i \right\rangle = \left\langle \Psi_f \middle| u \middle| \Psi_i \right\rangle \cdot v_{ak}^0 + \left\langle \Psi_f \middle| u^3 \middle| \Psi_i \right\rangle \cdot \sum_k v_{ak'}^0 \frac{1-f(\varepsilon_{k'})}{\varepsilon_k - \varepsilon_{k'}} v_{k'a}^0 \frac{1}{\varepsilon_k - \varepsilon_a} v_{ak}^0 + \cdots. \quad (21)$$

It is noted that the translational energy difference of moving particles between the initial state and the intermediate state disappears in the energy denominator of the second term of (21) as compared with that of (18), and the same factor, $1-f(\varepsilon_{k'})$, appears in (21) as in (18). As a result, a logarithmic divergence comes out at $\delta E = -\varepsilon_a$ in $F(\delta E)$ just like the Kondo effects in this case[1,14].

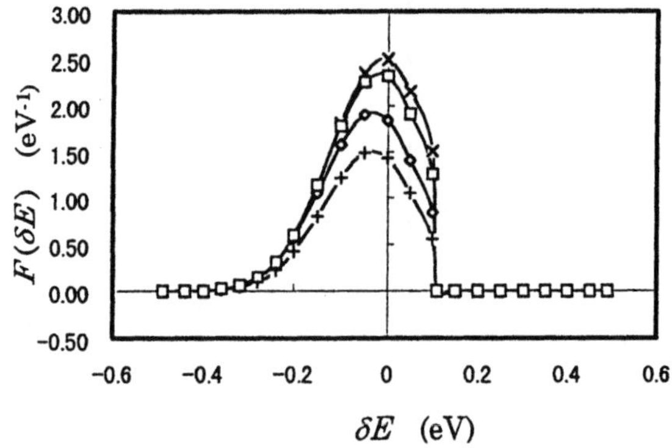

Fig.2. Same as fig.1, but for the case where an intra-atomic Coulomb interaction term is taken into account within the results obtained up to the fourth order with respect to v_{ak}^0. It is assumed that $\varepsilon_a - \varepsilon_F = -0.1$eV and $U = 0.2$eV (◇), 0.3eV (+), 0.5eV (□), $+\infty$ (×).

Here, we present the results for the case where an intra-atomic Coulomb interaction term is included in the Hamiltonian. It is generally considered that the many body effects induced by an intra-atomic Coulomb interaction appear in the results of more than sixth order term. In the present scattering case, however, it is not expected that the divergence does come out in $F(\delta E)$, even if an intra-atomic Coulomb interaction is included, as stated in section 1. When an intra-atomic Coulomb interaction is included the three charge-states of moving particles, viz. the positive state, the neutral state and the negative state, can be expressed in the calculations. In fig.2, the numerical results obtained up to the fourth order with respect to v_{ak}^0 are shown for the translational energy distribution, $F(\delta E)$, of moving particles after the scattering process with the charge-state change from the positive state to the neutral state. As the value of U is increased from $U = -2(\varepsilon_a - \varepsilon_F)$ (the symmetric case), the probability distribution decreases at first and then increases, which is approaching the results given in fig.1(a). This non-monotonous behaviour of the probability distribution can simply be explained as follows. The neutralization through the scattering process, in which the charge-state of moving particles is the negative state at the intermediate stage of the electronic transitions, is suppressed with a slight increase of U at fast. And, with a further increase of U, the neutralization is enhanced since the probability of the double occupied state (the negative state) is decreased, tending to the single level case.

4. Time Evolution Method

The above-mentioned effects of the interplay between the state change of the particle motion and that of the electron system during the scattering process can also be taken up approximately by the time evolution method[10]. The final energy distribution of the electron system, for example, is given in this method as,

$$P(\varepsilon)=\left\langle\phi_i\middle|\tilde{U}(-\infty,+\infty)\cdot\delta(\varepsilon-H_{\rm electron}(Z(+\infty))+E_{\rm electron,i})\cdot\tilde{U}(+\infty,-\infty)\middle|\phi_i\right\rangle, \tag{22}$$

where $\tilde{U}(t,t')$ is the time evolution operator and $E_{\rm electron,i}$ is the energy for the electron system in the initial state. In the wide band limit, we can calculate $P(\varepsilon)$ numerically in closed form with respect to v_{ak}^0 when we neglect an intra-atomic Coulomb interaction term[8]. We describe, here, the characteristics of the perturbation term with respect to v_{ak}^0. For simplicity, the trajectory of the moving particle is assumed to be $Z(t)=v_\perp\cdot|t|$ and $u(z(t))=\exp(-\alpha\cdot|Z(t)|)$. Then, as a matter of course, an additional imaginary part, $i\alpha\cdot v_\perp$, appears automatically in the energy denominator of the perturbation terms of more than third order, e.g. $(i\alpha\cdot v_\perp+\varepsilon_k-\varepsilon_{k'})^{-1}$ (c.f. (18) and (21)), which corresponds to the inverse of the average interaction time between the moving particle and the metal surface (the order of $(\alpha\cdot v_\perp)^{-1}$) [10]. This fact means that the time evolution method is expected to give nearly the same results qualitatively as compared with the quantum mechanical method given in the previous section. That is, this method has advantage in that one can calculate $P(\varepsilon)$ numerically in closed form with respect to v_{ak}^0 as mentioned above and that one can calculate other physical quantities more easily [7-9].

5. Summary

The effects of electronic excitations on the motion of particles (atoms or molecules) scattered by metal surfaces are investigated in the framework of a model consisting of the Newns-Anderson Hamiltonian for the electron system and the translational energy of particles. The state change of the particle motion is directly connected to the state change of the electron system, which gives rise to an increase of the number of the electron states involved in the intermediate stage of the scattering process. This is the reason why there is no divergence of the final translational energy distribution of scattered particles, in so far as the particle velocity is finite. On the basis of the obtained results, it can be said generally that there is no singular response of the metal electrons near the Fermi surface in the case of the finite particle velocities. In our view, this should be accepted as fundamental in the development of the theory of dynamical process of particles at metal surfaces.

Note that we just examined the above effects in the scattering case of particles which are accompanied with the charge-state change (from positive to neutral). It can be expected that the divergence does not come out, even in the case where scattered particles are not accompanied with the charge-state change.

Furthermore, it is said that the time evolution method is useful in order to investigate dynamical processes of particles at metal surfaces qualitatively regardless of the particle velocity, fast or slow. It seems not so easy, however, to extend this method to a multi-dimensional version which involves various degrees of freedom of particles. On the other hand, on the premise that the divergence does not come out in so far as the particle velocity is finite, it can be expected that the multi-dimensional PES picture approach gains an advantage over the time evolution method at least in the case of the particle motion being slow. As stated above, however, we must take care that the divergence might come out in the slow motion limit of the moving particle.

References

[1] A. Okiji: in *Fermi Surface Effects*, Springer Series Solid-State Sciences, Vol. 77 (Springer, Berlin Heidelberg 1987), p.63

[2] A. Okiji and N. Kawakami, Phys. Lett. **109**A, 65 (1985)

[3] A. Okiji and H. Kasai, Phys. Rev. B**38,** 8102 (1988)

[4] W. A. Diño, H. Kasai and A. Okiji, J. Phys. Soc. Jpn. **64,** 2478 (1995)

[5] H. Kasai, A. Okiji and W. A. Diño, in this issue.

[6] H. Kasai and A. Okiji, Surf. Sci. **225,** L33 (1990)

[7] H. Kasai and A. Okiji, Surf. Sci. **242,** 394 (1991)

[8] H. Nakanishi, H. Kasai and A. Okiji, Surf. Sci. **216,** 249 (1989)

[9] H. Nakanishi, H. Kasai and A. Okiji, Surf. Sci. **197,** 515 (1988)

[10] H. Nakanishi, H. Kasai and A. Okiji, Surf. Sci. **242,** 410 (1991)

[11] D. M. Newns, Surf. Sci. **171,** 600 (1986)

[12] The other mechanisms working for the vibrational excitations of NO are investigated, e.g. in A. Gross, Y. Chiba and W. Brenig, Prog. Theor. Phys. Suppl. **106,** 335 (1991).

[13] H. Kasai, A. Okiji, Prog. Surf. Sci. 44, 101 (1993)

[14] H. Kasai, N. Okamoto and A. Okiji, J. Phys. Soc. Jpn. **64,** 4308 (1995)

[15] A. Kato, A. Okiji and Y. Osaka, Prog. Theor. Phys. **44,** 287 (1970); A. Yoshimori and A. Okiji, Phys. Rev. B**16,** 3838 (1977)

Do the Highly Correlated Kondo and Mixed-Valent States Influence Charge Transfer and Energy Transfer in the Motion of Atoms near Metallic Surfaces?

T. Brunner[1], *D.C. Langreth*[1], *P. Nordlander*[2], *and H. Shao*[1,2]

[1]Department of Physics and Astronomy, Rutgers University, Piscataway, NJ 08855-0849, USA

[2]Department of Physics and Rice Quantum Institute, Rice University, Houston, TX 77251-1892, USA

Abstract. Physical arguments and non-equilibrium calculations based on the time dependent Anderson model are given that the highly correlated Kondo and mixed valent states are under certain circumstances relevant for an atom in motion outside a metal surface. It is predicted that anomalous temperature, energy, or velocity dependence may as a result occur in physically measured properties.

1. Introduction

The collective many-body resonances that occur in the context of impurity states in metals under certain conditions cause anomalous temperature dependence in the thermal and transport properties. An analogous situation exists with atoms outside metallic surfaces, where under similar conditions these Kondo or highly correlated mixed valent states should also exist. The surface case, where one normally deals with atoms in motion, is more difficult both theoretically and experimentally because of the time dependence. On the theoretical side we have solved the problem of time dependence in the experimentally interesting temperature range, by extending the so-called non-crossing approximation to the non-equilibrium situation, and then solving it exactly numerically. Here we will discuss the solution in the context of (a) the non-equilibrium spectral functions and a novel source of non-adiabaticity, (b) linear friction coefficients and a novel source of energy transfer and (c) novel temperature and velocity dependence of the charge transfer to an atom moving away from the surface. In the latter case we present both physical arguments and quantitative calculations which indicate that these dramatic effects should be directly seen experimentally when the system parameters such as the work function are appropriately tuned. Finally experimental systems are suggested where these phenomena might be observed. In fact, because the atom–surface system is so rich in respect to the possible variations of the appropriate variables such as work function and atom velocity, it may be the ideal experimental ground for studying these collective many-body states.

Springer Series in Solid-State Sciences, Volume 121
Elementary Processes in Excitations and Reactions on Solid Surfaces
Eds.: A. Okiji, H. Kasai and K. Makoshi

2. Method of Calculation

For the calculational aspects we describe the dynamical interaction between an atom and a surface using the Anderson model [1], which was first applied to the surface problem by Newns and collaborators [2, 3]. Here we use an N fold degenerate version with time dependent parameters:

$$\begin{aligned} H(t) &= \sum_{\kappa} \epsilon_{\kappa} n_{\kappa} + \sum_{a=1}^{N} E(t) n_a + \tfrac{1}{2} \sum_{a \neq a'} U_{a,a'}(t) n_a n_{a'} \\ &\quad + \sum_{\kappa,a} [V_{\kappa a}(t) c_a^{\dagger} c_{\kappa} + V_{\kappa a}^{*}(t) c_{\kappa}^{\dagger} c_a]. \end{aligned} \tag{1}$$

Here κ represents the quantum numbers of the states in the substrate band and a represents the quantum numbers of the atomic states. The time-dependent parameters are the level position E the level width

$$\Gamma = 2\pi N \sum_{\kappa} |V_{a\kappa}|^2 \delta(\varepsilon_{\mathrm{F}} - \varepsilon_{\kappa}), \tag{2}$$

and the intra-atomic Coulomb interaction U. The effect of the latter is crucial in this type of problem, as has been recognized in pioneering work [4, 5, 6, 7]. Here we discuss a simplified but realistically occurring situation where $E + U - \varepsilon_{\mathrm{F}} \gg \Gamma$. In this case U may as well be infinite, thus allowing at most one of the degenerate atomic levels to be occupied at the same time. This is a useful assumption, because there exists a well established approximation method for this case, known as the non-crossing approximation (NCA) [8, 9].

The NCA was originally justified in terms of a $1/N$ expansion, but is a conserving self-consistent approximation in its own right which preserves the qualitative features expected of the exact solution including the Kondo and mixed valent resonances [9, 10]. It has recently been compared [11] with exact numerical renormalization group calculations and shown to be creditable even for $N = 2$ at low temperatures for the type of application envisaged here. Although it has a well known spurious Fermi level resonance [9, 10] at very low temperatures (well below the Kondo temperature), it is believed not to be a concern for the temperature ranges of interest here. In any case we found that it occurs only at lower temperatures than those for which we reported results.

We have generalized the NCA [12] to the time-dependent non-equilibrium case by combining Coleman's auxiliary boson implementation of the NCA with the Kadanoff-Baym equations [13]. An exact numerical solution of these time-dependent equations was later given [14], which is used in the time-dependent curves reported here.

3. Equilibrium and Non-equilibrium Spectral Functions

One of the simplest ways to understand when electronic non-adiabaticity becomes an issue is through the behavior of the spectral functions of the atomic

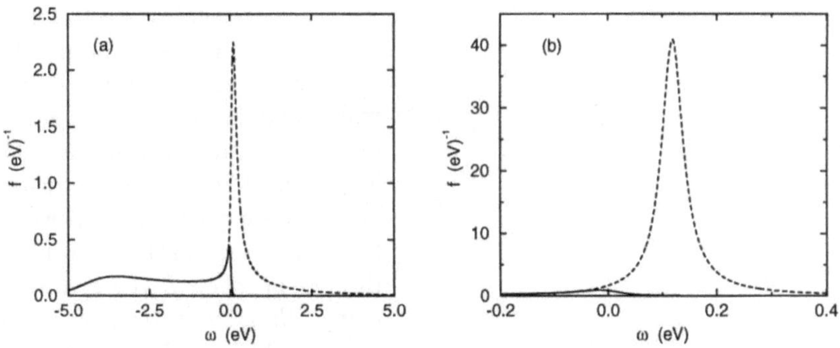

Fig. 1. Thermal equilibrium spectral functions in (a) the Kondo region and (b) the mixed valence region. Solid lines: f_{occ}; dashed lines: f_{tot}.

level. Although there are some technical intricacies in their definition with respect to the operators of an auxiliary boson Hamiltonian, they are defined in the usual way in terms of the time dependent operators $c(t_1)$ and $c^\dagger(t_2)$ of the original Hamiltonian Eq. (1). We will denote the occupied spectral weight by $f_{\text{occ}}(\omega, t)$ and the total spectral weight by $f_{\text{tot}}(\omega, t)$, where $t = \frac{1}{2}(t_1 + t_2)$ and ω is the Fourier transform variable with respect to $(t_1 - t_2)$.

In Fig. 1(a) are shown the thermal equilibrium spectral functions for $E = -2.4$ eV, $\Gamma = 5.8$ eV, $T = 200$ K, $N = 4$, and $D = 5$ eV. [Throughout this article we use a parabolic band with half-width D. Because the figures in this article have been collected from several projects with somewhat different purposes, the degeneracy N takes on a number of different values; we find, however that the results for different $N \geq 2$ are not qualitatively different from one another]. The broad peak at energy $\sim E$ and width $\sim \Gamma$ is qualitatively similar to what occurs for $U = 0$, while the so called Kondo peak occurring a small amount $\sim k_B T_l$ above the Fermi level, is produced by the intra-atomic correlation U. Here $k_B T_l \sim 0.1$ eV and its ratio with the Kondo temperature T_K is a number of order unity. As T is raised above T_K the Kondo peak gradually disappears. As E is increased, on the other hand, the two peaks merge into a single mixed valent peak when $E \sim 0$ [i.e., at the Fermi level]. This is illustrated in Fig. 1(b). This single peak is narrower and more strongly temperature dependent than the corresponding peak for the case where intra-atomic Coulomb interaction U is negligible.

What happens then in the time dependent case? If U were 0 then there would be a single characteristic rate Γ or inverse time-scale, and we would expect local thermal equilibrium to be maintained (no dissipation) if $|\dot{\Gamma}|/\Gamma \ll \Gamma$ and $|\dot{E}|/\Gamma \ll \Gamma$. However, the presence of the narrow peak above the Fermi level provides a second much slower characteristic rate $\sim \gamma$, where γ is the width of the narrow peak. Thus one might guess that to maintain thermal equilibrium one must in addition satisfy the additional requirements $|\dot{\gamma}|/\gamma \ll \gamma$ and $|\dot{\varepsilon}|/\gamma \ll \gamma$, where ε is the position (energy) of the narrow peak.

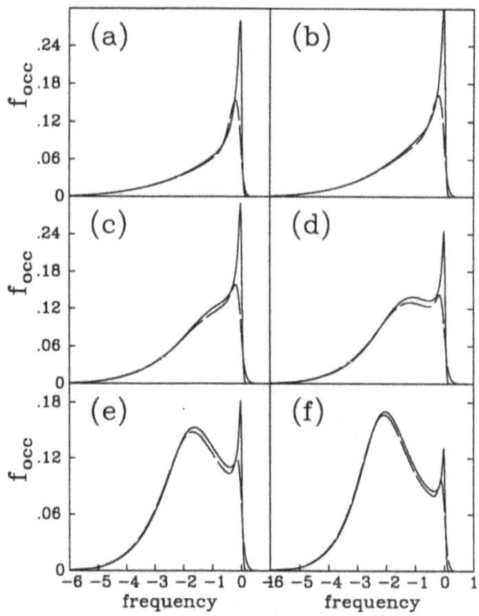

Fig. 2. Occupied spectral functions in and out of equilibrium. The solid curves are the occupied spectral weight functions of the system in local thermal equilibrium. The long dashed curves are the instantaneous occupied spectral weight functions (see text). Panels (a) through (f) show $f_{\rm occ}(\omega, t)$ at times t when $E(t)/\Gamma$ equals -0.05, -0.20, -0.35, -0.50, -0.65 and -0.80 respectively. The energy or frequency on the abscissa is measured from the Fermi level and is in units of $\Gamma/2$. $T = 0.013\Gamma$, $D = 2.5\Gamma$, and $N = 2$. Reprinted from Ref. [14].

The above is illustrated by a model calculation of $f_{\rm occ}(\omega, t)$ made several years ago [14] where Γ was held fixed and E was decreased at a rate $\dot{E} = -0.0625\Gamma^2$. This is a very slow rate, and the criterion $|\dot{E}|/\Gamma \ll \Gamma$ was well satisfied, but the more severe criterion $|\dot{\varepsilon}|/\gamma \ll \gamma$ was not. As shown in Fig. 2, the peak at E follows within a few percent of thermal equilibrium, but the Kondo peak does not. There are two lessons to be learned: first that electronic non-adiabaticity occurs much more frequently than would be supposed from estimates using traditional $U = 0$ models, and second, that one is going to have to achieve very slow rates of parameter variation to measure *directly* the effect of the highly correlated Kondo or mixed valent state.

4. Linear Response

As an example of how the presence of U can enhance the dissipation or electronic non-adiabaticity in the Kondo region, and the dramatic temperature dependence that occurs as one raises T above T_K, we present a tidbit from our work on linear response theory in the NCA [15]. Suppose that $E(t) = E_0 + \delta E \exp(-i\Omega t)$ with Γ held fixed, where $\delta E \ll E_0$. Then the atomic occupation $n \equiv \sum_a^N \langle c_a^\dagger c_a \rangle$ will also have a small oscillating component. That is, $n(t) = n_0 + \delta n \exp(-i\Omega t)$, were n_0 is the value of n implied in thermal equilibrium by $E = E_0$. The quantities δn and δE are related by a linear response function or susceptibility: $\delta n = -\chi \delta E$, where χ is dependent on frequency Ω. Note that this is not the more trivial magnetic susceptibility, whose diagrammatic expansion within the NCA of the Kubo formula terminates at the first

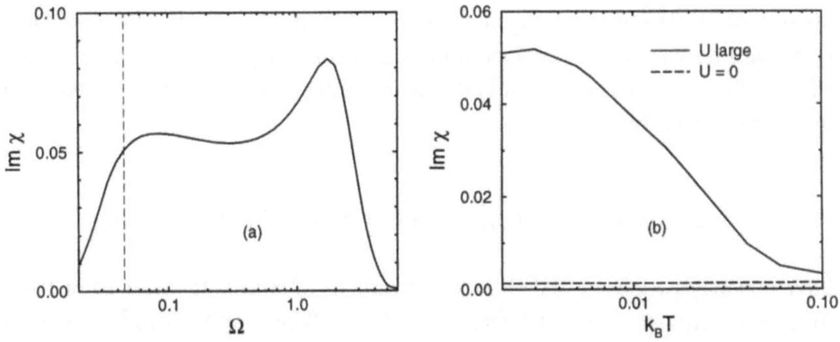

Fig. 3. Linear response function $\mathrm{Im}\,\chi$ plotted against (a) Ω with $k_BT = 0.002$ and against (b) k_BT with $\Omega = 0.045$ (position of dashed line in left panel). $E_0 = -1.5$, $D = 5$, and $N = 20$, where all energies are in units of $\Gamma/2$.

term, but rather what has been called the charge susceptibility in the literature [16, 17, 18, 19]. The two susceptibilities have vast qualitative and quantitative differences. Although χ could presumably be obtained by using the Kubo formula and summing the appropriate infinite series of NCA diagrams [16], our approach was simply to linearize our previous time dependent equations [12], which were exact within the NCA, and then solve the resulting equations.

The piece of χ of most interest to us is the imaginary part, which is a measure of the dissipative processes. In Fig. 3(a) we show $\mathrm{Im}\,\chi(\Omega)$ vs. Ω for a temperature somewhat below T_l. These first results were done at $N = 20$ where (a) the numerics were simpler and (b) they could be compared (favorably) with large N mean field results at $T = 0$ [17]; the results for physical N's, however, are similar [15]. We note that there is a peak on each of the two different energy scales of the problem. A simplistic interpretation would be that the high frequency peak represents electronic transitions from the broad peak around E in the spectral density [See Fig. 1(a)] to states above the Fermi level, and that the low frequency peak represents transitions within the Kondo peak or from states around the Fermi level to the Kondo peak. In any case the low frequency peak in χ has no counterpart in the $U = 0$ theory, and in the present theory goes away as the temperature is raised above the Kondo temperature. This is illustrated in Fig. 3(b), which shows $\mathrm{Im}\,\chi$ vs. T for $\Omega = 0.045(\Gamma/2)$. Also plotted is the analogous function for $U = 0$. This shows clearly that many-body correlation effects have not only enhanced the electronic non-adiabaticity by more than an order of magnitude over what simple theory predicts, but also provided a dramatic temperature dependence. Note that if $\Gamma \sim 1$ eV, the temperature dependence would occur on a scale $k_BT \sim .005$ eV, that is on a scale that one might normally associate with vibrational processes.

Recent work [20, 21, 22] suggests that the inclusion or electronic friction is an important element in the theory of motions of atoms and molecules outside metallic surfaces. Charge susceptibilities as defined above are an ingredient in the theory of electronic friction coefficients. The present work suggests that under certain conditions the inclusion of intra-atomic correlation U will have a significant effect on the results, and efforts are being directed toward developing a more complete theory.

5. Charge Transfer to an Atom Moving Away from a Surface

We have recently proposed [23] that measurement of charge transfer to or from an atom scattering or desorbing from a metal surface may be an ideal way to observe the dramatic temperature dependent effects associated with the highly correlated mixed valent state Fig. 1(b). As argued earlier, one needs to find a case where the quantities ε and γ vary slowly. This can be accomplished by playing off the image potential shift of E with distance z from the surface against the increase in Γ as z gets smaller. To be specific, we make a simple model for $E(z)$ and $\Gamma(z)$:

$$E = E(z) = E(\infty) - \frac{1}{4(z-1)} \tag{3}$$

and

$$\Gamma = \Gamma(z) = 1.5e^{-0.65z}, \tag{4}$$

where all quantities are expressed in atomic units (a.u.). Such a parameterization approximates the results of electronic structure calculations[24, 25], for typical atom-surface systems; we have found that the qualitative nature of our results is not changed by reasonable changes in the above assumptions. We take $E(\infty) = 1/27.2$ a.u., so that these equations describe an atom whose affinity level is 1 eV above the Fermi level when the atom is at $z = \infty$. As the atom slowly approaches the surface the affinity level moves below the Fermi level, first the mixed valent state [Fig. 1(b), $z = 8$ a.u.] forms, and then the Kondo

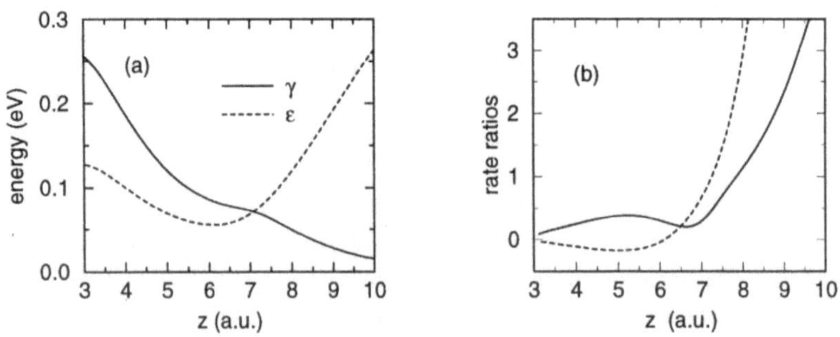

Fig. 4. (a) Peak position ε and width γ in thermal equilibrium at 200 K vs. atomic position z from the surface. (b) The corresponding ratios $\dot{\varepsilon}/\gamma^2$ and $-\dot{\gamma}/\gamma^2$.

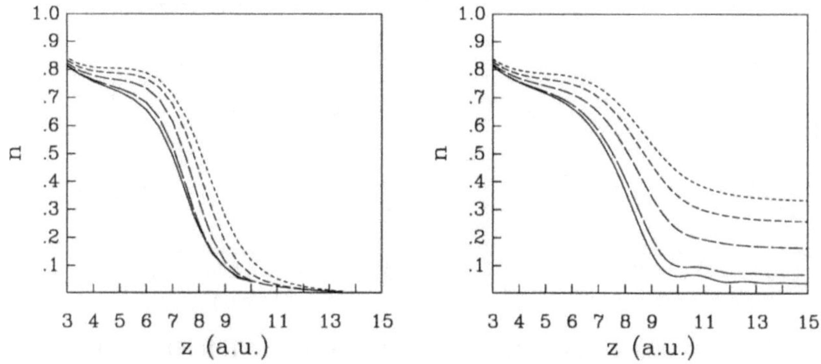

Fig. 5. Occupation probabilities for the atomic level vs. z for (a) thermal equilibrium ($v \approx 0$) and (b) $v = 0.004$ a.u. Temperatures of 100, 200, 400, 600, and 800 K.

state [Fig. 1(a), $z = 3.5$ a.u.] forms. On the outward trajectory, as the level moves upward again, there will be a competition for whether the affinity level will remain occupied or whether the extra electron will return to the metal leaving the original neutral atom as z approaches infinity again. We argue that the highly-correlated mixed valent state will make a substantial contribution to the outcome of this competition.

The crucial question is how slow the velocity v of the atom must be for the highly correlated state to form. We answer this via a two-fold approach. First we make equilibrium calculations of ε and γ vs. z, along with there derivatives with respect to z so that the local equilibrium conditions $|\dot{\gamma}|/\gamma \ll \gamma$ and $|\dot{\varepsilon}|/\gamma \ll \gamma$ can be tested. Second we make full time-dependent non-equilibrium calculations of the occupation of the atomic state. In Fig. 4(a) we show $\varepsilon(z)$ and $\gamma(z)$. In this section we take $N = 4$ unless otherwise indicated. It is clear that the peak at ε remains as if suspended just a little above the Fermi level, with fairly constant width, over a wide range of z values. Contrast this with the $U = 0$ case where the single peak's position ($\sim E$) and width ($\sim \Gamma$) vary rapidly with z. Thus in the highly correlated case, the mixed valent state should remain in some semblance of thermal equilibrium as z reaches the place where E crosses the Fermi level ($\sim$ 8 a.u. in this example). We illustrate this in Fig. 4(b), where the crucial ratios $-\dot{\gamma}/\gamma^2$ and $\dot{\varepsilon}/\gamma^2$ are plotted for a typical experimental velocity of 0.004 a.u. These ratios remain less than unity until z reaches around 8 a.u., that is the position of Fig. 1(b), at which point they undergo a fairly rapid switching to the the fast limit. This means that there is a good chance that the final distribution at $z = \infty$ will have a memory of the regime $z < 8$ a.u.

However, the occupation in the the mixed valent state which occurs in this regime is highly temperature dependent, as shown in Fig. 5(a), where we plot $n(z)$ for a range of temperatures. The spread in n over this temperature range is as much as 0.3 in the mixed valence region $z \sim 8$ a.u. This temperature

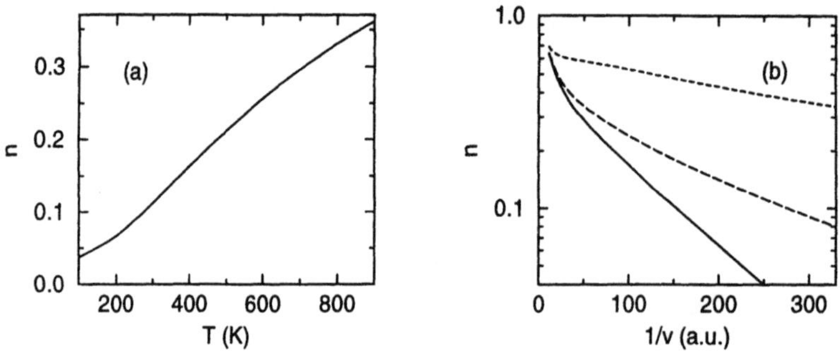

Fig. 6. Final average populations (a) vs. temperature T for $v = 0.004$ a.u. and (b) *vs.* $1/v$ for temperatures of 100, 300, and 1000 K.

spread can be understood from the presence of the strong peak in the unoccupied spectral weight just above the Fermi level [Fig. 1(b)] which becomes occupied as the temperature is raised, plus the fact that the width of this peak increases sharply with temperature, thus moving its effective lower edge to lower energies. A plot analogous to Fig. 5(a) for the $U = 0$ case shows essentially no temperature dependence of n.

Because of the rapid switching from the nearly adiabatic regime (in the $U = \infty$ case) when z crosses ~ 8 a.u. [see Fig. 4(b)], we expect a memory of this spread of n vs. T to be preserved. That this is so is shown in Fig. 5(b), where the results for $n(z)$ for different temperatures from our full time-dependent non-equilibrium calculation is illustrated. The final population probabilities ($z = \infty$) are shown in Fig. 6(a). Finally in Fig. 6(b) we plot the results vs. $1/v$ for several different temperatures. The fact that these curves are not straight lines (note the log scale) is due to the second time scale [26] provided by the highly correlated state. One can also see clearly from this figure that the anomalous temperature dependence persists over a fairly wide range of velocity.

6. Experimental Considerations

The unusual temperature dependences were found to occur over varying strengths and temperature ranges for atoms with affinity levels from about 0.7 to 2.0 eV above the Fermi level, with reasonable variation of the parameters used in Eq. (3) and (4). It may be necessary to tune the work function by varying the crystal face to get the temperature dependence into a range that is convenient. The effect also was found to occur over a wide range of velocities from 0.001 to 0.05 a.u. To eliminate Doppler broadening of the mixed valent resonance, which will weaken the effect, one should detect atoms escaping as close to the surface normal as possible. Careful estimates [23] show that substrate phonons do not significantly affect the results.

In addition to negative ion systems discussed explicitly here, our predictions are equally valid for appropriate atoms whose *ionization* levels are ~ 1 eV *below* the Fermi level, and where *positive* ion yield is measured [one merely has to interchange the roles of electrons and holes in the theoretical treatment]. The cleanest recommended example of an $N = 2$ system which we recommend is of this type, consisting of alkaline earth atoms over noble metal surfaces. In Ref. [23] we discuss further systems and the possibility of larger N.

Acknowledgments

This work is supported in part by the National Science Foundation under grants DMR 94-07055 (DCL) and DMR 95-21444 (PN).

References

[1] P. W. Anderson, Phys. Rev. **124**, 41 (1961).

[2] S. Edwards and D. M. Newns, Phys. Lett. **24A**, 236 (1967).

[3] D. M. Newns, Phys. Rev. **178**, 1123 (1969).

[4] R. Brako and D. M. Newns, Surf. Sci. **108**, 253 (1981).

[5] A. Yoshimori, H. Kawai, and K. Makoshi, Prog. Theor. Phys. Suppl. **80**, 203 (1984).

[6] H. Kasai and A. Okiji, Surf. Sci. **183**, 147 (1987).

[7] R. Brako and D. M. Newns, Rep. Prog. Phys. **52**, 655 (1989).

[8] P. Coleman, Phys. Rev. **B 29**, 3035 (1984).

[9] N. E. Bickers, Rev. Mod. Phys. **59**, 845 (1987).

[10] A. C. Hewson, *The Kondo Problem to Heavy Fermions* (Cambridge University Press, Cambridge, 1993).

[11] T. A. Costi, J. Kroha, and P. Wölfle, Phys. Rev. B **53**, 1850 (1996).

[12] D. C. Langreth and P. Nordlander, Phys. Rev. **B 43**, 2541 (1991).

[13] L. P. Kadanoff and G. Baym, *Quantum Statistical Mechanics* (Benjamin, New York, 1962).

[14] H. Shao, D. C. Langreth, and P. Nordlander, Phys. Rev. **B 49**, 13929 (1994).

[15] T. Brunner and D. C. Langreth, to be published (1996).

[16] Y. Kuramoto, Z. Phys. B **53**, 37 (1983).

[17] P. Coleman, Phys. Rev. B **35**, 5072 (1987).

[18] B. Jin, T. Matsuura, and Y. Kuroda, J. Phys. Soc. Jpn. **60**, 580 (1991).

[19] Y. Kuroda *et al.*, Progr. Theor. Phys. Suppl. **108**, 173 (1992).

[20] D. M. Newns, T. F. Heinz, and J. A. Misewich, Prog. Theor. Phys. Suppl. **106**, 411 (1991).

[21] C. Springer, M. Head-Gordon, and J. C. Tully, Surf. Sci. Lett. **320**, L57 (1994).

[22] M. Head-Gordon and J. C. Tully, J. Chem. Phys. **103**, 10137 (1996).

[23] H. Shao, P. Nordlander, and D. C. Langreth, Phys. Rev. Lett. (in press).

[24] P. Nordlander and J. C. Tully, Phys. Rev. **B 42**, 5564 (1990).

[25] A. G. Borisov, D. Teillet-Billy, and J. P. Gauyacq, Phys. Rev. Lett. **68**, 2842 (1992).

[26] H. Shao, P. Nordlander, and D. C. Langreth, Phys. Rev. B **52**, 2988 (1995).

Second Order Self-Energy and Interpolation Formula for the Charge Transfer Probability in Time-Dependent Models

K. Makoshi[1] **and** *T. Mii*[2]

[1]**Faculty of Science, Himeji Institute of Technology, Kamigoricho, Ako-gun, Hyogo 678-12, Japan**
[2]**Department of Material Physics, Faculty of Engineering Science, Osaka University, Toyonaka, Osaka 560, Japan**

Abstract. A simple approximation for charge transfer probability in time-dependent Hamiltonians is proposed with the use of the Mahanthappa-Baksi-Keldysh Green function. The Auger process and the resonant tunneling process are both considered. We show that the approximation interpolates between the limit of fast ion motion and that of slow ion motion. Also shown is that the range of validity is quite wide in Auger processes. The similar procedure is applied for the resonant tunneling process, leading to the interpolation formula for the two limits. By using model time-dependences, we also discuss the validity of our approximation.

1. Introduction

We look for an interpolation scheme for the charge fraction in atom-surface collisions, based on the time-dependent Hamiltonian:

$$H(t) = H_0 + H'(t). \tag{1}$$

Here, H_0 is an unperturbed part which is assumed to be diagonalized by an appropriate set of states. The time-dependence comes in through the trajectory approximation, when the position of an ion is classically given at each instance of time. A difficulty emerging from this type of Hamiltonian is due to the non-commutativity of $H'(t)$'s at different times. Therefore the time evolution operator must be written in an appropriate time-ordered exponential.

If the ionic motion is very fast compared to the electronic motion, the second order perturbation theory obviously works, since the electron transfer probability is small. When the ionic motion is slow enough, on the contrary, it has been believed that the description by the rate equation for the electron transfer works. The situation is quite clear in the Auger process such as neutralization of He^+ and deexcitation of metastable $He(1s2s)^*$,[1)] since the ground state products no longer make any contribution to the electronic states in the solid. We simply consider the survival probability of the ions. In such a situation, the survival probability of the ground state hole of excited species or ionized species is described by integrating the rate equation as[1)]

$$n(\infty) = n(-\infty)\exp\left[-\int_{-\infty}^{\infty} R(t)dt\right]. \tag{2}$$

Springer Series in Solid-State Sciences, Volume 121
Elementary Processes in Excitations and Reactions on Solid Surfaces
Eds.: A. Okiji, H. Kasai and K. Makoshi

Here, the transition rate at each instance of time t is obtained by the golden rule:

$$R(t) = \frac{2\pi}{\hbar} \sum_{\rm f} |\langle {\rm f}|H'(t)|{\rm i}\rangle|^2 \, \delta(E_{\rm f} - E_{\rm i}). \tag{3}$$

The summation is made over all possible final states satisfying the energy conservation expressed in terms of the last δ-function.

Therefore we expect to obtain an interpolation scheme between the fast and slow ionic motion limits, when we employ an appropriate function instead of the golden-rule product. We first consider the Auger process in the next section and in §3, proceed to the resonant tunneling case.

2. Auger process

The Auger process is essentially the two electron process as shown in Fig. 1. The time-dependent model delt with is

$$H_0 = \sum_{\mu} \varepsilon_\mu C_\mu^\dagger C_\mu + \sum_q \varepsilon_q A_q^\dagger A_q, \tag{4}$$

$$H'(t) = \sum_{q\mu\nu} \left(V_{q\mu\nu} u(t) C_{\rm g}^\dagger A_q^\dagger C_\mu C_\nu + {\rm h.c.} \right). \tag{5}$$

Here we assume that the ground state (denoted as g) is occupied by a hole in the incident beam and the double occupancy of the ground state hole is excluded. Also LEED states (q) are assumed to be independent from the states in the solid or in the excited states of the particle (μ and ν).[2,3] We assume, for simplicity, that the time-dependence of the matrix elements can be separated from those state-dependences as often assumed in literature. Then $V_{q\mu\nu}$ is the value at the closest approach of the ion to the surface.

To calculate the survival probability, we employ the Mahanthappa-Baksi-Keldysh[4-8] Green function formalism which allows us to deal with the nonequi-

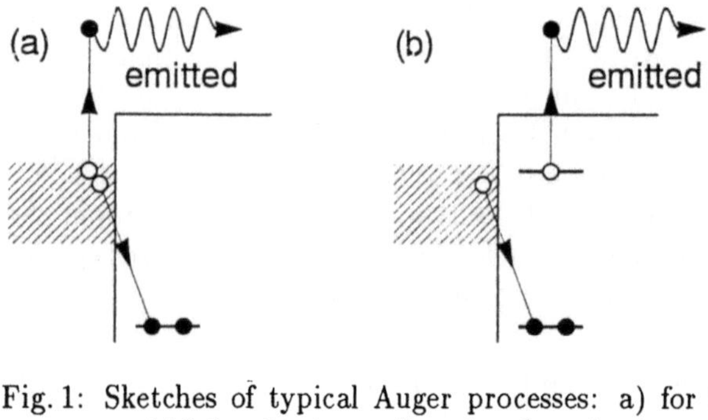

Fig. 1: Sketches of typical Auger processes: a) for the Auger neutralization process and b) for the Auger-Penning deexcitation process

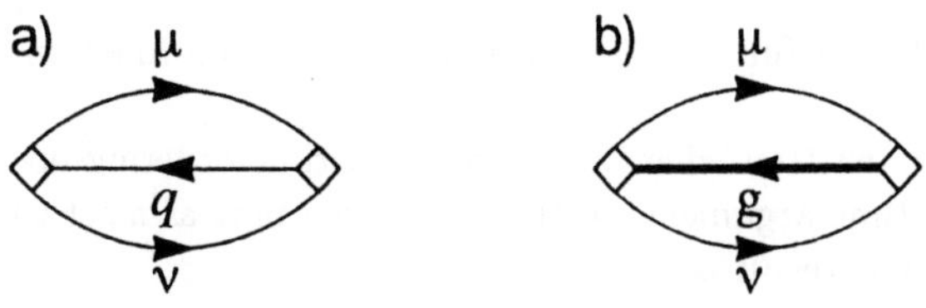

Fig. 2: The second order self-energy for a) ground state (g) Green function, and b) LEED (q) state Green function.

librium system. The form of the Green functions and the self-energy part is in 2×2 matrix:

$$\boldsymbol{G} = \begin{pmatrix} 0 & G^{\mathrm{r}} \\ G^{\mathrm{a}} & F \end{pmatrix}, \qquad \boldsymbol{\Sigma} = \begin{pmatrix} \Omega & \Sigma^{\mathrm{a}} \\ \Sigma^{\mathrm{r}} & 0 \end{pmatrix}. \tag{6}$$

The Dyson equation is

$$\boldsymbol{G} = \boldsymbol{G}_0 + \boldsymbol{G}_0 \boldsymbol{\Sigma} \boldsymbol{G}. \tag{7}$$

Here, the matrix multiplication includes the integration over time as well as the sum over quantum indeces. After obtaining the retarded Green function, G^{r}, G^{a} is given by the Hermitian congugate of G^{r} and F by

$$F = G^{\mathrm{r}}(\Omega - i\mathcal{N})G^{\mathrm{a}}, \tag{8}$$

where $\mathcal{N}$ is the initial ($t = -\infty$) occupation matrix defined as

$$\mathcal{N}_{\alpha\beta} = (1 - 2n_\alpha^0)\delta_{\alpha\beta}\delta(t - t')\delta(t + \infty), \tag{9}$$

with the initial occupation number of α-state, n_α^0.

Our first task is to look for a suitable approximation to obtain the retarded function. In our restriction of subspace, the Dyson equation for the g-Green function with the second order self-energy reads

$$G_{\mathrm{gg}}^{\mathrm{r}}(t_1, t_2) = G_{\mathrm{gg}}^{\mathrm{r}0}(t_1, t_2) + \int d^2\tau G_{\mathrm{gg}}^{\mathrm{r}0}(t_1, \tau_1)\tilde{\Sigma}_{\mathrm{gg}}^{\mathrm{r}(2)}(\tau_1, \tau_2)G_{\mathrm{gg}}^{\mathrm{r}}(\tau_2, t_2). \tag{10}$$

The diagrammatic expression of the second order self-energy is shown in Fig. 2. The analytic expression is given by

$$\tilde{\Sigma}_{\mathrm{gg}}^{\mathrm{r}(2)}(\tau_1, \tau_2) = -i\theta(\tau_1 - \tau_2)u(\tau_1)u(\tau_2)\sum_{q\mu\nu} \tilde{V}_{q\mu\nu}(\tau_1)\tilde{V}_{q\mu\nu}^*(\tau_2)n_\mu^0 n_\nu^0. \tag{11}$$

Here, θ is the unit step function expressing the retarded nature of functions. We moved explicit phase factors of the Green functions to the matrix elements which are redefined as

$$\tilde{V}_{q\mu\nu}(t) = V_{q\mu\nu} \exp\left[i(\varepsilon_q + \varepsilon_\mu - \varepsilon_\nu - \varepsilon_{\mathrm{g}})t\right]. \tag{12}$$

In this convention, the free Green function becomes simply the step function: $\tilde{G}_0^{\mathrm{r}}(t_1,t_2) = -i\theta(t_1 - t_2)$.

The formal solution can be written when we introduce a time-ordering operator $\tilde{T}_+$ arranging the two time arguments in the self-energy part as a set to put the future to the left.[2] The result is

$$\tilde{G}^{\mathrm{r}}_{\mathrm{gg}}(t_1,t_2) = -i\theta(t_1 - t_2)\tilde{T}_+ \exp\left[-\int d^2\tau \tilde{\Sigma}^{\mathrm{r}(2)}(\tau_1,\tau_2)\right], \tag{13}$$

where $d^n\tau$ means that the integration is of n-dimensional. We note that $\tilde{T}_+$ also exclude the possibility such a case as $\tau_1 > \tau_3 > \tau_2 > \tau_4$.

Our approximation is to remove the time ordering $\tilde{T}_+$:[2]

$$\tilde{G}^{\mathrm{r}}_{\mathrm{gg}}(t_1,t_2) = -i\theta(t_1 - t_2)\exp\left[-\int d^2\tau \tilde{\Sigma}^{\mathrm{r}(2)}(\tau_1,\tau_2)\right]. \tag{14}$$

In the second order self-energy, we have the intermediate sum over the LEED states:

$$\sum_q |V_{q\mu\nu}|^2 \exp\left[i\varepsilon_q(\tau_1 - \tau_2)\right] = \int d\varepsilon \rho_{\mathrm{L}}(\varepsilon)\langle |V_{q\mu\nu}|^2\rangle_\varepsilon e^{i\varepsilon(\tau_1-\tau_2)}. \tag{15}$$

Here, $\rho_{\mathrm{L}}(\varepsilon)$ is the density of states of the LEED states and the brakets mean the equienergy average at $\varepsilon_q = \varepsilon$.

For the Auger process. the equienergy average of $|V|^2$ is expected to extend in a wide energy range. Then we may apply the wide band approximation leading to the self-energy local in time:

$$\tilde{\Sigma}^{\mathrm{r}(2)}_{\mathrm{gg}}(\tau_1.\tau_2) \quad\propto\quad -iu(\tau_1)^2\theta(\tau_1 - \tau_2)\delta(\tau_1 - \tau_2). \tag{16}$$

In this case, the lift of the time ordering is harmless. Therefore, if the velocity of the incident beam is small compared to the electronic motion, the rate equation is justified for the Auger process. We emphasize that the electronic time scale is given by the "effective" band width of the "LEED" states which is quite wide. This tells us why the description by of the rate equation can be applied to the Auger cases with a rather wide range of applicability.

As for the spectrum of emitted electrons, the consistent approximation can be achieved by the second order self-energy as is shown in Fig. 2-b).[9] we employ the renormalized Green function for the g-state instead of the free one, to take the effect of the ground state hole life-time into account.

In the quasi-static case and an exponential variation of $u(t) = \exp(-\alpha v \mid t \mid)$ with the ion velocity v, we obtain the spectrum of the metastable deexcitation if most of ions are neutralized on the incoming leg, and find that the broadening of the spectrum due to the ion motion as well as to the life time effect is obtained by convoluting the broadening function:[9]

$$B(\varepsilon) = \left[2\alpha v \cosh\frac{\pi\varepsilon}{2\alpha v}\right]^{-1}. \tag{17}$$

3. Resonant tunneling

This case has recently been studied using the nonequilibrium Green function formalism.[10] We study the noninteracting Newns-Anderson model:[11, 12]

$$\mathcal{H}_0(t) = \sum_k \varepsilon_k C_k^\dagger C_k + \varepsilon_a(t) C_a^\dagger C_a \tag{18}$$

$$\mathcal{H}'(t) = \sum_k \{V_k u(t) C_a^\dagger C_k + \text{h.c.}\}. \tag{19}$$

Here, we again assume that the time-dependence of the admixture matrix element is separated. The original self-energy in terms of $\{a, k\}$-space is schematically written, in 2×2 form as

$$\boldsymbol{\Sigma}(\tau_1, \tau_2) = \delta(\tau_1 - \tau_2) \begin{pmatrix} 0 & V \\ V & 0 \end{pmatrix}. \tag{20}$$

We rewrite the equation for diagonal a-state, eliminating the k-states. Then we find

$$\tilde{\Sigma}_{\mathrm{aa}}^{\mathrm{r}}(t_1, t_2) = -i\theta(t_1 - t_2) u(t_1) u(t_2) \int_{-\infty}^{\infty} \frac{d\varepsilon}{\pi} \Delta(\varepsilon) \exp\{i \int_{t_2}^{t_1} [\varepsilon - \varepsilon_{\mathrm{a}}(\tau)] d\tau\} \tag{21}$$

is the self-energy. Here, $\Delta(\varepsilon)$ is defined as

$$\Delta(\varepsilon) = \pi \sum_k |V_k|^2 \delta(\varepsilon - \varepsilon_k). \tag{22}$$

Again the formal solution of $\tilde{G}_{\mathrm{aa}}^{\mathrm{r}}$ is obtained with the use of pair time-ordering. We apply the same approximation as the Auger case that lifts the pair time-ordering operator:

$$\tilde{G}_{\mathrm{aa}}^{\mathrm{r}}(t, t') = -i\theta(t - t') \exp\left[-i \int_{t'}^{t} d^2\tau \tilde{\Sigma}_{\mathrm{aa}}^{\mathrm{r}}(\tau_1, \tau_2)\right]. \tag{23}$$

This approximation leads to the same result for the charge fraction, as that given by Tsukada and Shima.[13]

In the wide band limit (ε-independent Δ), the approximation leads to the exact result since the self-energy becomes local in time. The average occupation after the scattering of the ion is given by that obtained by Brako and Newns:[14]

$$\begin{aligned} n_{\mathrm{a}}(\infty) = {} & n_{\mathrm{a}}^0 \exp\left[-\frac{2\Delta_0}{\alpha v}\right] + \frac{\Delta_0}{\pi} \int_{-\infty}^{\infty} d\varepsilon n_\varepsilon^0 \\ & \times \left| \int_{-\infty}^{\infty} d\tau u(\tau) \exp\left[i \int_{\tau}^{\infty} d\tau' \{\varepsilon - \varepsilon_{\mathrm{a}}(\tau') + i\Delta(\tau')\}\right] \right|^2. \end{aligned} \tag{24}$$

Even if the band width is finite, when the atom level and the band cross linearly with each other, we can obtain at least the leading term of order of

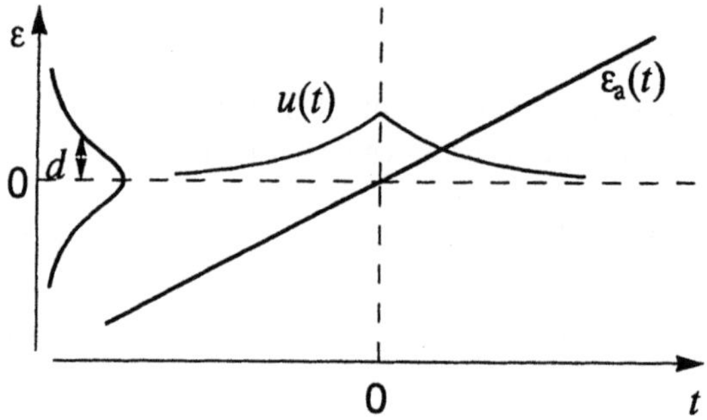

Fig. 3: Sketch of energy dependence of $\Delta(\varepsilon)$ and time dependences of $u(t)$ and $\varepsilon_a(t)$ in our model.

$(1/v)^n$ in the $(2n)$-th order retarded Green function. For this purpose, we model the finite band by the Lorentzian form:

$$u(t) = e^{-\alpha v|t|}, \qquad \varepsilon_{\mathrm{a}}(t) = \beta v t, \qquad \Delta(\varepsilon) = \frac{\Delta_0}{\varepsilon^2 + d^2}. \tag{25}$$

The model is sketched in Fig. 3. Then we find the leading order term of $(2n)$-th order retarded Green function reads

$$\tilde{G}^{\mathrm{r}(2n)}_{\mathrm{aa}}(\infty, -\infty) \propto \left(\frac{2}{\beta v}\right)^n \int_{-\infty}^{0} d^{2n-1}\zeta \exp\left[i \sum_{k=1}^{n} (\zeta_{2k} + 2 \sum_{l=2}^{2k-1} \zeta_l)\zeta_{2k}\right]. \tag{26}$$

Here, we simply changed time-variables as $\zeta_k = -\tau_{k-1} + \tau_k$ for $k \geq 2$ and $\zeta_1 = \tau_1 + \tau_2$, and performed the integration over ζ_1. This expression does not involve the band width d. Therefore the leading term is the same as the wide band limit if the ion velocity is small enough.

In the surface molecule limit (zero band width), the Landau-Zener formula is well known in the slow ionic motion limit. We assume a similar model to that in the previous argument, now with $d \to 0$ in Δ:

$$u(t) = e^{-\alpha v|t|}, \qquad \varepsilon_{\mathrm{a}}(t) = \beta v t, \qquad \Delta(\varepsilon) = \Delta_0 \delta(\varepsilon). \tag{27}$$

We obtain, in our approximation, following result for the charge fraction at $t = \infty$:

$$n_{\mathrm{a}}(\infty) = \exp\left[-\frac{4}{\pi}\Delta_0 |A|^2\right] \tag{28}$$

with

$$A = \frac{\alpha v i}{2\beta v}\left\{-2F(1, \frac{3}{2}; \frac{i\alpha^2 v^2}{2\beta v}) + \left(\frac{i\alpha^2 v^2}{2\pi\beta v}\right)^{-\frac{1}{2}} F(\frac{1}{2}, \frac{1}{2}; \frac{i\alpha^2 v^2}{2\beta v})\right\}. \tag{29}$$

Here, F is Kummer's hypergeometric function.

This expression interpolates two limits. When $\alpha^2 v^2 \ll \beta v$ (slow limit), we obtain the Landau-Zener formula:

$$n_a(\infty) \quad = . \ \exp\left(-\frac{2\Delta_0}{|\beta| v}\right). \tag{30}$$

When the ion motion is fast, i.e. $\alpha^2 v^2 \gg \beta v$, we obtain by employing the asymptotic formula for F,

$$n_a(\infty) \quad = \quad \exp\left(-\frac{4\Delta_0}{\pi \alpha^2 v^2}\right). \tag{31}$$

When the a-level does not cross the level in the solid, we obtain an oscillatory behavior of the charge fraction depending on the value of αv, as has been obtained by Tully.[15] The period cannot be correctly given by our approximation, however, unless the admixture matrix element is small. A simple example is given by assuming that (the atom level)$\equiv \varepsilon_0 \neq \varepsilon_a$ and $u(t) = \theta(t)$. This case is well known as an excercise of the quantum mechanics. When the value of V is large enough, the oscillation period is given by $\sqrt{(\varepsilon_0 - \varepsilon_a)^2 + 4V^2}$. Our approximation is based on the second order perturbation, thus cannot give rise to such a period.

4. Concluding remarks

We presented a simple interpolation formula for the charge fraction of ion scattered from metal surfaces. We have shown that the survival probability of ion, in the Auger process, can be well described by the rate equation, since the effective "LEED" bandwidth is expected to be large. Even if the velocity is not so small, our approximation leads to a good description of the survival probability, it interpolates the fast and slow motion limits at least.

In the resonant tunneling case, our approximation for the Green functions leads to the same result as Tsukada and Shima obtained by dealing with the time-evolution operator.[13] In the Green function formalism, it is achieved rather easily and thus able to investigate in some models for the resonant width and time-dependences. The approximation gives fairly satisfactory result when the atom level and the electronic band cross with each other. The formula leads to the interpolation between the fast and slow motion limits. In the final case without level crossing, however, our approximation is rather poor.

References

1) H.D. Hagstrum: in *Chemistry and Physics of Solid Surfaces VII* eds. R. Vanselow & R. Howe (Springer-Verlag, Berlin, 1988) p. 341.

2) H. Kaji, K. Makoshi and A. Yoshimori, Surf. Sci. **227** (1990) 138.

3) K. Makoshi and H. Kaji, Prog. Theor. Phys. Suppl. no. 106 (1991) 327.

4) K.T. Mahanthappa, Phys. Rev. **126** (1961) 329.

5) P.M. Baksi and K.T. Mahanthappa, J. Math. Phys. **4** (1963) 1.

6) P.M. Baksi and K.T. Mahanthappa, J. Math. Phys. **4** (1963) 12.

7) L.V. Keldysh, Sov. Phys. JETP **20** (1965) 1018.

8) A. Blandin, A. Nourtier and D.W. Hone, J. Phys. (Paris) **37** (1976) 369.

9) K. Makoshi, Surf. Sci. **254** (1991) 281.

10) T. Mii and K. Makoshi, to be published in Surf. Sci.

11) P.W. Anderson, Phys. Rev. **124** (1961) 41.

12) D.M. Newns, Phys. Rev. **178** (1969) 1123.

13) M. Tsukada and N. Shima, in *Dynamical Processes and Ordering on Solid Surfaces*, eds. A. Yoshimori and M. Tsukada (Springer-Verlag, Berlin, 1985) p. 34.

14) R. Brako and D.M. Newns, Rep. Prog. Phys. **52** (1989) 655.

15) J.C. Tully, Phys. Rev. B **16** (1977) 4324

Surface Reactions Activated by High Electron Temperatures

T.F. Heinz[1], *J.A. Misewich*[1], *D.M. Newns*[1], *M. Brandbyge*[2], *and P. Hedegård*[3]

[1]IBM Thomas J. Watson Research Center, P.O. Box 218, Yorktown Heights, NY 10598, USA

[2]Physics Department, Technical University of Denmark, DK-2800 Lyngby, Denmark

[3]Ørsted Laboratory, Niels Bohr Institute, DK-2100 Copenhagen, Denmark

Abstract. Reaction processes are normally activated by collisional energy transfer from the surrounding medium. In metallic environments, e.g. on metal surfaces, reaction processes activated by direct coupling to electronic degrees of freedom are theoretically possible. Femtosecond pulse laser desorption is proposed as a canonical example of this new class of electronically-activated reaction process. The theory of femtosecond pulse laser desorption is outlined starting from the concept of electronic friction.

1. Introduction

A reaction process such as we have in mind here involves a microscopic system such as a molecule or molecules undergoing a chemical process along some reaction coordinate x. For example a molecule might dissociate, when x starts off finite and goes to infinity. The presumption is that the process is activated, i.e. the free energy as a function of x has a maximum, say at an activation energy E_a above the bottom of the initial energy minimum, either at finite or infinite x.

The rate R of an activated reaction is written in the Arrhenius form:

$$R = Pe^{-E_a/T}, \tag{1}$$

in $k_B = 1$ units. The usual assumption is that the ion coordinates propagate on the Born-Oppenheimer energy surface, which is defined as the ground electronic state of the system. The prefactor P in Eq.(1) is then taken to represent a vibration frequency and to be of order $10^{12} sec^{-1}$.

The metallic environment provides many low-lying excited states, which can be coupled to a local object such as a molecule. We shall see that it may then be appropriate to drop the assumption of ionic motion on the ground Born-Oppenheimer surface, and to look at the electronic excitations as an activation channel capable of competing with the more conventional vibrational one [1].

Friction for the degree of freedom x is introduced via the frictional force F

$$F = -\eta \dot{x} M, \tag{2}$$

where the mass M has been incorporated so as to make the dimension of friction η that of frequency. For a weakly damped harmonic oscillator, the vibrational linewidth is η.

At finite temperatures Brownian motion occurs due to noise introduced into the x-degree of freedom. For white noise this comes in via the force-force correlation function $< F_t F_{t'} >= 2\eta M T \delta(t-t')$.

Springer Series in Solid-State Sciences, Volume 121
Elementary Processes in Excitations and Reactions on Solid Surfaces
Eds.: A. Okiji, H. Kasai and K. Makoshi

Brownian motion may excite the particle over the activation barrier. Kramers gave the expression for the rate in the low-friction limit (appropriate for electronic friction) as [1]

$$R = \eta \frac{E_a}{T} e^{-E_a/T}, \tag{3}$$

containing an explicit expression for the prefactor P as of order η.

Electronic friction is a phenomenon [1-5] which arises when an atom is moving through the electron gas environment, for example at a metal surface. The finite scattering cross-section of the atom results in electrons being scattered from occupied to unoccupied states near the Fermi surface, with consequent transfer of energy from the atom to the electron gas, which appears as electronic friction.

We can make an order of magnitude estimate of the effect by calculating the electrical dissipation on the basis of the quantized conductance [7] of the channel presented by the atom to the electron gas. The energy change dE for a current I flowing into the atom during a short time dt will be

$$dE = (I^2/G_0)dt \tag{4}$$

where G_0 will be of order of the quantum of conductance $G_0 \sim e^2/h$. Expressed in terms of electronic occupation n and velocity v

$$dE \sim \frac{h}{2}(\frac{dn}{dt})^2 dt \sim \frac{h}{2}(\frac{dn}{dx})^2 v dx, \tag{5}$$

Then comparing with Eq.(2) we have for the electronic friction η

$$\eta = \frac{\hbar}{\pi M}(\frac{d\delta^F}{dx})^2, \tag{6}$$

δ^F is phase shift at the Fermi level, related to $n(t)$ by $n = \delta^F/\pi$; the coefficient in (6) turns out to be unity. We see that what comes into our result (6) is the change in charge on the atom (or molecule) as it moves around in the inhomogeneous surface electron gas, i.e. the electronic friction is related to the chemistry of the atom-surface interaction.

The result derived in this heuristic manner is in fact a special case of the complete result [1]. However it is an extremely useful one which will be adequate for use in the present paper once it is generalized to finite temperatures. The generalization is [2, 6]

$$\eta = \frac{\hbar}{\pi M} \int d\epsilon (\frac{d\delta(\epsilon)}{dx})^2 [-\frac{df(\epsilon)}{d\epsilon}]. \tag{7}$$

where $f(\epsilon)$ is Fermi function and $\delta(\epsilon)$ is the phase shift at energy ϵ.

If we put in reasonable assumptions, for say the phase shift expected for a π^* orbital in a diatomic such as NO, into Eq.(6), we find that a friction η of order $10^{12} sec^{-1}$ is feasible. Estimates for the magnitude of the electronic friction can be derived from the vibrational linewidths η, as has been shown in extensive work by Persson, coworkers and others [5, 8]. A few values quoted from a recent review article are quoted in Table I.

They are indeed of the order 10^{12} sec^{-1}; hence the reaction rates expected on the basis of the Kramers rate (3) are predicted to be comparable with vibrationally activated ones!

2. The DIMET process

Can electronic friction indeed activate a reaction process? We think that well-studied examples exist, for systems in which desorption was observed under conditions with the electron system thermally excited but with the lattice relatively cold. Hence only the electronic activation channel is open, and the observation of reaction under these conditions proves that it is effective.

The experiment [9-12] involves exciting an adsorbed monolayer of CO, NO or O_2 on a metallic substrate with a short laser pulse lasting approx. $100fs$.. The laser pulse absorption process has been extensively modelled, and is believed to involve direct transfer of the incident energy into the electron gas within $\sim$ a skin depth of the surface. The excited region of the electron gas can be described thermally, with a temperature rising to several thousand $^{\circ}K$ for a duration of order $200fs$. [13].

The desorption yield (which lies in the range of 10^{-3} to 10^{-4} per pulse) is a very rapidly increasing function of the laser fluence, interpreted as a very nonlinear function of the surface electron gas temperature - which strongly resembles a thermally activated reaction (in contradistinction to the linearity in fluence of the single photon desorption (DIET) process).

A measure for the time scale of the desorption process is obtained by looking at the yield in a two-pulse experiment as a function of the time interval between the pulses [12]. A very short characteristic time scale of order $400fs$. is determined in this manner.

The non-vibrational interpretation for the activation process in this reaction comes essentially from (a) the low lattice temperature achieved after $\sim$1-2 ps., a few hundred $^{\circ}K$ which is insufficient to desorb in the available time frame given the $\sim 1eV$. adsorption energy, and (b) the long time scale expected from modelling for the rise in lattice temperature, $\sim$ 1-2 ps., relative to the short reaction time deduced from the 2-pulse experiment.

Support for this interpretation comes from the discrepancy between desorbate translational (a few hundred $^{\circ}K$) and vibrational (one to two thousand $^{\circ}K$) energies. The high degree of vibrational excitation is expected for the electronic mechanism because the adsorbate stretch mode also experiences relatively high electronic friction (see Table I) [14, 15]. The relatively low translational excitation energy is consistent with our modelling (see below).

The approach we have taken to understanding the theory behind the femtosecond pulse laser desorption process is twofold. On the one hand we have built a simple theoretical model [6], essentially soluble analytically, and on the other hand we have numerically modelled desorption trajectories using the Langevin formalism [6, 16]. Since the treatments agree, we emphasize the modelistic approach [6] in the present description.

In the model treatment we describe the evolution in the adsorbate center of mass degree of freedom, and assume a truncated harmonic oscillator form for the well in which the adsorbate is bound, of depth V_B. In this well we define a probability distribution $W(\epsilon, t)$ for the adsorbate to have energy ϵ at time t. Treating motion in the well classically, the distribution $W(\epsilon, t)$ obeys the Fokker-Planck equation

Table I

system	vib. mode	linewidth (sec.)	freq. (cm^{-1})
CO/Cu(100)	C-O stretch	2.10^{-12}	2084
CO/Cu(100)	CO-Cu stretch	2.10^{-12}	285
H/metal	all modes	$\sim 10^{-12}$	$\sim$ 1000

$$\frac{\partial W}{\partial t} = \eta(t)\frac{\partial}{\partial \epsilon}\left(\epsilon\left[1 + T(t)\frac{\partial}{\partial \epsilon}\right]W\right). \tag{8}$$

Here the time-dependent friction η arises because of the time dependence of the reservoir (electron gas) temperature T in (7).

For the desorption problem a solution is required with the boundary conditions (a) at $t = 0$ $W(\epsilon, t) = \delta(\epsilon)$ (distribution initially concentrated at bottom of well), and (b) $W(V_B, t) = 0$ at all times, corresponding to the Kramers low friction limit [17].

The solution is obtained in terms of an auxiliary function, the "adsorbate temperature", $T_A(t)$, which obeys Newton's Law of Cooling:

$$\frac{dT_A}{dt} + \eta(t)T_A = \eta(t)T(t). \tag{9}$$

Previously[6] we showed that, by introducing an "image" solution with negative sign in the region $\epsilon > V_B$, it was possible to satisfy the boundary condition (b). In the regime appropriate for the present problem, in which temperatures are small compared with the well depth, the desorption probability obtained is

$$P_{des} = V_B \int_0^\infty dt \frac{\eta(t)}{T_A(t)} e^{-V_B/T_A(t)}. \tag{10}$$

Note that this result is just the time integral over the Kramers low-friction rate Eq.(3), except that it is noteworthy that the temperature appearing is the adsorbate temperature T_A, rather than the electron gas temperature T.

We may use this model to get an estimate of the desorption behavior. For the π^* we include a spin degeneracy of 2, together with an orbital degeneracy of 2, which introduces an additional factor of 4 in front of Eq. (7). We need a model to describe the π^* resonance in the adsorbate, characterizing it by an energy ϵ_a and a lifetime broadening Δ. We model the dependencies of these quantities on distance x from the well minimum by

$$\Delta = \Delta^0 e^{-\beta x}, \tag{11}$$

$$\epsilon_a = \epsilon_\infty - Ce^{-\alpha x}. \tag{12}$$

The phase shift $\delta = atan[\Delta/(\epsilon - \epsilon_a)]$, can be inserted into (7) to obtain the friction. We take η at $x = 0$ as the friction appearing in (9) and (10), and hence employ (9) to obtain T_A and (10) to get P_{des}.

We illustrate in Fig's 1 and 2 the behavior of various quantities as a function of time, based on a function for the electronic temperature T derived from numerical calculations.

In Fig. 1 we have chosen Δ^0 to be comparable with the separation of ϵ_a above the Fermi level, so that the DOS at the Fermi level is quite large and hence the zero-temperature friction significant. Since Δ^0 is also large compared with T, the friction is relatively T-, and therefore time- independent, with a value of order 10^{12} sec^{-1}. The desorption probability is of the same order as that found in experimental data.

In Fig. 2 we have altered only Δ^0 relative to Fig. 1, making it much smaller so that the DOS at the Fermi level, and hence the zero-temperature friction, is negligible. Now there is a significant thermally activated friction, lasting only for the duration of the peak in the electronic temperature T, which again results in a reasonable value for the desorption probability.

A limiting classical picture becomes reasonable in the latter type of case where the friction is primarily activated. We introduce the ground state neutral PE curve for the adsorbate (π^* unoccupied) and an excited, ionized PE curve, (π^* occupied). The de-excitation rate from excited to ground state curves will be Δ, while there will be an excitation

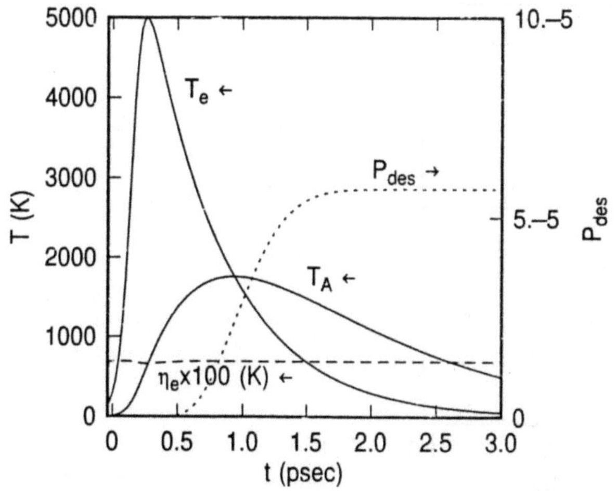

Figure 1: Time dependence of various quantities in the femtosecond pulse laser desorption problem. Electronic friction $\eta(t)$, adsorbate temperature $T_A(t)$, and fraction desorbed up to time t, are shown. The regime illustrated is non-activated friction, parameters are C = 40000K, ϵ_∞ = 58000K, $\Delta^0 = \epsilon_a(0)$, V_B = 1.4eV, β = 1.0au, α = 0.45au.

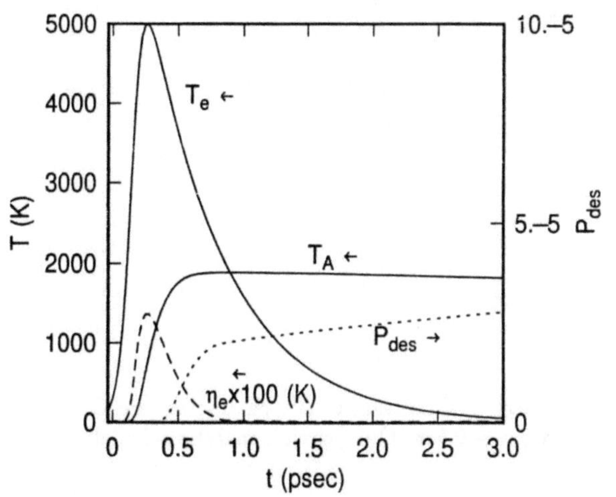

Figure 2: Time dependence of various quantities in the femtosecond pulse laser desorption problem, DIMET case, parameters as previous figure except for $\Delta^0 = 0.08\ \epsilon_a(0)$.

rate $\Delta exp(-\epsilon_a/T)$. The picture of Desorption which results is one Induced by Multiple Electronic Transitions between classical trajectories on the two surfaces (DIMET), a term which has become generic for the whole femtosecond pulse laser desorption regime.

In Fig. 3 we compare the desorption probabilities for large and small Δ regimes in the 2-pulse experiment. There is a clear trend for the desorption process to be faster in the activated friction or DIMET regime, because the activated friction is only large

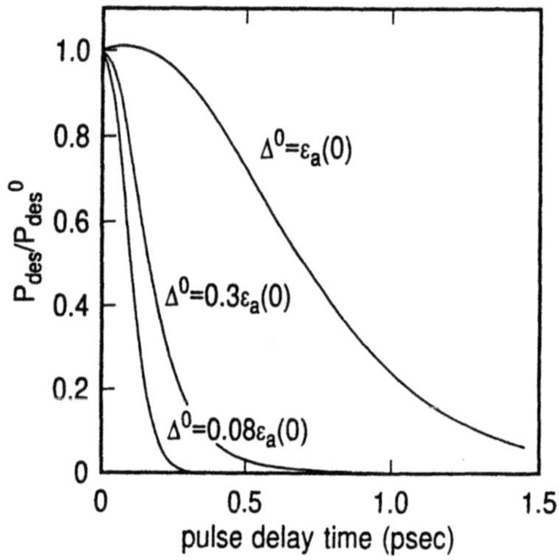

Figure 3: Two pulse experiment, in which a laser pulse corresponding to the parameters of Fig's 2 and 3 is split into two equal components separated by a pulse delay time. The desorption probability is plotted as a function of delay for three values of Δ^0, other parameters as Fig's 2 and 3.

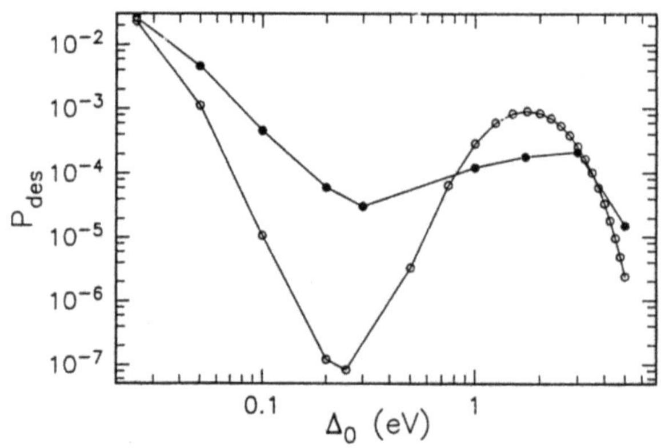

Figure 4: Plot of desorption probability as a function of Δ^0, parameters are C = 3.02 eV,ϵ_∞ = 5eV, V_B = 1.4eV, β = 1.0au, α = 0.5au. Open circles, analytic calculation, full circles, Langevin simulation.

for the short duration of the T-pulse. The observed time scale of order 0.4*ps.* is in an intermediate regime with some degree of activation of the friction.

It would be desirable to check the foregoing calculation with a model allowing for spatial dependence of the friction and a more realistic well shape. This has been done numerically by running large numbers of trajectories in the Langevin formalism. A comparison between the results of the two approaches is seen in Fig. 4. The approaches differ in the intermediate regime between the constant friction (right side of figure) and DIMET (left side) regimes. This is where the model approach using friction only at the well minimum gives a very low value for the desorption probability, while in the model allowing for spatial dependence of the friction is able to avail of significant friction values elsewhere, giving a shallower minimum in desorption probability.

Measurement of the desorbate kinetic energy in the Langevin approach yields values below $1000K$, consistent with experimental observations. Treatment of desorbate vibrational excitation as electronic-friction induced [14], following earlier lines [15], has yielded results consistent with the exceptionally large vibrational excitation seen experimentally.

In conclusion, we think that the early suggestion by Suhl and coworkers that electronic friction may be playing a significant role in activated processes at surfaces has some merit, on the basis that the DIMET process, occurring under controlled conditions with a hot electron system and a cold lattice, provides a laboratory in which its action seems to be clearly observed.

References

[1] K.P. Bohnen, M. Kiwi, and H. Suhl, Phys. Rev. Lett. **34**, 1512 (1975); A. Nourtier, J. de Physique **38**, 479 (1977).

[2] A. Yoshimori and J.L. Motchane, J. Phys. Soc. Japan **51**, 1826 (1982); A. Okiji and H. Kasai, Surf. Sci. **B188**, L717 (1987).

[3] K. Makoshi, J. Phys. C **16**, 3617 (1983)

[4] J.C. Tully, M. Gomez and M. Head Gordon, J. Vac. Sci. Technol. **11**, 1914 (1993); J.C. Tully and M. Head Gordon, Phys. Rev. **B46**, 1853 (1992);C. Springer, M.Head-Gordon, and J.C. Tully, Surf. Sci. **320**, L57 (1994)

[5] B.N.J. Persson and M. Persson, Solid State Commun. **36**, 175 (1980); Phys. Rev. Lett. **54**, 2119 (1985); S. Gao, B.I. Lundqvist, and W. Ho, Surf. Sci (in press).

[6] D. M. Newns, T. F. Heinz and J. A. Misewich, Prog. Theor. Phys.(Kyoto), **106**, 411 (1991); M. Brandbyge, P. Hedegård, T.F. Heinz, J.A. Misewich and D.M. Newns, Phys. Rev. **B52**, 6042 (1995).

[7] D. M. Newns, T. F. Heinz and J. A. Misewich, M. Brandbyge and P. Hedegård, "Canonical Surface Reaction Processes Driven by Electronic Friction", submitted as chapter in "Electronic Processes at Solid Surfaces", publisher World Scientific, eds. E. Ilisca and K. Makoshi.

[8] A.I. Volokitin and B.N.J. Persson, "Energy Transfer in Vibrational Spectroscopy of Adsorbates", preprint; R.R. Cavanagh, E.J. Heilweil, and J.C. Stephenson, Surf. Sci. **299-300**, 643 (1994).

[9] J.A. Prybyla, T.F. Heinz, M.M.T. Loy and J.H. Glownia, Phys. Rev. Lett. **64**, 1537 (1990); J.A. Misewich, A. Kalamarides, T.F. Heinz, U. Hofer, and M.M.T. Loy, J. Chem. Phys. **100**, 736 (1994).

[10] J.A. Prybyla, H.W.K. Tom and G.D. Aumiller, Phys. Rev. Lett. **68**, 503 (1992).

[11] F.J. Kao, D.G. Busch, D. Cohen, D.G.D. Costa and W. Ho, Phys. Rev. Lett. **71**, 2094 (1993); F.J. Kao, D.G. Busch, D.G.D. Costa and W. Ho, Phys. Rev. Lett. **70**, 4098 (1993); D.G. Busch, S. Gao, R.A. Pelak, M.F. Booth and W. Ho, Phys. Rev. Lett. **75**, 673 (1995); S. Deliwala, R.J. Finlay, J.R. Goldman, T.H. Her, W.D. Mieher and E. Mazur, Chem. Phys. Lett. **242**, 617 (1995)

[12] F. Budde, T.F. Heinz, M.M.T. Loy, J.A. Misewich, F. de Rougemont, and H. Zacharias, Phys. Rev. Lett. **66**, 3024 (1991).

[13] P.B. Allen, Phys. Rev. Lett. **59**, 1460 (1987).

[14] F.Budde, T.F. Heinz, A. Kalamarides, M.M.T. Loy and J.A. Misewich, Surf. Sci. **283**, 143 (1993).

[15] D.M. Newns, Surf. Sci. **171**, 600 (1986); H. Kasai and A. Okiji, Surf. Sci. **225**, L33 (1990).

[16] J. A. Misewich, T. F. Heinz, D. M. Newns, Phys. Rev. Lett. **68**, 25, 3737 (1992).

[17] W. Brenig, H. Muller, and R. Sedlmeier, Phys. Lett. **54A**, 109 (1975)

Density Functional Theory Study of Self-Diffusion on the (111) Surfaces of Ni, Pd, Pt, Cu, Ag and Au

J.J. Mortensen[1], *B. Hammer*[1,2], *O.H. Nielsen*[3], *K.W. Jacobsen*[1], *and J.K. Nørskov*[1]

[1]Center for Atomic-scale Materials Physics and Physics Department, Technical University of Denmark, DK 2800 Lyngby, Denmark
[2]Joint Research Center for Atom Technology (JRCAT), 1-1-4 Higashi, Tsukuba, Ibaraki 305, Japan
[3]UNI•C Technical University of Denmark, DK 2800 Lyngby, Denmark

Abstract: The energy barriers for self-diffusion on the (111) surfaces of Ni, Pd, Pt, Cu, Ag, and Au are calculated using density functional theory. Also the effect of strain on the diffusion barriers is studied. Energy barriers calculated within the effective-medium theory are found to agree reasonably with the first principles calculations for the three noble metals. Including non-local corrections to the local density approximation for exchange and correlation is found to lower energy barriers by typically 0.04 eV.

1 Introduction

The growth of a metal by deposition from the gas phase has been extensively studied over the last few years[1, 2, 3, 4, 5, 6, 7, 8]. The experimental observations show that even fairly simple systems where only one kind of metal atoms are involved can produce a wealth of different behaviors. The growth mode and the shapes of the islands are sensitively dependent on temperature and deposition rate[4]. If more than one kind of atom is involved even more complicated processes become important involving for instance alloying in the first and deeper layers[9, 10].

The growth process consists of a number of elementary processes which can all be described as diffusion processes: motion over the flat surface, nucleation into islands, detachment of an atom from an island, diffusion along the island edge, exchange with the substrate, diffusion down from an island edge etc. To understand why a given system grows in a particular fashion it is therefore necessary to gain information about the elementary diffusion processes involved. The most important experimental source of information about diffusion rates has been Field Ion Microscopy (FIM)[11, 12, 13, 14], but recently Scanning Tunneling Microscopy (STM) has also provided such information[8, 6]. Another source has been electron microscopy[15].

Due to the many different processes involved it may also be useful with guidance from theory. The complexity of the systems considered has made it necessary to use approximate total energy methods, often based on the effective-medium theory[16, 17, 18] or the related embedded atom method[19, 20]. Such

Springer Series in Solid-State Sciences, Volume 121
Elementary Processes in Excitations and Reactions on Solid Surfaces
Eds.: A. Okiji, H. Kasai and K. Makoshi

studies has made it possible to study the dynamics of the diffusion process to understand, for instance, whether the energy barrier can be put equal to the activation energy found from an Arrhenius-plot of the diffusion constant[21, 22, 23], or to guide the choice of barriers in Monte Carlo simulations of the full growth process[24, 25, 26].

Recently, density functional calculations of diffusion barriers have also begun to appear[13, 27, 28, 29]. They are all based on the local density approximation (LDA), and from the calculations done so far it is hard to judge the accuracy of the LDA approximation in this context. A calculation by Feibelman, Nelson and Kellogg for Pt/Pt(111)[13] with some relaxations in the substrate, shows a barrier of 0.38 eV whereas the experimentally determined value is 0.25 eV[13], the calculation by Stumpf et al. for Al/Al(111) shows a very low barrier of 0.04 eV and the recent calculation by Boisvert et al.[29] for Ag, Au and Ir shows a barrier for Ag on Ag(111) of 0.14 eV which should be compared with the experimentally determined values of 0.15 ± 0.1[15] and 0.097 ± 0.01 eV[6].

In this paper we present a new set of density functional calculations of the barrier for diffusion of Ni, Pd, Pt, Cu, Ag, and Au on the (111) surfaces of these metals. In cases where comparison is possible our results compare well with those previously published, and they supplement them in four respects: i) We consider Cu, Ni and Pd in addition to the metals already considered. ii) We study the effect of nonlocal corrections to the local density approximation, and show that the generalized gradient approximation (GGA)[30] tends to lower the calculated barriers somewhat. This makes the comparison to experiment better in some cases (Pt/Pt(111)) but not all. iii) We make a detailed and systematic comparison to the approximate effective medium theory (EMT) results and show that for the noble metals the EMT gives the same trends and even the same magnitude of the barriers as the GGA. For Ni, Pd and Pt the EMT barriers are 10% higher, 20% lower and 50% lower than the GGA-barriers respectively. iv) Finally we study the effect of strain on the diffusion barriers. By studying the barrier as a function of the lattice constant we show that the effect of strain can be substantial and in reasonable agreement with what can be deduced from experiment[8]. This effect seems to be reasonably well described in the EMT.

2 Methods

The energy barrier, that is relevant for surface diffusion, is the energy of the adatom in the transition state minus the energy of the atom in the most energetically favorable site. For all metals dealt with here, we find that the atom in the fcc-site has the highest binding energy, and for the transition state we take the bridge position midway between the fcc- and hcp-sites.

Using EMT we calculate the energy barriers in two ways. One way with all relaxations included, and one way where only the height of the adatom over the substrate is relaxed and the distance between the layers is fixed at the theoretical bulk values. Using DFT we only calculate the second type of barrier, where only the height of the adatom is relaxed over a static substrate.

In the density functional theory calculations the surfaces are realized as unit cells consisting of 4 layers of atoms and 5 layers of vacuum, repeated periodically in three dimensions. Each layer is a $\sqrt{3} \times \sqrt{3}$ R-30° unit cell containing 3 atoms, and the adatom is placed on one side of the slab. The difference in work-function between the two slab-surfaces is corrected for by adding a dipole layer in the middle of the 5 vacuum layers[31].

To solve the Kohn-Sham equations, we use the methods of Payne, Teter and co-workers[32] and of Gillan[33]. The core electrons are treated using scalar-relativistic pseudopotentials constructed using a method of Troullier and Martins[34], and the pseudo-wavefunctions are expanded in plane-waves with a cutoff at 40 Ry for Pd, Pt, Ag and Au, and 50 Ry for Ni and Cu. For Brillouin-zone integrations we use 15 and 6 special k-points[35] in the irreducible part of the Brillouin-zone for the bridge and fcc configurations respectively, which corresponds to 54 points in the full zone. Occupation numbers are Fermi distributed with $kT = 0.1$ eV, in order to stabilize the numerical procedures. Total energies are extrapolated to an electronic temperature of 0 K[33].

The LDA XC-functional we use is that of Ceperly and Alder[36] as parameterized by Perdew and Zunger[37]. All GGA-energies reported here are obtained by applying the XC-functional of Perdew and Wang[30] to the self-consistent LDA density. Because of the variational property of the density functional, a small error in the density will only give rise to errors of second order in the energy, and we believe that the difference between the present approach and a fully self-consistent GGA-calculation is insignificant. This has been confirmed in GGA-calculations by Kong *et al.*[38] and by Hammer *et al.*[39].

For LDA the theoretical lattice constants for Ni, Pd, Pt, Cu, Ag and Au are 3.47, 3.88, 3.93, 3.57, 4.04 and 4.07Å respectively, which is 0-2% smaller than experimental values. The GGA-functional, which tends to favor large values of the scaled density gradient $|\vec{\nabla} n|/n^{4/3}$, gives lattice constants around 1-2% larger than experiment.

The EMT calculations are performed using the method described in Refs. [16, 17, 18]. The EMT is based on density functional theory, but the constructed potentials also contain experimental input about lattice constants, bulk moduli, shear moduli, and cohesive energies for the metals considered[17, 18]. The potentials include a description of many-atom interactions, and are in their mathematical form rather similar to the embedded atom method (EAM)[19, 20] or other N-body[40, 41, 42] or glue-type potentials[43].

3 Results and discussion

The results of the calculations are shown in Table 1. The accuracy of the calculated energy barriers can be found by increasing the number of plane-waves in the calculations. This has been done by raising the cut-off energy by 5-20 Ry for Pt, Cu and Ag, keeping the geometry found with the lower cut-off energies (40, 50 and 40 Ry respectively). From these calculations, it can be estimated that the energy barriers reported here are converged to within ±0.02 eV with

Table 1: Calculated energy barriers in eV using LDA and GGA. The EMT results are from Ref. [17] and the results from the two embedded atom methods, AFW (Adams, Foiles, and Wolfer) and VC (Voter and Chen), are from Ref. [45]. The effect of relaxations in the substrate can be estimated from the two EMT-columns.

	Static substrate			Relaxed substrate			
	LDA	GGA	EMT	EMT	AFW	VC	Exp.
Ni	0.16	0.11	0.12	0.07	0.06	0.06	0.33[12]
Pd	0.21	0.17	0.13	0.10	0.03	0.06	
Pt	0.42	0.39	0.19	0.16	0.01	0.08	0.25 ± 0.02[13] 0.26 ± 0.02[44]
Cu	0.16	0.12	0.09	0.05	0.03	0.04	
Ag	0.14	0.10	0.09	0.06	0.06	0.04	0.15 ± 0.1[15] 0.10 ± 0.01[8]
Au	0.20	0.15	0.13	0.10	0.02	0.04	

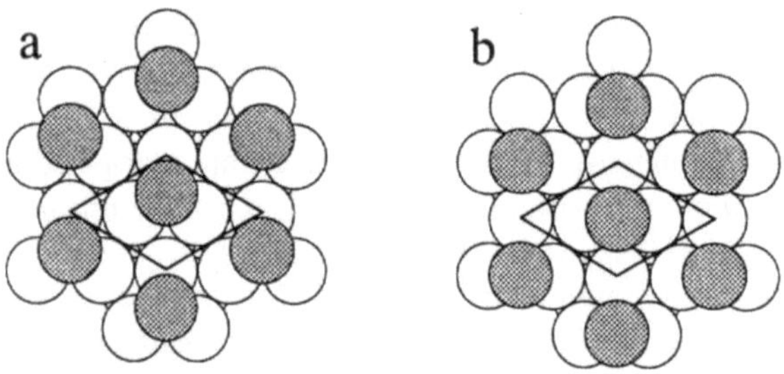

Figure 1: Top view of surfaces used in the calculations. a) fcc position, b) bridge position.

respect to cut-off energy. Another error could come from the small size of the unit cell used. However, it should be noted that even though the shortest distance between adatoms is only $\sqrt{3}$ times the bulk-nearest neighbor distance, the overlayer of adatoms never have common nearest neighbors in the surface layer (see Fig. 1).

To estimate the size of the finite size error we calculated the energy barrier for Pt in a 3×3 unit cell with the LDA-lattice constant. The result was 0.40 eV and 0.33 eV for LDA and GGA respectively. This should be compared with the corresponding numbers from the smaller unit cell (0.42 eV and 0.36 eV) where again both values are calculated at the LDA-lattice constant. The LDA values are in good agreement with the 0.41 eV that Feibelman et al.[13] find with one adatom on an infinite static substrate.

We have not tested the dependence of energy barriers on slab thickness. However Boisvert et al.[29] find that this dependence is weak, and our LDA results for Ag and Au agree well with their LDA results.

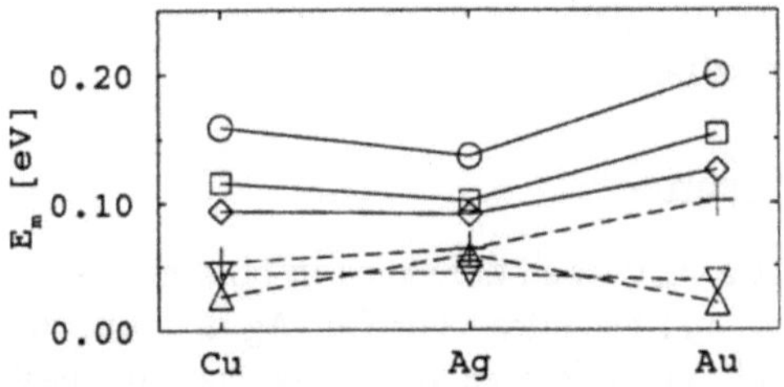

Figure 2: Energy barriers for self-diffusion on (111) surfaces of Cu, Ag and Au. Calculations with a static substrate: ○ LDA, □ GGA and ◇ EMT. Relaxed substrate: + EMT, △ AFW[45] and ▽ VC[45].

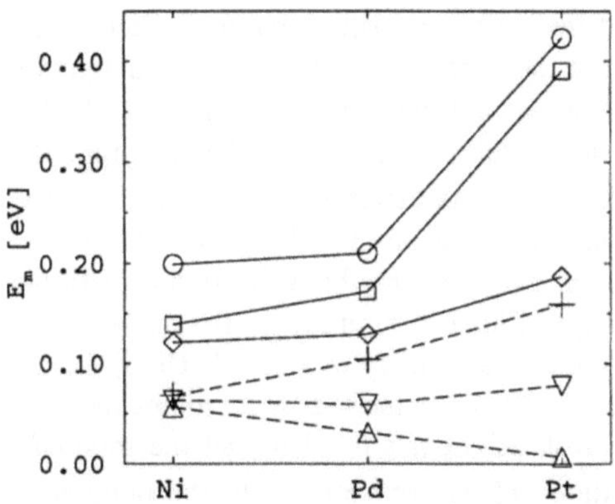

Figure 3: Energy barriers for self-diffusion on (111) surfaces of Ni, Pd and Pt. Calculations with a static substrate: ○ LDA, □ GGA and ◇ EMT. Relaxed substrate: + EMT, △ AFW[45] and ▽ VC[45].

3.1 Energy barriers

Several observations can be made from Table 1 and Fig. 2 and 3. First, the effect of GGA is to lower the energy barriers by typically 0.04 eV which is a rather significant shift in light of the small absolute sizes of the diffusion barriers. The LDA is well known to generally give too large cohesive energies for metals with the main error coming from the calculation at low coordination (i.e. isolated atoms). The GGA does at least to some extent remedy this problem at low coordination, and the decrease of the diffusion barriers in GGA compared to LDA can be viewed as a consequence of this.

Second, the agreement between the GGA values and the EMT values is very good for the the noble metals. The deviations between the GGA and the EMT calculations are more pronounced for Ni, Pd and Pt. This can be understood in terms of the so-called one-electron correction energy[16], which is only included via a pair-potential term in the simple EMT calculations used here. It is likely that the directional bonding caused by the partially filled d-bands for Ni, Pd, and Pt is the main reason for this deviation.

3.2 FCC-HCP energy difference

The EMT-calculations always predict a very small difference in binding energy between an adatom positioned in fcc- and hcp-sites. For Cu, Ag and Au, the DFT calculations also show no significant difference (i.e. less than 10 meV difference) between the two types of sites. For Ni, Pd and Pt the binding energy in the fcc-site is 0.05, 0.13, and 0.14 eV higher than in the hcp-site respectively. The fcc-hcp energy differences reported here are GGA-energies found by placing the adatom in the hcp-site at a height above the substrate taken to be the same as for the adatom in the fcc-site.

We will now address the origin of the fcc-hcp energy difference. Comparing the two configurations, we see that both the adatom and the substrate neighbors next to it have the same number of nearest and next-nearest neighbors. In the simple EMT-model, where there is no angular character of the bonding, the energy difference between these configurations is therefore very small, and that is also the case for the noble metals as confirmed by the DFT calculations. For Ni, Pd, and Pt the electrons in the open d-shells contribute through directional bonding to the fcc-hcp energy difference. This contribution to the one-electron energy is determined by the change in the local density of states (LDOS) for the atoms, and the LDOS of the i'th atom can be characterized by its moments $\mu_n^i = \int_{-\infty}^{\infty} \epsilon^n \rho(\epsilon) d\epsilon$. In a tight-binding nearest neighbor s-band model the moments have a simple geometrical interpretation: μ_n^i is proportional to the number of paths with n legs starting and ending at atom i.

It is sufficient to include only contributions up to the fourth moment to understand the trends in stacking fault energies of late transition metals[46] and the fcc-hcp difference can also be estimated from this approach. The contributions from the s-orbitals will be the same in the fcc- and hcp-configurations. However, for the d-orbitals the angular dependence of the hopping matrix elements leads to a difference in the fourth moment. If we use the screened two-center integrals from Andersen et al.[47], we estimate an fcc-hcp difference of 0.19 eV for Pt. In this simple model the energy difference will scale as the d-bandwidth, and for metals with a completely filled d-band there will be no contribution from the one-electron correction term. Relaxations of the adatom complicate the situation, and we should therefore not expect the d-band scaling to be exact.

4 Comparison with experiment

Experimental information about the energy barriers for self-diffusion has been obtained for Ni, Pt, and Ag. In the comparison between our calculated energy barriers and the experimental results, several factors have to be taken into account. The DFT calculations do not include relaxations of the substrate but at least it is possible to get an estimate of this effect from the EMT calculations. We now discuss the three systems separately.

For the Ag self-diffusion the energy barrier is determined experimentally from the Arrhenius behavior of the island density using electron microscopy[15] or STM[8]. In the former case the uncertainty is estimated to be as high as 70% while the STM experiments give a value of 0.1eV within a 10% uncertainty. The EMT calculations find the lowering of the energy barrier due to relaxations to be about 30 meV. The barriers calculated with LDA and GGA can both be considered to be within the combined experimental and theoretical error bar.

The self-diffusion barrier on Pt has been determined rather accurately in two independent experiments using onset of mobility observed with FIM[13] (assuming a "typical" prefactor in the rate expression), and from both onset of mobility and saturation island density with STM[44]. The experiments agree on a value close to 0.25 eV. The calculated values for the $\sqrt{3} \times \sqrt{3}$ unit-cell are 0.42 eV and 0.39 eV for LDA and GGA, respectively and in the larger 3×3 unit cell the LDA value is only 0.02 eV lower. The relaxation effect on the energy barrier is estimated from the EMT to result in a further lowering of the barrier of around 0.03 eV, in agreement with what is found in LDA including some relaxations of nearest neighbors[13]. Even taking the remaining uncertainties in the calculation due to finite energy cut-off, unit-cell size, and slab thickness into account, the calculated DFT values, however, still seem to be on the high side. The gradient corrections do in this case improve on the comparison by lowering the barrier.

The calculated DFT diffusion barriers for Ni are found to be significantly lower than the experimentally observed value[12] of 0.33 eV even without considering relaxation effects in the calculations. However, in this case the FIM experiments are very difficult to perform because of the low desorption field for adatoms[12] and the experimental value obtained from the onset of mobility at 120K is therefore rather uncertain. Also in this case a "typical" prefactor has to be assumed in order to deduce an energy barrier from the onset temperature.

4.1 Strain

In heteroepitaxy one can have a situation where a complete layer forms in registry with the substrate (Frank-van der Merwe or Stranski-Krastanov mode of growth). The energy barrier for diffusion on this layer will mainly be controlled by three things: the kind of atom diffusing, the kind of atoms in the top layer and the amount of stress in the top layer.

Recently Brune et al.[8] have measured the activation energy for Ag diffusion on a pseudomorphically grown Ag monolayer on Pt(111). It is found that the energy barrier is 0.060 eV compared to 0.097 eV for diffusion on Ag(111) and by comparison to EMT calculations it is concluded that this reduction in activation energy is mainly caused by the 4.2% compressive strain of the Ag layer.

EMT calculations show that with no relaxations of the substrate atoms, the dependence of E_m on the lattice constant is approximately linear for $\Delta a/a$ in the range ±4%, where a is the lattice constant. As we have done the calculations of

Table 2: Derivative of E_m with respect to strain in units of meV/%.

	LDA	GGA	EMT	EMT(relaxed)
Ni	5	1	4.9	2.8
Pd	12	8	8.6	3.5
Pt	19	21	15.7	3.6
Cu	8	3	4.5	3.6
Ag	9	7	5.0	3.8
Au	13	10	10.9	2.8

energy barriers both with the LDA lattice constant and with the 1 to 3% larger GGA lattice constant, we can use our results to estimate the derivative of E_m with respect to strain: $\Gamma = \Delta E_m/(\Delta a/a)$. The results are shown in Table 2 together with the EMT values with no relaxations. From the data by Brune et al.[8] we get a rough estimate ($\Gamma \sim 10$ meV/%) in reasonable agreement with the calculated values for Ag. It should be noted, however, that relaxations can be very important for the strain effects. For comparison we have included in Table 2 the values obtained with EMT if the crystal is only strained in the directions parallel to the surface with all other relaxations included.

5 Conclusions

Energy barriers for adatom self-diffusion on the (111)-surfaces have been calculated for Ni, Pd, Pt, Cu, Ag, and Au using density functional theory. Gradient corrections are seen to lower the energy barriers by about 0.04 eV. The semi-empirical EMT is seen to agree well with the DFT calculations for the noble metals, while the deviations are larger for the late transition metals.

The comparison with experiment is not very conclusive. For Ag the calculated energy barrier is in agreement with the experimental one within the error bars, while for Pt the DFT calculated values are on the high side. In the case of Ni the calculated energy barrier is clearly lower than the one found in the FIM experiments, however, because of the experimental complications for this system further experiments would be of interest.

We would like to acknowledge computing facilities at JRCAT Supercomputer System as financially supported by New Energy and Industrial Technology Development Organization (NEDO). The present work was in part financed by The Danish Research Councils through grant #9501775. Center for Atomic-scale Materials Physics is sponsored by the Danish National Research Foundation.

References

[1] E. Bauer, Appl. Surf. Sci. **11/12**, 479 (1982).

[2] J. A. Venables, G. D. T. Spiller, and M. Hanbücken, Rep. Prog. Phys. **47**, 399 (1984).

[3] R. Q. Hwang, J. Schröder, C. Günter, and R. J. Behm, Phys. Rev. Lett. **67**, 3279 (1991).

[4] T. Michely, M. Hohage, M. Bott, and G. Comsa, Phys. Rev. Lett. **70**, 3943 (1993).

[5] E. Hahn *et al.*, Surf. Sci. **319**, 277 (1994).

[6] H. Brune, H. Röder, C. Boragno, and K. Kern, Phys. Rev. Lett. **73**, 1955 (1994).

[7] S. Günther *et al.*, Phys. Rev. Lett. **73**, 553 (1994).

[8] H. Brune *et al.*, Phys. Rev. B **52**, R14380 (1995).

[9] L. P. Nielsen *et al.*, Phys. Rev. Lett. **71**, 754 (1993).

[10] L. P. Nielsen *et al.*, Phys. Rev. Lett. **74**, 1159 (1995).

[11] G. L. Kellogg, Surf. Sci. Rep. **21**, 1 (1994).

[12] T. Tung and W. R. Graham, Surf. Sci. **97**, 73 (1980).

[13] P. J. Feibelman, J. S. Nelson, and G. L. Kellogg, Phys. Rev. B **49**, 10548 (1994).

[14] S. C. Wang and G. Ehrlich, Phys. Rev. Lett. **68**, 1160 (1992).

[15] G. W. Jones, J. M. Marcano, J. K. Nørskov, and J. A. Venables, Phys. Rev. Lett. **65**, 3317 (1990).

[16] K. W. Jacobsen, J. K. Nørskov, and M. J. Puska, Phys. Rev. B **35**, 7423 (1987).

[17] P. Stoltze, J. Phys. Condensed Matter **6**, 9495 (1994).

[18] K. W. Jacobsen, P. Stoltze, and J. K. Nørskov, to be published.

[19] M. S. Daw and M. I. Baskes, Phys. Rev. Lett. **50**, 1285 (1983).

[20] S. M. Foiles, M. I. Baskes, and M. S. Daw, Phys. Rev. B **33**, 7983 (1986).

[21] G. Boisvert and L. J. Lewis, in *Mechanisms of Thin-Film Evolution*, Vol. 317 of *MRS Symposia Proceedings*, edited by S. M. Yalisove, C. V. Thomson, and D. J. Eaglesham (Materials Research Society, Pittsburgh, 1994), p. 71.

[22] P. Blandin and C. Massobrio, Surf. Sci. **279**, L219 (1992).

[23] D. E. Sanders and A. E. DePristo, Surf. Sci. Lett. **264**, L169 (1992).

[24] M. Villarba and H. Jónsson, Surf. Sci. **317**, 15 (1994).

[25] J. Jacobsen, K. W. Jacobsen, P. Stoltze, and J. K. Nørskov, Phys. Rev. Lett. **74**, 2295 (1995).

[26] S. Liu, Z. Zhang, G. Comsa, and H. Metiu, Phys. Rev. Lett. **71**, 2967 (1993).

[27] R. Stumpf and M. Scheffler, Surf. Sci. **307-309**, 501 (1994).

[28] R. Stumpf and M. Scheffler, Phys. Rev. Lett. **72**, 254 (1994).

[29] G. Boisvert, L. J. Lewis, M. J. Puska, and R. M. Nieminen, Phys. Rev. B **92**, 9078 (1995).

[30] J. P. Perdew *et al.*, Phys. Rev. B **46**, 6671 (192).

[31] J. Neugebauer and M. Scheffler, Phys. Rev. B **46**, 16067 (1992).

[32] M. C. Payne *et al.*, Rev. Mod. Phys. **64**, 1045 (1992).

[33] M. J. Gillian, J. Phys. **1**, 689 (1989).

[34] N. Troullier and J. L. Martins, Phys. Rev. B **43**, 1993 (1991).

[35] D. J. Chadi and M. L. Cohen, Phys. Rev. B **8**, 5747 (1973).

[36] D. M. Ceperly and B. J. Alder, Phys. Rev. Lett. **45**, 566 (1980).

[37] J. P. Perdew and A. Zunger, Phys. Rev. B **23**, 5048 (1981).

[38] X. J. Kong, C. T. Chan, K. M. Ho, and Y. Y. Ye, Phys. Rev. B **42**, 9357 (1990).

[39] B. Hammer, K. W. Jacobsen, and J. K. Nørskov, Phys. Rev. Lett. **70**, 3971 (1993).

[40] M. W. Finnis and J. E. Sinclair, Philos. Mag. A **50**, 45 (1984).

[41] A. P. Sutton and J. Chen, Philos. Mag Lett. **61**, 139 (1990).

[42] S. B. Sinnot, M. S. Stave, T. J. Reaker, and A. E. DePristo, Phys. Rev. B **44**, 8927 (1991).

[43] F. Ercolessi, E. Tosatti, and M. Parinello, Phys. Rev. Lett. **57**, 719 (1986).

[44] M. Bott *et al.*, to be published.

[45] C. L. Liu, J. M. Cohen, J. B. Adams, and A. F. Voter, Surf. Sci. **253**, 334 (1991).

[46] K. Stokbro and K. W. Jacobsen, Phys. Rev. B **47**, 4916 (1993).

[47] O. K. Andersen, O. Jepsen, and D. Glötzel, in *Highlights of Condensed Matter Theory*, Corso Soc. Italiana di Fisica, edited by F. Bassani, F. Fumi, and M. P. Tosi (North-Holland, New York, 1985), Vol. LXXXIX, pp. 59–176.

Surface Resistivity and Related Phenomena on Metal Surfaces

H. Ishida

College of Humanities and Sciences, Nihon University,
Sakura-josui, Tokyo 156, Japan

A microscopic expression of the surface resistivity for semi-infinite metallic systems is derived by linear response theory. It coincides with the semi-classical one based on the Boltzmann equation, if the probability of diffuse scattering at the surface was independent of the direction of scattered waves. For adsotbates on flat jellium surfaces, it is shown that the resistivity is related with the electronic friction force and the "electron-wind" force exerted on the adsorbate by simple equations. Results of ab-initio calculations are presented of the resistivity of isolated atoms and molecules on semi-infinite jellium substrates.

1 Introduction

The resistivity of a carefully prepared thin metallic film is very sensitive to surface conditions. At low temperatures (~10 K), the resistivity of metallic films with thickness $l_f = 10^2 - 10^3$ Å can change by ~10% upon adsorption of atoms and molecules at less than monolayer coverages [1]. Surface defects such as adsorbates and steps scatter conduction electrons incident from the interior of the metal off-specularly, causing momentum changes of the electrons in the plane. As the surface is a 2-dimensional (2-D) scattering center, the surface (surface-induced) resistivity, ρ_s, scales as $1/l_f$. In this paper, we consider only the elastic scattering of conduction electrons at the surface and ignore the inelastic scattering due to phonon excitations. This approximation may be valid if the temperature is low enough.

Most of previous theories of the surface resitivity utilized semi-classical models based on the Boltzmann equation [2, 3]. In Sec. 2 of this paper, we derive a microscopic expression of the surface resistivity by linear response theory, and compare it with the corresponding one in the semi-classical theory. In our model, the interior of the metal is treated as jellium for simplicity, whereas the surface region may have any 3-D structures such as isolated adsorbates and ordered overlayers.

In Sec. 3, the relation between the surface resistivity and other quantities appearing in dynamical processes at the surface is discussed. Recently, Persson [4] derived a simple relation between ρ_s and the electron-hole pair contribution to the life-time of the parallel vibrational motion of adsorbates. We point out that his relation can be recast in the same form as the relation between ρ_s and the "electron-wind" force exerted on adsorbates located in current densities [5].

In Sec. 4, results are presented of ab-initio calculations of the adsorbate-induced surface resistivity. We calculate the electronic structure of isolated atoms and molecules adsorbed on semi-infinite jellium surfaces using the impurity Green function method and investigate the dependence of ρ_s on the adatom–substrate distance and the adatom–adatom bond length.

Springer Series in Solid-State Sciences, Volume 121
Elementary Processes in Excitations and Reactions on Solid Surfaces
Eds.: A. Okiji, H. Kasai and K. Makoshi

2 Surface Resistivity

Let us consider a semi-infinite metallic surface. We choose the z−axis as the surface normal pointing toward the vacuum. In the interior of the metal ($z \leq 0$), where the effective one-electron potential $V_{eff}(\mathbf{r})$ is assumed to be constant (jellium), the wave function of the one-electron state incident from the interior of the metal with energy ϵ and the parallel wave vector $\mathbf{k} = (k_x, k_y)$ ($\epsilon >\mid \hbar\mathbf{k} \mid^2 /2m$) is written as

$$\begin{aligned} \psi(\epsilon, \mathbf{k}, \mathbf{r}) &= \frac{1}{\sqrt{S}}\Big\{\frac{1}{ik_z(\epsilon, \mathbf{k})}\exp[i\mathbf{k}\cdot\mathbf{x} + ik_z(\epsilon, \mathbf{k})z] \\ &+ \sum_{\mathbf{k}'} r(\epsilon, \mathbf{k}, \mathbf{k}')\exp[i\mathbf{k}'\cdot\mathbf{x} - ik_z(\epsilon, \mathbf{k}')z]\Big\}, \end{aligned} \tag{1}$$

Here $\mathbf{r} = (\mathbf{x}, z)$, $k_z(\epsilon, \mathbf{k}) = \sqrt{2m\epsilon/\hbar^2 - \mid \mathbf{k} \mid^2}$, and S denotes the area of the surface. The current conservation implies $\sum_{\mathbf{k}'} p(\epsilon, \mathbf{k}, \mathbf{k}') = 1$, where

$$p(\epsilon, \mathbf{k}, \mathbf{k}') = \mathrm{Re}k_z(\epsilon, \mathbf{k})\mathrm{Re}k_z(\epsilon, \mathbf{k}') \mid r(\epsilon, \mathbf{k}, \mathbf{k}') \mid^2 . \tag{2}$$

The probability of specular reflection, $p_s(\epsilon, \mathbf{k})$, is given by the diagonal element $p(\epsilon, \mathbf{k}, \mathbf{k})$.

We apply a uniform electric field with frequency ω parallel to the surface. The α (x or y) component of the current density induced in the system is given to the lowest order as

$$J_\alpha(\mathbf{r}) = \frac{ie^2 n_0(\mathbf{r})}{m\omega}E_\alpha + \frac{ie}{m\omega}\sum_\beta \chi_0\big(\hat{j}_\alpha(\mathbf{r}), \hat{p}_\beta, \omega\big)E_\beta, \tag{3}$$

where n_0 is the ground-state charge density, $\hat{j}_\alpha$ is the current density operator, and the bare polarization function χ_0 is defined by

$$\chi_0(\hat{A}, \hat{B}, \omega) = 2\sum_{i,j}\frac{f_i - f_j}{\epsilon_i - \epsilon_j + \hbar\omega + i\delta}\langle\psi_i \mid \hat{A} \mid \psi_j\rangle\langle\psi_j \mid \hat{B} \mid \psi_i\rangle. \tag{4}$$

Here $f_i = \theta(\epsilon_F - \epsilon_i)$ is the occupation for state i with energy ϵ_i, where ϵ_F denotes the Fermi energy. The macroscopic current density in the bulk, $J_\alpha = in_e e^2 E_\alpha/(m\omega)$, originates from the first term of (3), while its second term produces a microscopic current density localized in the surface (n_e is the electron density in the bulk). The energy dissipation in the system per unit time is

$$P = \sum_{\alpha,\beta}\mathrm{Re}\Big[\frac{ie^2}{m^2\omega}\chi_0(\hat{p}_\alpha, \hat{p}_\beta, \omega)E_\alpha E_\beta\Big]. \tag{5}$$

Since the bulk metal has no resistivity in the present model, P may also be written by introduing the surface resistivity in tensor form as,

$$P = Sl_f\sum_{\alpha,\beta}\rho_s^{\alpha\beta}(\omega)J_\alpha^* J_\beta. \tag{6}$$

From (5) and (6), one has

$$\begin{aligned} (Sl_f)\ \rho_s^{\alpha\beta}(\omega) &= -\frac{\omega}{n_e^2 e^2}\mathrm{Im}\chi_0(\hat{p}_\alpha, \hat{p}_\beta, \omega) \\ &= \frac{2\pi\omega}{n_e^2 e^2}\sum_{i,j}\delta(\epsilon_j - \epsilon_i - \hbar\omega)(f_i - f_j)\langle\psi_i \mid \hat{p}_\alpha \mid \psi_j\rangle\langle\psi_j \mid \hat{p}_\beta \mid \psi_i\rangle. \end{aligned} \tag{7}$$

For the present model, (7) can be evaluated most easily by substituting $\psi(\epsilon, \mathbf{k}, \mathbf{r})$ for state i, and $\psi^*(\epsilon + \hbar\omega, -\mathbf{k}', \mathbf{r})$ for state j. Using the relation $r(\epsilon, \mathbf{k}, \mathbf{k}') = r(\epsilon, -\mathbf{k}', -\mathbf{k})$, one has at small ω,

$$\int_{-\infty}^{0} dz \int d\mathbf{x}\, \psi(\epsilon+\omega, -\mathbf{k}', \mathbf{r}) \hat{p}_\beta\, \psi(\epsilon, \mathbf{k}, \mathbf{r}) = \frac{\hbar^2}{im\omega} \Delta k_\beta\, r(\epsilon, \mathbf{k}, \mathbf{k}') + o(\omega^0), \tag{8}$$

where $\Delta k_\beta = k'_\beta - k_\beta$. The integration in the surface region ($z \geq 0$) does not contribute to the divergence of the matrix element. As a result, by inserting (8) into (7), one obtains in the limit of $\omega \to 0$ [6],

$$(Sl_f)\, \rho_s^{\alpha\beta} = \frac{\hbar}{2\pi n_e^2 e^2} \sum_{\mathbf{k},\mathbf{k}'} [\Delta k_\alpha \Delta k_\beta] p(\epsilon_F, \mathbf{k}, \mathbf{k}'), \tag{9}$$

Hereafter we omit ω for the dc resistivity. Alternatively, the diagonal element of (9) is written as

$$l_f\, \rho_s^{xx} = \frac{3\hbar k_F}{16 n_e e^2} \langle 1 - p_s \rangle - \frac{\hbar}{\pi S n_e^2 e^2} \sum_{\mathbf{k}} k_x \sum_{\mathbf{k}'(\neq \mathbf{k})} k'_x p(\epsilon_F, \mathbf{k}, \mathbf{k}'), \tag{10}$$

where $\langle 1 - p_s \rangle$ denotes the probability of off-specular scattering averaged with a weight factor k_x^2,

$$\langle 1 - p_s \rangle = \frac{\sum_{\mathbf{k}} k_x^2 [1 - p_s(\epsilon_F, \mathbf{k})]}{\sum_{\mathbf{k}} k_x^2} = \frac{16\pi}{S k_F^4} \sum_{\mathbf{k}} k_x^2 [1 - p_s(\epsilon_F, \mathbf{k})]. \tag{11}$$

If the probability of off-specular scattering was independent of the direction of $\mathbf{k}'$, the second term of (10) vanishes. With this assumption, (10) coincides exactly with ρ_s derived by semi-classical approaches [2, 3].

3 Friction force and electron-wind force

Recently, Persson [4] derived a simple relationship between ρ_s and the electron-hole pair contribution to the life-time of the parallel vibrational motion (frustrated translation) of adsorbates, τ. His relation is

$$\frac{1}{\tau} = \frac{n_e^2\, e^2}{M n_a} [l_f\, \rho_s], \tag{12}$$

where n_a is the 2-D density of adsorbates, and M is the adsorbate mass. With $F^f = -Mv/\tau$, the friction force on the adsorbate oscillating with velocity v, and $J^f = -n_e e v$, the apparent current density in the metal seen from a reference frame fixed on the adsorbate, (12) is read as

$$\frac{F^f}{J^f} = \frac{n_e e}{n_a} [l_f\, \rho_s]. \tag{13}$$

Changing to a reference frame fixed on the metal, the force acting on adsorbates located in a (real) current density J^w is called the "electron-wind" force F_w. For an impurity in the electron gas, it is well-known that F^w is related with the impurity resistivity by a simple relation [5, 7]. For the case of adsorbates, the relation is

$$\frac{F^w}{J^w} = \frac{n_e e}{n_a} [l_f\, \rho_s]. \tag{14}$$

It is plausible that both equations have the same form, since the reference frame fixed on the metal and that fixed on the adsorbate may be equivalent as far as the adsorbate velocity is low enough

Let us consider a single adatom on a semi-infinite *flat* jellium substrate vibrating in the x direction. The displacement of the atom is assumed to be given as $\Delta x \exp(-i\omega t)$. The friction force is calculated from the imaginary part of the dynamical matrix. Noting that $J^f = i\omega n_e e \Delta x \exp(-i\omega t)$, we obtain, to the lowest order of ω,

$$\frac{F^f}{J^f} = \frac{-1}{n_e e \omega} \mathrm{Im}\chi_0 \left[\frac{\partial V_{eff}}{\partial R_x}, \frac{\partial V_{eff}}{\partial R_x}, \omega \right], \tag{15}$$

where R_x denotes the x-coordinate of the adatom. As the substrate is translationally invariant,

$$\frac{\partial V_{eff}}{\partial R_x} = -\frac{\partial V_{eff}}{\partial x} = -\frac{i}{\hbar}[\hat{p}_x, \hat{H}], \tag{16}$$

where $\hat{H} = -\hbar^2 \Delta / 2m + V_{eff}$ is the one-electron Hamiltonian. By substituting (16) in (15) and comparing the resultant expression with (7), one immediately finds that the relation (13) holds true. In the same way, one can prove (13) [8]. From (15), one has in the static limit,

$$\begin{aligned} (Sl_f)\ \rho_s^{xx} &= M/(n_e^2 e^2 \tau) = \frac{2\pi\hbar}{n_e^2 e^2} \sum_{i,j} \delta(\epsilon_i - \epsilon_F)\delta(\epsilon_j - \epsilon_F) \mid \langle \psi_i \mid \partial V_{eff}/\partial x \mid \psi_j \rangle \mid^2 \\ &= \frac{2\hbar}{\pi n_e^2 e^2} \int d\mathbf{r} d\mathbf{r}'\ \mathrm{Im}G(\mathbf{r}, \mathbf{r}', \epsilon_F + i\delta) \frac{\partial V_{eff}}{\partial x} \mathrm{Im}G(\mathbf{r}', \mathbf{r}, \epsilon_F + i\delta) \frac{\partial V_{eff}}{\partial x'}, \end{aligned} \tag{17}$$

where the Green function is defined by $\hat{G}(\lambda) = 1/(\lambda - \hat{H})$. (17) applies also for adsorbed molecules.

In the same way, the friction force in the surface normal direction can be calculated by replacing $\partial V_{eff}/\partial x$ by $\partial V_{eff}/\partial R_z$ in (17) where R_z denotes the z-coordinate of the adsorbate. Such calculations were performed in the past for H and H_2 on jellium in the context of atom–surface scattering [9].

4 Adsorbate resistivity

We evaluate the surface resistivity of isolated adsorbates on jellium substrates. We make use of the fact that the potential change ΔV_{eff} upon adsorption is localized in the neighborhood of the adsorbate, and the Green function in this region is expanded by a nonorthogonal basis set $\{\phi_n\}$ as $G(\mathbf{r}, \mathbf{r}', \lambda) = \sum_{n,n'} G_{nn'}(\lambda)\phi_n(\mathbf{r})\phi_{n'}^*(\mathbf{r}')$. $G_{nn'}$ satisfies the Dyson equation (λ is ommited for simplicity and summation is implied for repeated indices) [10],

$$G_{nn'} = G_{nn'}^0 + G_{nm}^0\ \Delta V_{mm'}\ G_{m'n'} \tag{18}$$

where $\Delta V_{nn'} = \langle \phi_n \mid \Delta V_{eff} \mid \phi_{n'} \rangle$, $G_{nn'} = S_{nm}^{-1} \langle \phi_m \mid \hat{G} \mid \phi_{m'} \rangle S_{m'n'}^{-1}$, and $G_{nn'}^0$ is the corresponding matrix for the clean jellium substrate (S^{-1} stands for the inverse of the overlap matrix $S_{nn'} = \langle \phi_n \mid \phi_{n'} \rangle$). With the definitions $G_{nn'}^I = [G_{nn'} - (G_{\bar{n}\bar{n}'})^*]/(2i)$, where $\phi_{\bar{n}} = \phi_n^*$, and $(\Delta V_\alpha)_{nn'} = \langle \phi_n \mid \partial(\Delta V_{eff})/\partial x_\alpha \mid \phi_{n'} \rangle$, we obtain from (17),

$$(Sl_f)\ \rho_s^{\alpha\beta} = \frac{2\hbar}{\pi n_e^2 e^2} \mathrm{Tr}\left[G^I(\epsilon_F + i\delta)\ \Delta V_\alpha\ G^I(\epsilon_F + i\delta)\ \Delta V_\beta \right]. \tag{19}$$

In the present work, $G_{nn'}$ is calculated self-consistently within the local density approximation in density functional theory [11], and the adsorbate ion cores are represented by a non-empirical pseudo potential [12]. We utilize a plane-wave-like basis set,

$$\phi_{n_x n_y n_z}(\mathbf{r}) = \sqrt{\frac{2}{L_x L_y L_z}} \exp\left(\frac{2\pi i n_x x}{L_x}\right) \exp\left(\frac{2\pi i n_y y}{L_y}\right) \sin\left(\frac{\pi n_z z}{L_z}\right). \tag{20}$$

In order to be able to represent the Green function whose boundary condition varies with energy, (20) should be used in a smaller unit cell: $\mid x \mid \leq L_x'/2$, $\mid y \mid \leq L_y'/2$, and $b_1 \leq z \leq b_2$, where $L_x' < L_x$, $L_y' < L_y$, and $0 < b_1 < b_2 < L_z$.

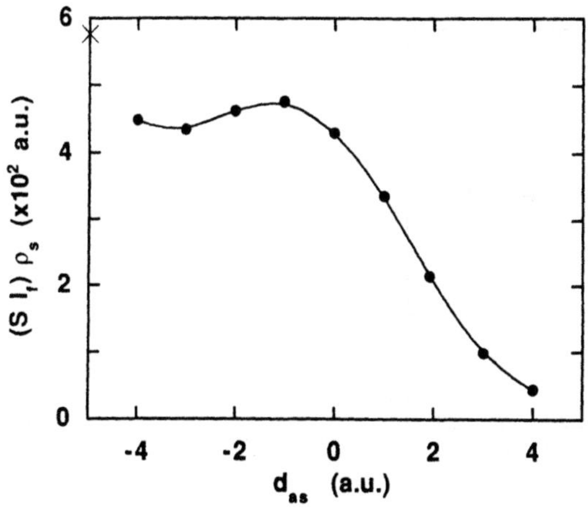

Fig. 1. The surface resistivity of a single Al atom on jellium with r_s=2.07 a.u. as a function of $d_{as} = R_z - z_j$, the z-coordinate of the adatom minus that of the edge of the positive jellium charge.

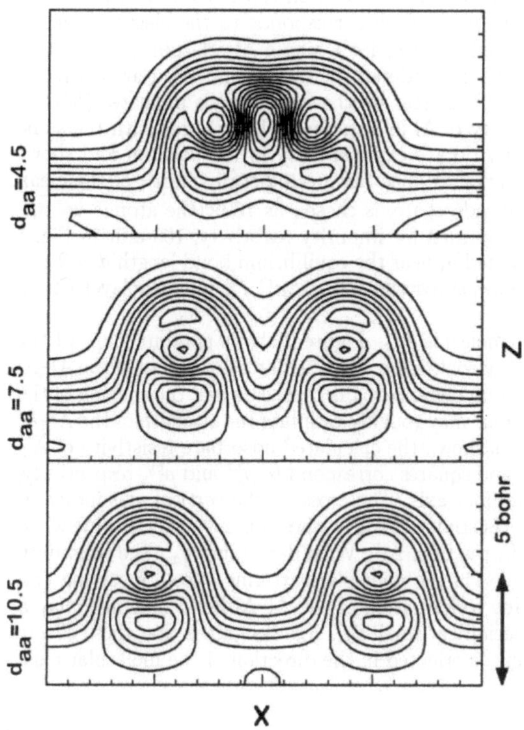

Fig. 2. Contour map of the charge density for Al_2 on jellium with r_s=2.07 a.u. on the vertical cut plane containing Al nuclei. The contour spacing is 0.003 a.u., and d_{as}=1.91 a.u.

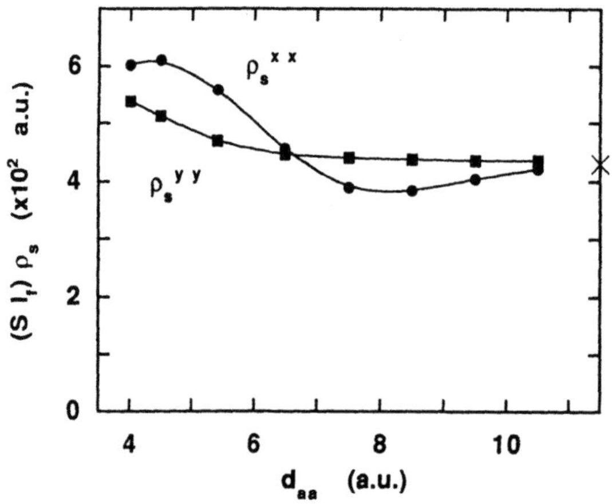

Fig. 3. The surface resistivity of Al_2 on jellium with r_s=2.07 a.u. as a function of the Al–Al bond length, d_{aa}. The solid circles and squares indicate ρ_s^{xx} and ρ_s^{yy}, respectively. The molecule is oriented in the x direction. d_{as}=1.91 a.u.

Numerical results are presented for Al atoms adsorbed on top of jellium with r_s=2.07 a.u. (Hartree atomic units with $m=e=\hbar=1$) which corresponds to the electron density of Al [13]. Figure 1 shows the calculated $(Sl_f)\ \rho_s$ for a single Al adatom as a function of $d_{as} = R_z - z_j$, where z_j denotes the z-coordinate of the jellium edge at which the positive background charge drops to 0. The cross on the vertical axis indicates $(Sl_f)\ \rho_s$ at $d_{as} = -\infty$, i.e., the impurity resistivity of Al in a uniform electron gas with the same n_e. The calculated adsorbate resistivity decays exponentially in the vacuum and shows an oscillatory behavior toward the interior as does the ground-state charge of the clean jellium surface. If Sl_f, the volume per adsorbate, is chosen as 100×(the atomic volume of Al in bulk fcc Al), following the convention for impurity resistivity, 100 a.u. in Fig. 1 corresponds to 0.19 $\mu\Omega/\%$. The calculated ρ_s near the equilibrium bond length $d \sim 2$ a.u. is seen to be of the same order as the measured one, 0.6±0.2$\mu\Omega$ /% for monolayer Cu on Cu(111) [1].

Next we study ρ_s of the diatomic molecule Al_2. Figure 2 shows contour maps of the ground-state charge on the vertical plane containing Al nuclei for a few values of the Al–Al bond length d_{aa}. Here the adatom–jellium distance d_{as} is chosen as half the spacing of the neighboring Al(001) planes, 1.91 a.u. One may roughly estimate the strength of the Al–Al interaction from this figure. In Fig. 3 is shown the calculated adsorbate resistivity of Al_2 as a function of d_{aa}. The solid circles and squares correspond to ρ_s^{xx} and ρ_s^{yy}, respectively, where the x axis is chosen as the molecular axis. The cross on the vertical axis indicates 2×(ρ_s of an isolated Al adatom). The resistivity of Al_2 reaches this asymptotic value when the Al–Al bond length is $\sim 2 \times$ (d_{aa} in the bulk Al). With decreasing d_{aa}, the resistivity of Al_2 starts to deviate from this value via the multiple scattering effects involving two atomic sites. The potential change due to formation of the Al–Al bond may also play a significant role in the change of ρ_s especially at small d_{aa}. Fig. 3 signifies that these effects are appreciably larger when the current is oriented in the direction of the molecular axis.

5 Conclusion

Measurements of the surface resistivity ρ_s provide us with useful information about the elementary processes at solid surfaces such as adsorption and diffusion. In the present paper, we derived a microscopic expression of ρ_s for semi-infinite metallic surfaces, and discussed its relation to the friction force and the electron-wind force. The sensitivity of ρ_s to the surface electronic structure has been demonstrated for isolated adsorbates on semi-infinite jellium by ab-initio calculations.

This work was supported by a Grant-in-Aid from the Ministry of Education, Scienece, and Culture of Japan.

References

[1] D. Schumacher, *Surface Scattering Experiments with Conduction Electrons*, Springer Tracts in Modern Physics, Vol.128 (Springer, Berlin, 1993).

[2] M.S.P. Lucas, J. Appl. Phys. **36**, 1632 (1965).

[3] D.L. Lessie and E.R. Crosson, J. Appl. Phys. **59**, 504 (1986).

[4] B.N.J. Persson, Phys. Rev. B **44**, 3277 (1991); B.N.J. Persson, J. Chem. Phys. **98**, 1659 (1993).

[5] For a review, see J. van EK and A. Lodder, *Defect and Diffusion Forum*, Vol.115–116 (Scitec, Switzerland, 1994).

[6] H. Ishida, Phys. Rev. B **52**, 10819 (1995).

[7] V.B. Fiks, Sov. Phys. -Solid State **1**, 16 (1959); H.B. Huntington and A.R. Grone, J. Phys. Chem. Solids, **20**, 76 (1961).

[8] H. Ishida, Surf. Sci., in press (1996). For impurities in a uniform electron gas, see H. Ishida, Phys. Rev. B **51**, 10345 (1995).

[9] B. Hellsing, M. Persson, and B.I. Lundqvist, Surf. Sci. **126**, 147 (1983); M. Persson and B. Hellsing, Phys. Rev. Lett. **49**, 662 (1982).

[10] G. Wachutka, A. Fleszar, F. Máca, and M. Scheffler, J. Phys. C **4**, 2831 (1992).

[11] W. Kohn and L.J. Sham, Phys. Rev. **140**, A1133 (1965).

[12] L. Kleinman and D.M. Bylander, Phys. Rev. Lett. **48**, 1425 (1982).

[13] H Ishida, unpublished.

Part III

New Methods and Structures

Use of Renormalized Radiation Green's Function in Microscopic Nonlocal Optical Response Theory: Applications to Surface Optical Problems

K. Cho and *J. Ushida*

Faculty of Engineering Science, Osaka University, Toyonaka 560, Japan

Abstract In many surface optical problems, we need to consider both contributions from resonant surface transitions and nonresonant substrate transitions. Treating the latter part as a semi-infinite local dielectrics, we solve the whole problem in two steps: We first calculate radiation Green's function including the effect of semi-infinite dielectrics, and then, in terms of this renormalized Green's function, we solve the nonlocal response due to a small number of resonant transitions. This paper describes the general formulation and its application to a model calculation of resonant SNOM (scanning near-field optical microscopy).

1 Introduction

In spatially confined matter systems, it is essentially important to retain the coherence of matter in the calculation of any physical quantities. By retaining the quantum mechanical coherence of wave functions, response functions generally exhibit nonlocality. For example, an optical polarization can be exerted on a position different from the point irradiated by an incident light, if the two points are within the range of the wave function excited resonantly.

Such a nonlocality has been studied very little, in comparison with the standard macroscopic response theory. The reason for the lack of research in the microscopic response theory would partly be due to the opinion such as " most physicists have felt in the past, and still feel, that the task of microscopic theory is to explain dielectric constants and magnetic permeabilities already measured macroscopically rather than to define something not already known" /1/.

This kind of understanding has faced to the necessity of a change by the appearance and popularity of mesoscopic systems, where measurements are no more restricted to be macroscopic, and microscopic details of a mesoscopic system may be reflected also on macroscopic measurements in such forms as size quantized resonant levels and resonant size dependence of internal field strength.

In view of such a current interest, we formulated a microscopic, nonlocal response theory which can be applied to both linear and nonlinear optical

Springer Series in Solid-State Sciences, Volume 121
Elementary Processes in Excitations and Reactions on Solid Surfaces
Eds.: A. Okiji, H. Kasai and K. Makoshi

responses of confined matter systems /2,3/. In this framework, we set up two relationships for vector potential $\boldsymbol{A}(\boldsymbol{r},t)$ in Coulomb gauge ($\nabla \cdot \boldsymbol{A} = 0$) and current density $\boldsymbol{j}(\boldsymbol{r},t)$, both of which are c-numbers. The first one

$$\boldsymbol{A}(\boldsymbol{r},t) = \boldsymbol{A}_0(\boldsymbol{r},t) + \mathcal{G}[\boldsymbol{j}] \tag{1}$$

represents the solution of microscopic Maxwell equations, where $\boldsymbol{A}_0$ is a free field and the functional $\mathcal{G}[\boldsymbol{j}]$ the source-induced field. The second relationship is given as

$$\boldsymbol{j}(\boldsymbol{r},t) = \mathcal{F}[\boldsymbol{A}] \;, \tag{2}$$

where the functional $\mathcal{F}$ represents various integrals containing linear and nonlinear susceptibilities, which are obtained by solving the quantum mechanics of a matter in a given electromagnetic field $\boldsymbol{A}$.

The selfconsistent motion of matter and field can be obtained by solving these two equations as simultaneous equations as functions of the initial conditions for field ($\boldsymbol{A}_0$) and matter (contained in susceptibilities). By noting the separability of integral kernels in $\mathcal{F}$ as a good approximation both for resonant and nonresonant cases /4/, these equations turn out to be a set of polynomial equations, which is certainly easier to solve (numerically). In this way, the framework of nonlocal response theory is quite general and is expected to be useful in many applications.

In numerical treatments, the dimension of the simultaneous equations is restricted to a finite number. Since the number of nonresonant levels is infinite, we need to find out an appropriate way to circumvent this difficulty. One method is to choose a finite number of nonresonant states, which can manage, in a good approximation, to describe the spatial variation of induced polarization for a given frequency range of interest. This approach allows us to retain the separable nature of susceptibility in the formulation, and has been used in several cases of our model calculations /3/.

However, there is a completely different type of approach to deal with the contribution of nonresonant levels. This is possible when the nonresonant levels are replaced by a background dielectrics with local susceptibility. Then, the total polarization can be divided into resonant (nonlocal) polarization $\boldsymbol{P}_r$ and nonresonant (local) polarization. If we use the latter to define the Green's function of radiation field, we can follow the logic of nonlocal response in terms of the matrix element of retarded interaction, via this renormalized radiation propagator, among the induced dipoles of resonant states. This approach was taken by Keller et al./5/ to calculate the linear response of point dipoles on a substrate. We are going to generalize it as a practical method to carry out the calculation of nonlocal response.

The formulation of the nonlocal response theory by such a renormalized radiation Green's function can be formally made for any systems. However, actual use of the formulation is restricted to simple geometries such as semi-infinite planar boundary or a sphere, because the calculation of the renormalized Green's function becomes quite complicated for less simple geometries.

2 Formulation

We start with the following equation obtained from microscopic Maxwell equations by eliminating the magnetic field component:

$$\nabla \times \nabla \times \boldsymbol{E} - q^2 \boldsymbol{E} = 4\pi q^2 (\boldsymbol{P}_r + \boldsymbol{P}_b) \ . \tag{3}$$

where the arguments $\boldsymbol{r}$ and ω are omitted for simplicity, $q = \omega/c$, and the suffices "r" and "b" stand for resonant and background part of induced polarization, respectively.

The background polarization is assumed to be described by a local susceptibility, i.e.,

$$\boldsymbol{P}_b(\boldsymbol{r},\omega) = \chi_b(\omega)\boldsymbol{E}(\boldsymbol{r})\Theta(\boldsymbol{r}) \tag{4}$$

where the factor $\Theta = 1$ or 0 for $\boldsymbol{r}$ inside or outside of the local background dielectrics. Then, we can rewrite eq.(3) as

$$\mathcal{L}\boldsymbol{E} = 4\pi q^2 \boldsymbol{P}_r \tag{5}$$

where the operator $\mathcal{L}$ is given as

$$\mathcal{L} = \nabla \times \nabla \times -q^2 - 4\pi q^2 \chi_b \Theta(\boldsymbol{r}) \ . \tag{6}$$

Let us define the Green's function for the operator $\mathcal{L}$ as

$$\mathcal{L}G_b(\boldsymbol{r},\boldsymbol{r}') = 4\pi\delta(\boldsymbol{r}-\boldsymbol{r}') \tag{7}$$

which describes the propagation of EM wave from $\boldsymbol{r}'$ to $\boldsymbol{r}$ in the presence of the background dielectrics. Then, the solution of eq.(5) is written as

$$\boldsymbol{E}(\boldsymbol{r}) = \boldsymbol{E}_0(\boldsymbol{r}) + q^2 \int d\boldsymbol{r}' G_b(\boldsymbol{r},\boldsymbol{r}')\boldsymbol{P}_r(\boldsymbol{r}') \ , \tag{8}$$

where $\boldsymbol{E}_0$ is the solution of

$$\mathcal{L}\boldsymbol{E}(\boldsymbol{r}) = 0 \ , \tag{9}$$

namely, the incident field in the presence of the background dielectrics.

Now we want to solve the above equation together with

$$\boldsymbol{j}_r(\boldsymbol{r}) = \int d\boldsymbol{r}' \chi_r(\boldsymbol{r},\boldsymbol{r}')\boldsymbol{A}(\boldsymbol{r}') \ , \tag{10}$$

$$= \sum_\nu{}' g_\nu(\omega)\boldsymbol{I}_{0\nu}(\boldsymbol{r})F_{\nu 0}(\omega) \tag{11}$$

where

$$g_\nu(\omega) = \frac{1}{E_{\nu 0} - \hbar\omega - i0^+} \tag{12}$$

$$F_{\nu 0}(\omega) = \int d\boldsymbol{r} < \nu \mid \hat{\boldsymbol{I}}(\boldsymbol{r}) \mid 0 > \cdot \boldsymbol{A}(\boldsymbol{r}) \tag{13}$$

and $\hat{I}(r)$ is current density operator /4/. The summation in (11) is restricted to a small number of resonant states.

Equation (11) is a resonant part of eq.(2), where $\boldsymbol{j}$ and $\boldsymbol{A}$ are related with $\boldsymbol{P}$ and $\boldsymbol{E}_t$, the transverse component of the Maxwell field $\boldsymbol{E}$, as

$$\begin{aligned} \boldsymbol{j}_r &= -i\omega \boldsymbol{P}_r \ , && (14)\\ iq\boldsymbol{A} &= \boldsymbol{E}_t \ . && (15) \end{aligned}$$

The transverse component of $\boldsymbol{E}$ is obtained by subtracting its longitudinal component, i.e., the depolarization field due to the induced polarization:

$$\boldsymbol{E}_t = \boldsymbol{E} - \nabla\nabla \cdot \int d\boldsymbol{r}' \frac{\boldsymbol{P}_r(\boldsymbol{r}') + \boldsymbol{P}_b(\boldsymbol{r}')}{\mid \boldsymbol{r} - \boldsymbol{r}' \mid} \qquad (16)$$

Using eq.(4) and eq.(8) in this expression, we obtain

$$\begin{aligned} \boldsymbol{E}_t(\boldsymbol{r}) &= \boldsymbol{E}_0(\boldsymbol{r}) - \chi_b \nabla\nabla \cdot \int d\boldsymbol{r}' \frac{\boldsymbol{E}_0(\boldsymbol{r}')\Theta(\boldsymbol{r}')}{\mid \boldsymbol{r} - \boldsymbol{r}' \mid} \\ &+ q^2 \int d\boldsymbol{r}' G_b(\boldsymbol{r}, \boldsymbol{r}') \boldsymbol{P}_r(\boldsymbol{r}') - \nabla\nabla \cdot \int d\boldsymbol{r}' \frac{\boldsymbol{P}_r(\boldsymbol{r}')}{\mid \boldsymbol{r} - \boldsymbol{r}' \mid} \\ &- q^2 \chi_b \nabla\nabla \cdot \int d\boldsymbol{r}' \int d\boldsymbol{r}'' \Theta(\boldsymbol{r}') \frac{G_b(\boldsymbol{r}', \boldsymbol{r}'') \boldsymbol{P}_r(\boldsymbol{r}'')}{\mid \boldsymbol{r} - \boldsymbol{r}' \mid} \ . \qquad (17) \end{aligned}$$

This result can be converted to the relationship between vector potential $\boldsymbol{A}$ and current density $\boldsymbol{j}_r$ by using (14) and (15). Then, eqs.(11) and (17) are regarded as simultaneous equations to determine $\boldsymbol{A}$ and $\boldsymbol{j}_r$ selfconsistently.

Noting that $\boldsymbol{P}_r$ in eq.(17) is a linear combination of $\{ F_{\nu 0} \}$, eq.(13), we obtain $\boldsymbol{A}$ from eq.(17) also as a linear combination of $\{ F_{\nu 0} \}$. Substituting this result in the definition of F, eq.(13), we get a set of selfconsistent equations for $\{ F_{\nu 0} \}$ as

$$F_{\nu 0} = F_{\nu 0}^{(0)} + \sum_{\mu}{}' B_{\nu\mu} g_\mu F_{\mu 0} \qquad (18)$$

where the summation over μ is restricted to a small number of resonant levels, and

$$\begin{aligned} F_{\nu 0}^{(0)} &= -\frac{i}{q} \int d\boldsymbol{r} \boldsymbol{I}_{\nu 0}(\boldsymbol{r}) \cdot \boldsymbol{E}_0(\boldsymbol{r}) \\ &- i\frac{\chi_b}{q} \int\int d\boldsymbol{r} d\boldsymbol{r}' \nabla \cdot \boldsymbol{I}_{\nu 0}(\boldsymbol{r}) \nabla \cdot \frac{\boldsymbol{E}_0(\boldsymbol{r}')\Theta(\boldsymbol{r}')}{\mid \boldsymbol{r} - \boldsymbol{r}' \mid} \qquad (19) \\ B_{\nu\mu}(\omega) &= \frac{1}{c} \int\int d\boldsymbol{r} d\boldsymbol{r}' \boldsymbol{I}_{\nu 0}(\boldsymbol{r}) \cdot G_b(\boldsymbol{r}, \boldsymbol{r}') \cdot \boldsymbol{I}_{0\mu}(\boldsymbol{r}') \\ &+ \frac{1}{cq^2} \int\int d\boldsymbol{r} d\boldsymbol{r}' \nabla \cdot \boldsymbol{I}_{\nu 0}(\boldsymbol{r}) \nabla \cdot \frac{1}{\mid \boldsymbol{r} - \boldsymbol{r}' \mid} \boldsymbol{I}_{0\mu}(\boldsymbol{r}') \qquad (20) \\ &+ \frac{\chi_b}{c} \int\int\int d\boldsymbol{r} d\boldsymbol{r}' d\boldsymbol{r}'' \nabla \cdot \boldsymbol{I}_{\nu 0}(\boldsymbol{r}) \nabla \cdot \frac{\Theta(\boldsymbol{r}') G_b(\boldsymbol{r}', \boldsymbol{r}'')}{\mid \boldsymbol{r} - \boldsymbol{r}' \mid} \boldsymbol{I}_{0\mu}(\boldsymbol{r}'') \ . \end{aligned}$$

The solution of eq.(18) for a given initial condition represented by $F_{\nu 0}^{(0)}$ determines the current density and vector potential uniquely. This result is very similar to that of the original version of the nonlocal theory /2-4/. The main difference is that the size of the equations is small in eq.(18), while it is essentially infinite in the original version. The effect of all the nonresonant levels is approximated as a local susceptibility and absorbed in the radiation Green's function in the definition of $B_{\nu\mu}$ in eq.(18), while the Green's function in the original version is the one in vacuum.

It should be noted that the matter Hamiltonian in the presence of a substrate should include the instantaneous Coulomb interaction among all the dipoles including their images. It can be explicitly shown that this type of interaction is forced to be zero in the expression of the radiative corrections (retarded interaction among excited dipoles) represented by $\{B_{\nu\mu}\}$, which can be seen most clearly in the non-relativistic limit. Therefore it should be included in the matter Hamiltonian.

The real problem in this formulation is how to calculate $G_b(\boldsymbol{r}, \boldsymbol{r}')$ explicitly for a given geometry. For the cases of planer, spherical, and cylindrical boundaries, including their multi-layer structures, the explicit forms of renormalized Green's functions are known /6/. Though there is only limited feasibility, it is quite useful in the cases of the above mentioned geometries.

3 Application to a surface problem

For surface problems, we often have a situation where polarizable part can be divided into two, i.e. resonant part on the surface and non-resonant background dielectrics as a substrate. The latter may be treated as a semi-infinite dielectrics. In this case, Θ in eq.(4) may be chosen, for (x, y, z) Cartesian coordinates, as

$$\Theta(\boldsymbol{r}) = 1 \qquad (z \leq 0) \tag{21}$$

$$= 0 \qquad (z > 0) \tag{22}$$

If the resonant part of the polarization is located on the surface, namely, on the side $z > 0$, the necessary components of the renormalized Green's function $G_b(\boldsymbol{r}, \boldsymbol{r}')$ in eq.(20) are those for $(z > 0, z' > 0)$ and $(z < 0, z' > 0)$. The explicit forms of these functions are given by Maradudin and Mills /7/.

The explicit form of $G_b(\boldsymbol{r}, \boldsymbol{r}')$ allows us to carry out the calculation of all the coefficients of eq.(18), and it is rather easy to solve this set of linear equations. Though we have explicitly treated linear response, it is straightforward to extend it to the case where the surface resonant contribution contains nonlinear effect such as resonant SHG effect.

As an example, we consider a problem concerning resonant SNOM (Scanning Near-field Optical Microscopy) in a simplified model. As a sample, we take a semiconductor sphere placed on a substrate. We consider an additional sphere as a probe tip. Each sphere has a degenerate resonant level

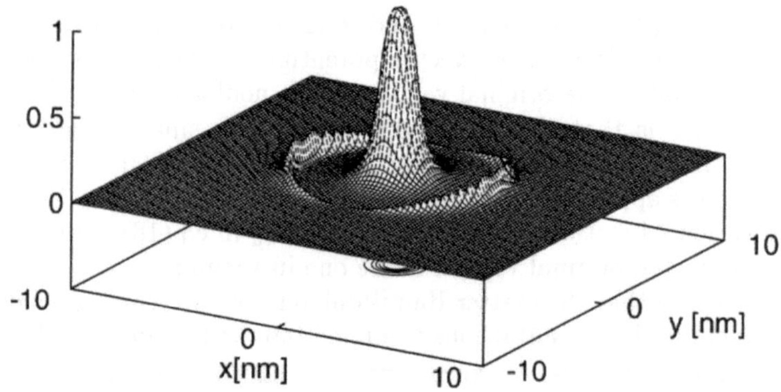

Fig.1: Map of signal intensity without substrate. The center of sample and probe spheres are at $(0,0,2)$ nm and $(x,y,8)$ nm, respectively. The intensity is normalized at the central peak.

(with x, y, z polarizations), and they interact with one another via dipole-dipole coupling. The signal intensity in this model is assumed to be the far field intensity produced by the induced polarization on the probe sphere alone. Though this way of defining signal intensity is not quite satisfactory, it is expected to relatively reflect the spatial variation of the real signal / 5,8,9/.

Both of the spheres have the same radius 1.5nm . The substrate surface is taken as the (x,y) plane $(z = 0)$, and the center of the sample sphere is located at (x,y,z)=(0, 0, 2) nm. The incident field is assumed to have x-polarization. The probe sphere is scanned on the surface z=8 nm.

The substrate is described as a semi-infinite local dielectrics having a dielectric constant of $\epsilon_b = 2.0$ and the sample sphere has a resonant level at $E_1 = 3.2772$ eV with its current density given in /10/. The non-resonant polarization of the sphere is approximated, as in /10/, by a single component in the complete set expansion of delta function, i.e., a uniform polarization with dielectric constant $\epsilon_{b1} = 5.6$. The probe sphere is assumed to have similar polarization characteristics as the sample sphere, but the resonant energy is set somewhat different from E_1 as $E_2 = E_1 + 0.01$ eV. This off-set energy is introduced to reduce the effect of configuration resonance /5/, which has a tendency to distort the image of SNOM /9/.

Fig.1 and Fig.2 show the signal intensity as a function of the probe sphere position (x,y) on the surface z=8 nm in the absence and presence of the substrate, respectively. The frequency of light is chosen at the highest resonance energy of the whole system for the probe position at $x = y = 0$, $z = 8$ nm. For this frequency the largest polarization occurs on the probe. In both Fig.1 and Fig.2 there occur two types of structure, a central peak and a "coral reef" surrounding it. The presence of the substrate enhances the coral reef structure.

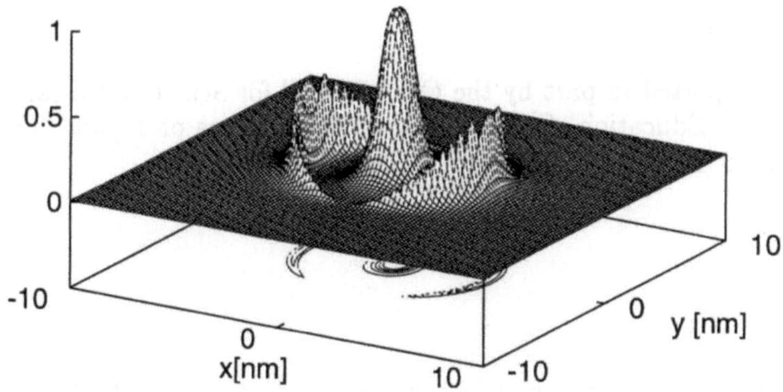

Fig.2: Same as Fig.1 with substrate.

The interpretation of the two types of structure is as follows: The central peak comes from the frequency used for this map which is resonant with the matter level (including radiative shift) for the probe position at $x = y = 0$. As the probe moves away from $x = y = 0$, the resonance condition becomes rapidly lost.

The origin of the coral reef structure is due to the resonance with an another level, which comes down in energy as the probe moves away from $x = 0$ along the x-axis. This level has pure z-polarization at $x = 0$, but, as x is increased, it acquires x-z mixed character. Because of this mixing, this level also can contribute to the resonant enhancement of the probe polarization. The stronger signal of coral reef structure in Fig.2 than in Fig.1 indicates the larger induced polarization on the probe sphere. This is due to the interplay of interactions among the dipoles on the sample and probe spheres and their images, which influences the resonant levels and the direction and the size of induced dipoles.

Though the coral reef structure has little connection with the shape of the sample, it contains important information about the level scheme of the sample (plus substrate) system.

The radiative lifetime is also expected to change by the presence of a substrate. Though it is not evident in Figs.1 and 2, a preliminary result shows an interesting dependence of the radiative lifetime on the polarization and the distance from the substrate. More systematic study in this respect is now being done. The details of this numerical result including other sample systems with internal degrees of freedom will be published elsewhere.

In summary, we have given a new type of nonlocal response theory by the use of a renormalized radiation Green's function, which takes care of the effect of a substrate with local dielectric constant. As a simple application, we have calculated the SNOM image of a model system for a fixed resonant energy. A large change in the image is brought about due to the different behavior of the matter levels by the presence of a substrate.

Acknowledgments

This work is supported in part by the Grant-in-Aid for Scientific Research from Ministry of Education, Science, Sports and Culture of Japan. The authors are grateful to Prof. Y. Ohfuti for useful discussions.

References

1. J.H.Van Vleck: *The Theory of Electric and Magnetic Susceptibilities,* (Oxford 1932) p.1
2. K. Cho: Prog. Theor. Phys. Suppl. **106** (1991) 225
3. K. Cho, H. Ishihara, and Y. Ohfuti: J. Luminescence **58** (1994) 95
4. Y. Ohfuti and K. Cho: Phys. Rev. B**52** (1995) 4828; ibid. Proc. Int. Conf. on *Dynamical Processes in Excited States of Solids,* Great Barrier Reef 1995, to be published in J. Luminescence
5. O. Keller, M. Xiao, and S. Bozhevolnyi: Surf. Sci. **280** (1993) 217
6. W. C. Chew: *Waves and Fields in Inhomogeneous Media,* (Van Nostrand Reinhold, New York, 1990)
7. A. A. Maradudin and D. L. Mills: Phys. Rev. B**11** (1975) 1392
8. C. Girard and D. Courjon: Phys. Rev. B**42** (1990) 9340
9. K. Cho, Y. Ohfuti and K. Arima: Jpn. J. Appl. Phys. **34** (1995) Suppl. **34-1,** pp.267
10. Y. Ohfuti and K. Cho: Phys. Rev. B**51** (1995) 14379

Theory for Nonlinear Magneto-Optics at Thin Films

T.A. Luce, U. Pustogowa, W. Hübner, and K.H. Bennemann
Institute for Theoretical Physics, Freie Universität Berlin,
Arnimallee 14, D-14195 Berlin, Germany

Abstract. We present a theoretical study of the nonlinear magneto-optical Kerr-effect, which demonstrates how various electronic material properties can be extracted from the nonlinear Kerr spectrum. We derive its dependence on exchange interaction and magnetization. We calculate the thickness dependence of the nonlinear Kerr spectra of thin Fe(001) films using an *ab initio* bandstructure calculation. We demonstrate how this dependence can be used for the characterization of the d bandwidth, the thickness, and the magnetic moment at the interfaces of thin films. The calculation of the nonlinear Kerr rotation demonstrates the strong enhancement of the Kerr rotation angle compared to the linear Kerr angle, thus allowing for the determination of the easy axis at the surface. The non-equilibrium electron temperature dependence of the SHG at noble metal surfaces is discussed. Calculations of quantum well oscillations in SHG for a xCu/1Fe/Cu(001) system are presented. A brief outlook on future prospects of nonlinear magneto-optics, in particular on investigations of the morphology of thin films, is given.

1. Introduction

Nonlinear magneto-optics for metals [1, 2] is a new expanding field of research. Extension of the symmetry analysis to an electronic theory [3, 4] made second harmonic generation (SHG) a powerful tool for investigating the electronic structure of surfaces, interfaces, thin films, and multilayers [5, 6, 7]. The surface sensitivity results, since the breakdown of inversion symmetry is necessary for SHG. Thus SHG allows for the determination surface magnetic moments, magnetic anisotropy, domains, magnetostriction, and properties of buried interfaces.

Using an electronic theory (with a bandstructure input) we calculate (i) the material specific nonlinear response function $\chi^{(2)}(\omega, \mathbf{M}, \lambda_{s.o.}, d, T_{el})$, depending on incident photon frequency ω, magnetization M, spin-orbit coupling $\lambda_{s.o.}$, film thickness d, and on the electron temperature T_{el} for a Fe film and for a Fe/Cu bilayer, (ii) the nonlinear Fresnel formula for the reflected nonlinear fields $E^{(2)}(\omega)$ and the complex nonlinear Kerr rotation angle $\Phi_K^{(2)}$ for a Fe monolayer, (iii) the dynamics of the SHG intensity $I^{(2)}(\omega, T)$ of a Cu surface layer and finally (iv) the layer number dependence of $I^{(2)}(\omega)$ for a xCu/1Fe/Cu(001) system,

Springer Series in Solid-State Sciences, Volume 121
Elementary Processes in Excitations and Reactions on Solid Surfaces
Eds.: A. Okiji, H. Kasai and K. Makoshi

resulting in quantum well oscillations. The presented results give an overview over the variety of material specific physical observables that can be probed with nonlinear magneto-optics. Some selected general results obtained previously are shown, but also interesting new results indicating important substrate effects.

2. Nonlinear susceptibility

In nonlinear response theory the second order optical polarization is given by

$$P_i^{(2)} = \chi_{ijm}^{(2)} E_j(\omega) E_m(\omega), \tag{1}$$

where $P_i^{(2)}$ is the i-th component of the nonlinear polarization, and the susceptibility tensor $\chi_{ijm}^{(2)}$ is determined by [1, 8]

$$\chi_{ijm}^{(2)}(2q_{\|}, 2\omega, \mathbf{M}) = \frac{e^3 C}{\Omega} \frac{\lambda_{s.o.}}{\hbar\omega} \times \sum_{\sigma} \sum_{\mathbf{k},l,l',l''} \Big\{ \langle \mathbf{k}+2\mathbf{q}_{\|}, l''\sigma | r_i | \mathbf{k}l\sigma \rangle \langle \mathbf{k}l\sigma | r_j | \mathbf{k}+\mathbf{q}_{\|}, l'\sigma \rangle \langle \mathbf{k}+\mathbf{q}_{\|}, l'\sigma | r_m | \mathbf{k}+2\mathbf{q}_{\|}, l''\sigma \rangle \times \frac{\dfrac{f(E_{\mathbf{k}+2\mathbf{q}_{\|},l''\sigma}) - f(E_{\mathbf{k}+\mathbf{q}_{\|},l'\sigma})}{E_{\mathbf{k}+2\mathbf{q}_{\|},l''\sigma} - E_{\mathbf{k}+\mathbf{q}_{\|},l'\sigma} - \hbar\omega + i\hbar\alpha_1} - \dfrac{f(E_{\mathbf{k}+\mathbf{q}_{\|},l'\sigma}) - f(E_{\mathbf{k}l\sigma})}{E_{\mathbf{k}+\mathbf{q}_{\|},l'\sigma} - E_{\mathbf{k}l\sigma} - \hbar\omega + i\hbar\alpha_1}}{E_{\mathbf{k}+2\mathbf{q}_{\|},l''\sigma} - E_{\mathbf{k}l\sigma} - 2\hbar\omega + i2\hbar\alpha_1} \Big\}. \tag{2}$$

Here $E_{\mathbf{k}l\sigma}$ are the electronic energy levels resulting from the bandstructure calculations, α_1 is the Lorentzian broadening and typically equals 0.1 eV in our calculation. $\langle \mathbf{k}l\sigma | r_i | \mathbf{k} + \mathbf{q}_{\|}, l'\sigma \rangle$ are the transition matrix elements, which control the symmetry and thus the interface sensitivity. The factor C [9] is a cut-off to guarantee the surface sensitivity, since we are assuming constant matrix elements M fitted to the linear dielectric function $\epsilon(\omega)$. This approximation is reasonable because the **k**-dependence of the matrix elements is expected to become less important in two dimensions due to the shrinking of the d-band width for the reduced coordination number and also due to the occurrence of additional allowed optical transitions. Typically, magnetic dipole effects are negligible if the electric dipole contribution does not vanish. $\lambda_{s.o.}$, the spin-orbit coupling, is the key parameter for nonlinear magneto-optics. In particular, for determining the Kerr effect it might be useful to decompose $\chi^{(2)}$ in even and odd contributions $\chi^{(2)} = \chi^{(2)o} + \chi^{(2)}(\mathbf{M})$.

For applications of the general theory thin films are of particular interest. Then, the layer-dependent contributions to $\chi^{(2)}$ reveal interesting film properties and substrate effects. For the thin film calculations we use $C = W_{\mathbf{k}+2\mathbf{q}_{\|},l''\sigma} W_{\mathbf{k}+\mathbf{q}_{\|},l'\sigma} W_{\mathbf{k}l\sigma}$, where W_α denotes the weight of the density of state $|\mathbf{k}l\sigma\rangle$ in the Wigner-Seitz cell of the first monolayer. Note, due to the factor C, SHG will always result from the surface layer or from the layer at the interface, however the electronic structure of this layer depends on the film thickness. The bandstructure $E_{\mathbf{k}l\sigma}$ is calculated employing the full-potential linear muffin-tin orbital method (FP-LMTO) [10] for thin Fe(001) films within a symmetric slab calculation.

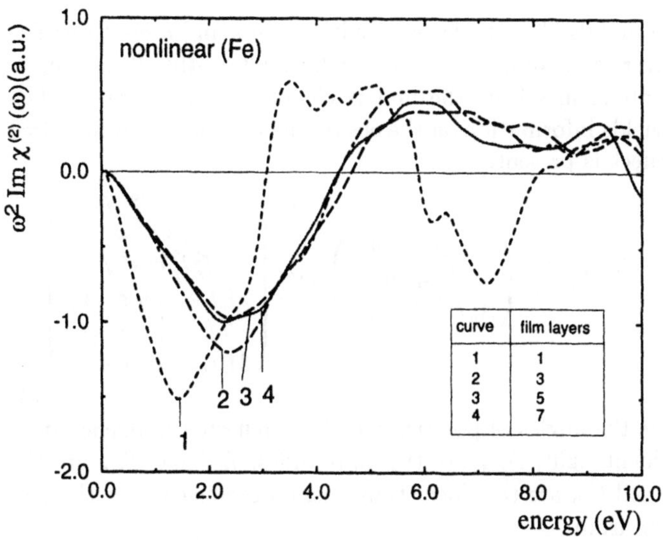

Figure 1: *Ab initio* calculated nonlinear magneto-optical Kerr spectra of Fe using Eq. (2) for a monolayer (dashed curve - 1), and films with 3 layers (dashed-dotted curve - 2), 5 layers (long-dashed - 3), and 7 atomic layers (dotted curve - 4). The second harmonic response results from the first atomic layer.

Results for one, three, five, and seven layers of Fe are presented in Fig. 1. The depth of the first minimum in the spectra scales linearly with the magnetic moment of the top layer while the energy of this minimum reflects the *d*-band width. As can be seen from calculations using different lattice constants, the nonlinear spectra reveal the structural changes induced by different substrates or temperatures. However, tight-binding calculations may be adequate if one uses as input parameters those determined from *ab initio* calculations for thin films.

From these results we learn that the nonlinear Kerr effect is a sensitive probe of the magnetic and electronic structure at interfaces. This might be of general significance also with respect to the tuning of interface properties, ($\lambda_{s.o.}$ in the substrate, **M** in films).

3. Symmetry, Kerr Rotation and Nonlinear Intensity

The pronounced symmetry sensitivity of SHG is furthermore demonstrated by its polarization dependence and by the remarkable enhancement of the Kerr rotation. The latter important result was first theoretically derived by Pustogowa *et al.* [11] and later observed in experiments by Böhmer *et al.* [12] and Koopmans *et al.* [13].

First, we note that due to the symmetry sensitivity SHG already exhibits a corresponding polarization dependence, which interestingly is very different for transition and noble metals [3]. The latter results, since s and d electrons feel the breakdown of inversion symmetry at the interface differently. This symmetry dependence mainly results from the matrix elements in $\chi^{(2)}$, see Eq. (2). Interestingly, SHG yields information on the electronic structure even for $\mathbf{M}=0$ and when no Kerr effect is present.

$$E^{(2\omega)}(\Phi,\varphi) = 2i\left(\frac{\omega}{c}\right)|E_0^{(\omega)}|^2 \begin{pmatrix} A_pF_c\cos\Phi \\ A_s\sin\Phi \\ A_pN^2F_s\cos\Phi \end{pmatrix} \chi^{(2)} \begin{pmatrix} f_c^2t_p^2\cos^2\varphi \\ t_s^2\sin^2\varphi \\ f_s^2t_p^2\cos^2\varphi \\ 2f_st_pt_s\cos\varphi\sin\varphi \\ 2f_cf_st_p^2\cos^2\varphi \\ 2f_ct_pt_s\cos\varphi\sin\varphi \end{pmatrix} . \tag{3}$$

Here, Φ and φ denote the angles of polarization of the reflected frequency doubled and of the incident light, $A_{p,s}$ are the amplitudes, $f_{c,s}$ and $F_{c,s}$ are the Fresnel coefficients, and $t_{s,p}$ are the linear transmission coefficients for the fundamental light. Then, from

$$\phi_K^{(2)} \approx Re\frac{E_\varphi^{(2\omega)}(s-SH)}{E_\varphi^{(2\omega)}(p-SH)} , \tag{4}$$

where $E_\varphi^{(2\omega)}(s-SH)$ and $E_\varphi^{(2\omega)}(p-SH)$ denote the s- and p-polarized SHG field from incident light with polarization angle φ, we determine the Kerr rotation for the polar and longitudinal configuration and for s- and p-polarized incident light.

Summarizing the results, we find that generally $\phi_K^{(2)M_\perp} < \phi_K^{(2)M_\parallel}$ and $\phi_{K,s}^{(2)} > \phi_{K,p}^{(2)}$, which is in agreement with the experiment by Koopmans *et al.* [13]. Numerical results of our theory using as input parameters the phase ratio of the various different tensor elements from Ni and linear complex indices of refraction are presented in Fig. 2. These results demonstrate clearly that the nonlinear Kerr angle depends sensitively on the direction of the magnetization and the incident beam, as well as on the light polarization. Depending on the experimental conditions the nonlinear Kerr angle might become even as large as 90°. Moreover, also the ellipticity will yield correspondingly large magnetic effects. We conclude that the enhanced nonlinear Kerr rotation allows for the determination of easy axis and canted-spin configurations due to large magnetic contrast of different interfaces.

The SHG yield for p-polarization within the dipole approximation is derived from Eq. (3) [3]. Since it is known from theory and also experiment that the $\chi_{zzz}^{(2)}$ tensor element dominates over $\chi_{xzx}^{(2)}$ and $\chi_{zxx}^{(2)}$ [3], we restrict our calculation to this single element of the nonlinear susceptibility for the SHG yield. Then the p-polarized SHG yield is approximately given by

$$I^{(2)}(p-SH) = |(2i\frac{\omega}{c})|^2 |E_0(\omega)|^4 |A_pN^2F_s\chi_{zzz}^{(2)}f_s^2t_p^2\cos^2\varphi|^2 . \tag{5}$$

Note, the nonlinear susceptibility tensor $\chi^{(2)}$ is material specific, and so are the linear dielectric function $\epsilon(\omega)$ and the indices of refraction n and N via the

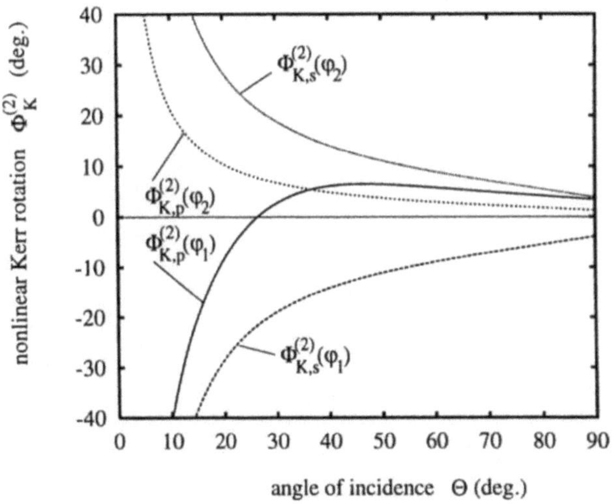

Figure 2: Nonlinear Kerr rotation angles for p polarized incident light $\phi^{(2)}_{K,p}$ (full and short-dashed curves) and for s polarized incident light $\phi^{(2)}_{K,s}$ (long-dashed and dotted curves) for Fe at 770 nm as a function of the angle of incidence θ in the longitudinal Kerr configuration. The relative phase between $\chi^{(2)}_{zxx} = \chi^{(2)}_{zyy}$ and $\chi^{(2)}_{zzz}$ is φ_1=0.505π in the full and long-dashed curves and φ_2=1.505π in the short-dashed and dotted curves.

electronic bandstructure. $\chi^{(2)}(\omega)$ is the susceptibility tensor already introduced in Eq. (2). The calculation of the dielectric function $\epsilon(\omega, T)$ is performed by including both the intra- and interband electronic transitions, using an electronic theory [15].

4. Dynamics of Electronic Systems and Temperature Effects

To gain insight into the dynamics of electronic systems, SHG may be used to probe the relaxation processes after optical excitations without background from unexcited bulk due to its surface sensitivity. With pump-probe experiments it is possible to investigate the time window when the electrons have thermalized, but the electron-phonon coupling not yet has become effective. Recent experiments exploiting the time dependence of SHG [16, 17] demonstrated the ability of SHG to analyze the dynamics of an electronic system during relaxation to the equilibrium state. The calculation of the SHG intensity with Eq. (6) for different electron temperatures, using a tight binding band structure of a Cu monolayer with parameters fitted to a Cu surface, shows that the main temperature effect for photon energies less than 3 eV comes from $\chi^{(2)}(\omega, T)$ and not from the Fresnel coefficients [18]. This is presented in Fig. 3, where we show the

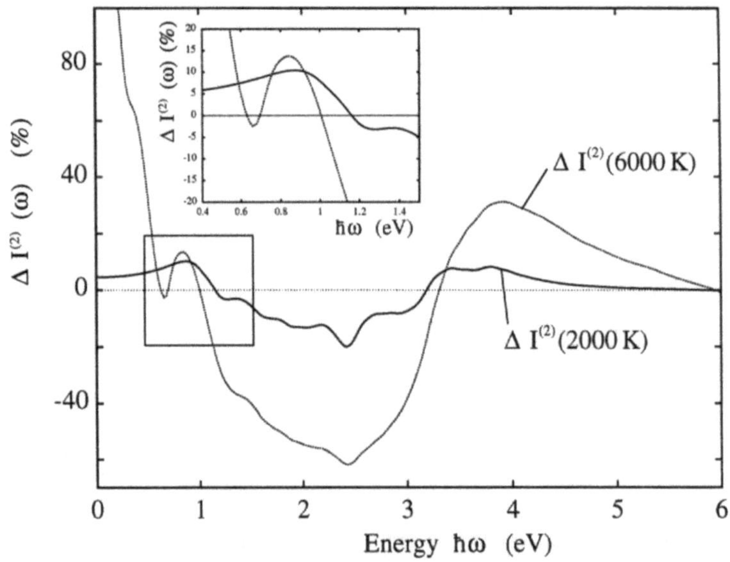

Figure 3: Change of the SHG intensity $\Delta I^{(2)}(\omega) = \frac{I^{(2)}(\omega,T_{el}) - I^{(2)}(\omega,300K)}{I^{(2)}(\omega,300K)}$ due to a rise of electron temperature as a function of the incident photon energy $\hbar\omega$ in percent for $T_{el} = 2000K$ and $T_{el} = 6000K$. The inset shows the energy range between 0.4 and 1.5 eV at an enlarged abscissa scale.

relative difference $\Delta I^{(2)}(\omega) = \frac{I^{(2)}(\omega,T_{el}) - I^{(2)}(\omega,300K)}{I^{(2)}(\omega,300K)}$ of the SHG yield as a function of frequency for different non-equilibrium electronic temperatures T_{el}. In particular, in the inset of Fig. 3 we show for the frequency range from 0.4 to 1.5 eV, that $I^{(2)}(\omega)$ may be reduced or enhanced due to increasing non-equilibrium temperature for the electrons, which is in fair agreement with experiment [17]. Our further analysis shows that there is a strong dependence of the SHG on the "mismatch" between (E_d - E_F) and the photon energy $\hbar\omega$, thus yielding a different behavior for other noble metals with a different position of the d-band edge, since the main contribution to the SHG yield comes from transitions with the initial and the intermediate state an occupied d state.

5. Nonlinear Response of Bilayers and from Quantum Well States

We expect generally interesting results for $\chi^{(2)}$ and $I^{(2)}$ at heterogeneous interfaces like Fe/Cu(001). For example, it is of interest to find out how the substrate interferes with the electronic structure of the thin film. For this reason we performed calculations of the Fe/Cu bilayer spectrum. The bandstructure has been obtained [19] within the combined interpolation scheme (CIS). The d-bands are parameterized in terms of Fletcher-Wohlfarth parameters [15], the parameters were obtained from a fit to literature band structures [9, 20]. Results for the

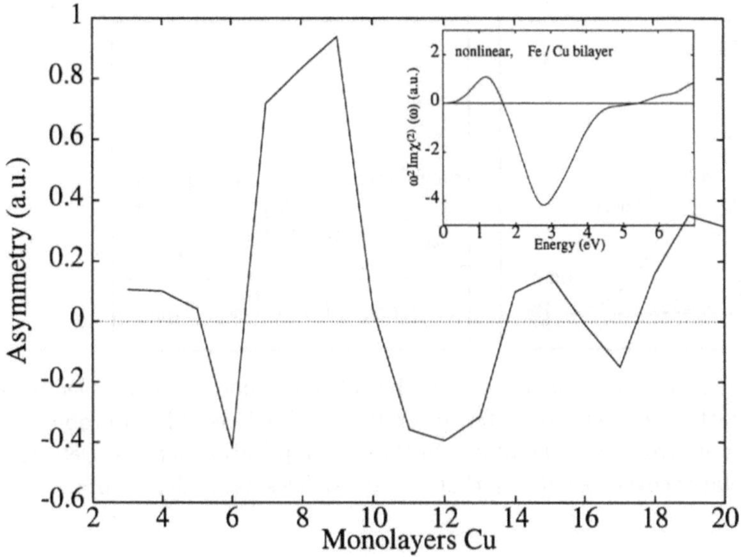

Figure 4: Quantum well oscillations of the asymmetry $\frac{I^{(2)}_{+\mathrm{M}}-I^{(2)}_{-\mathrm{M}}}{I^{(2)}_{+\mathrm{M}}+I^{(2)}_{-\mathrm{M}}}$ of a xCu/1Fe/Cu(001) layer system. The variation of the Cu cover layer thickness leads to oscillation periods of approx. 6 and 12 ML for the difference signal. The inset shows the nonlinear susceptibility spectrum of a Fe/Cu bilayer. The bilayer spectrum looks rather different from the corresponding one for a free-standing monolayer indicating a strong effect of Cu on the optical transitions.

nonlinear magnetic susceptibility of the Fe/Cu bilayer $\chi^{(2)}(\omega)$ are shown in the inset of Fig. 4. Compared with the spectrum of the Fe monolayer, due to the presence of the (nonmagnetic!) Cu layer there is a strong additional peak at higher photon energy, resulting from the amplification of the nonlinear response by additional Cu d-states serving as initial states. Further support for the importance of occupied d-states for the NOLIMOKE-signal comes from the calculation of the SHG yield for a xCu/1Fe/Cu(001) system. Experimentally there have been observed quantum well oscillations in similar systems in photoemission [21, 22] as well as in the SHG response and in the asymmetry $\frac{I^{(2)}_{+\mathrm{M}}-I^{(2)}_{-\mathrm{M}}}{I^{(2)}_{+\mathrm{M}}+I^{(2)}_{-\mathrm{M}}}$ [7], when the Cu cover-layer number is varied. We modeled the xCu/1Fe/(001)Cu system using a Cu bulk Hamiltonian (thus depending on k_x, k_y, k_z), and a Fe monolayer Hamiltonian that has no dispersion perpendicular to the in-plane $k^{||}$-directions. The Fletcher-Wohlfarth parameters for the Cu bulk bandstructure are taken from [23]. Since we are evaluating the SHG response at $(k_x,k_y) = (0,0)$ (k_z is perpendicular to the layers), the k-summation is performed over k-points along the Γ - X $(= k^{\perp})$ direction. Due to the confinement of electronic states in thin films, the quantization of $k^{\perp}$ results in a splitting of the continuous band into discrete points equally distributed along the $k_{\perp}$ direction[21].

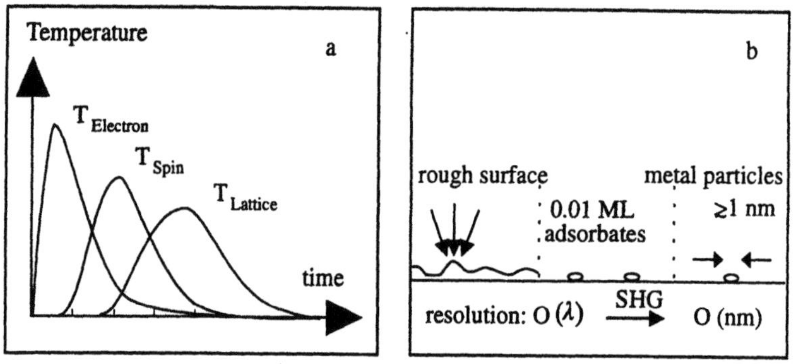

Figure 5: a) Schematic illustration for the different time scales for electron-, spin-, and lattice-temperature after an excitation. SHG is able to probe even the ultrafast electron-spin relaxation. b) Possible application of SHG: Detection of submonolayer coverages of adsorbates and particles as small as 1 nm.

For the formation of quantum well oscillations of the nonlinear response, transitions at those $k_\perp$-vectors are decisive, where (for Cu bulk) the s-band crosses the Fermi surface (at $k_{E_F}^{\perp}$). If the layer number increases, such $k_\perp$-states at the Fermi surface will occur if the layer number equals $n * \frac{k_{BZ}}{k_{BZ}-k_{E_F\perp}}$, with $n = 1, 2, 3....$ Then these new unoccupied s-states enable additional transitions, and the SHG yield increases. Our analysis shows that the Fe monolayer d-states are responsible for the enhancement of the SHG intensity compared with a Cu multilayer since they act as intermediate states, thus enhancing the SHG yield as in the case of the Cu/Fe bilayer. The spin polarization of the Fe d states results in a different amplification of the SHG for different direction of $\mathbf{M}$, thus leading to a nonvanishing $\mathbf{M}$-asymmetry. Fig. 4 shows results of our calculation of the layer dependence of the SHG yield. Periods of 6 and 12 layers can be deduced (the latter being more pronounced), in good agreement with experiment, which observes mostly twice the period of linear optics (12 ML) [24].

This selective discussion demonstrates the value of nonlinear magneto-optics as a promising spectroscopy. Our results show that SHG is a new sensitive tool for determining magnetic interface properties such as the interplay of structure and magnetism, magnetic anisotropy, and magnetic reorientation. In contrast to linear optics, which probes characteristic film-averaged features, the nonlinear Kerr effect originates essentially from the surface and interfaces and thus allows the analysis of buried interfaces.

Future theoretical investigations of the nonlinear magneto-optical Kerr effect will include its dynamics and its dependence on the surfaces morphology. Fig. 5a) shows different relaxation times for the electron, spin, and lattice temperatures of an excited system, which may allow for ultrafast magnetic switching without involving the lattice. A drastic SHG signal increase on rough surfaces is observed, due to the much larger local-field enhancement of SHG compared with linear optical techniques. Thus we will study the possibility to measure submonolayer coverages of adsorbates with NOLIMOKE, and even to detect particles of the size of 10^{-9}m with SHG and NOLIMOKE (see Fig. 5b)). Another promising feature of SHG, the enhanced sensitivity for lateral inversion symmetry breaking may be exploited to give information about the lateral structure of surfaces and thin films on the microscopic length scale.

References

[1] W. Hübner and K. H. Bennemann, Phys. Rev. B **40**, 5973 (1989).

[2] R.P. Pan, H. D. Wei, and Y. R. Shen, Phys. Rev. B **39**, 1229 (1989).

[3] W. Hübner, K. H. Bennemann, and K. Böhmer, Phys. Rev. B **50**, 17597 (1994).

[4] W. Hübner and K. H. Bennemann, accepted for publication in Phys. Rev. B.

[5] J. Reif, J. C. Zink, C. M. Schneider, and J. Kirschner, Phys. Rev. Lett. **67**, 2878 (1991).

[6] H. A. Wierenga, M. W. J. Prins, D. L. Abraham, and Th. Rasing, Phys. Rev. B **50**, 1282 (1994).

[7] H. A. Wierenga, W. de Jong, M. W. J. Prins, Th. Rasing, R. Vollmer, A. Kirilyuk, H. Schwabe, and J. Kirschner, Phys. Rev. Lett. **74**, 1462 (1995).

[8] W. Hübner, Phys. Rev. B **42**, 11553 (1990).

[9] U. Pustogowa, W. Hübner, K. H. Bennemann, submitted for publication.

[10] M. Methfessel, Phys. Rev. B **38**, 1537 (1988); M. Methfessel, C. O. Rodriguez, and O. K. Andersen, *ibid.* **40**, 2009 (1989); M. Methfessel and M. Scheffler, Physica B **172**, 175 (1991).

[11] U. Pustogowa, W. Hübner, and K. H. Bennemann, Phys. Rev. B **49**, 10031 (1994); Appl. Phys. A **59**, 611 (1994).

[12] K. Böhmer, J. Hohlfeld, and E. Matthias, Appl. Phys. A **60**, 203 (1995).

[13] B. Koopmans, M. Groot Koerkamp, Th. Rasing, and H. van den Berg, Phys. Rev. Lett. **74**, 3692 (1995).

[14] U. Pustogowa, T. A. Luce, W. Hübner, and K. H. Bennemann, accepted for publication in J. Appl. Phys.

[15] U. Pustogowa, W. Hübner, and K. H. Bennemann, Phys. Rev. B **48**, 8607 (1993)

[16] J. Hohlfeld, D. Grosenick, U. Conrad, and E. Matthias, Appl. Phys. A **60**, 137 (1995)

[17] J. Hohlfeld, U. Conrad, and E. Matthias, to appear

[18] T. A. Luce, W. Hübner, and K. H. Bennemann, submitted for publication to Phys. Rev. B

[19] H. Ehrenreich and L. Hodges, Methods Comput. Phys. **8**, 149 (1968)

[20] H. Krakauer, M. Posternak, and A. J. Freeman, Phys. Rev. B **19**, 1706 (1979).

[21] J. E. Ortega and F. J. Himpsel, Phys. Rev. Lett. **69**, 844 (1992); J. E. Ortega, F. J. Himpsel, G. J. Mankey, and R. F. Willis, Phys. Rev. B **47**, 1540 (1993)

[22] C. Carbone, E. Vescovo, O. Rader, W. Gudat, and W. Eberhardt, Phys. Rev. Lett. **71** 2805 (1993)

[23] N. V. Smith, Phys. Rev. B **19** 5019 (1978)

[24] M. Straub, R. Vollmer, and J. Kirschner, private communication

Size Dependent CO Dissociation on Rh Particles Supported on Thin Alumina Films

J. Libuda[1], M. Frank[1], A. Sandell[1], S. Andersson[2], P.A. Brühwiler[2], M. Bäumer[1], N. Mårtensson[2], and H.-J. Freund[1]

[1]Physical Chemistry 1, Ruhr-Universität Bochum, D-44780 Bochum, Germany

[2]Department of Physics, Uppsala University, Box 530, S-75121 Uppsala, Sweden

Small Rh particles are grown onto a well ordered alumina film which covers a NiAl(110) single crystal surface. Adsorption and dissociation of carbon monoxide is studied by means of photoelectron spectroscopy (XPS). The tendency to dissociate CO increases with increasing particle size. This statement is supported for the first time by a direct in-situ measurement, and is in partial contrast to literature data. It is speculated that the low tendency for dissociation on the small particles is connected with the spatial constraints of a dissociation reaction, i.e. it needs several separated sites to occur. Since the smallest particles contain less than ten Rh atoms these constraints may play an important role.

1. Introduction

In many catalytic processes at present dispersed metal catalysts are being used [1,2]. Often the metal particle size plays an important role for the activity and selectivity of the reaction in question. For so called structure sensitive reactions the size dependent distribution of specific adsorption sites on small metal particles can affect the reaction [3]. In order to study the details of the interaction of molecules such as CO, as a prototype system, with small metal particles as a function of temperature we resort to the study of model systems. These model systems consist in our case of a thin well ordered Al_2O_3 film grown on top of a NiAl(110) single crystal surface, onto which metal particles are deposited from the gas phase [4-13]. In the past we have studied various metal deposits, such as Pt, Pd, Ag, and Rh [4, 5, 8-13]. The different metals exhibit quite different growth behaviour which is due to the varying interaction strength between the metal and the oxide substrate. CO adsorption has been studied for almost all the different metal deposits. We have reported indications that e.g. for Pt particles dissociation occurs on very small particles [5]. The evidence was deduced from x-ray photoelectron spectroscopic measurements carried out for this system. However, it could not be finally excluded that this dissociation was photoinduced due to the rather extensive exposure times to x-rays in conventional x-ray spectrometers or was due to contaminations accumulating during the rather long accumulation times. Investigations carried out subsequently applying synchrotron radiation to record the corresponding spectra in short times indicated that this may in fact have been

Springer Series in Solid-State Sciences, Volume 121
Elementary Processes in Excitations and Reactions on Solid Surfaces
Eds.: A. Okiji, H. Kasai and K. Makoshi

the case [4]. It was therefore important to investigate a system where there are strong indications from the literature that CO dissociation does take place and that this dissociation proceeds via a thermal mechanism. Rh is known to induce CO dissociation, and CO adsorption and reaction has been studied extensively in the literature for small deposited metal particles [14-17] as well as single crystal surfaces [18-25]. While for the single crystal systems XPS results have been reported, such reports for small particle systems have been very scarce. Whereas previous SIMS and TPD studies have shown that dissociation is more important for smaller than for larger metal particles [15], thermal desorption data recently published gives very indirect indication for lower activity for dissociation on smaller particles [14]. On the single crystal surfaces it appears that while the close packed surfaces do not seem to induce C-O bond breaking [18-21], the more open or stepped faces are very well capable for CO dissociation [22-25].

In the present study we report on a synchrotron based XPS study of dissociation as a function of temperature of small Rh particles interacting with CO., CO dissociation is observed and its dependence on the particle size is discussed.

2. Experimental Aspects

The experiments have been performed in several ultrahigh vacuum systems located in Bochum and at beamline 22 located at Maxlab in Lund/Sweden. The setup contains a modified Zeiss SX-700 monochromator in conjunction with a large hemisperical electron energy analyser for photoemission and a multichannel plate with a retarding grid for x-ray absorption measurements by detection of secondary electrons. The C1s spectra were recorded with a total resolution of 0.4eV. All the XPS binding energies are referred to the Fermi level of the NiAl substrate. The absolute photonenergies were determined using photoemission spectra excited by first and second order radiation. The SPA-LEED measurements were carried out in a seperate ultrahigh vacuum system. From the SPA-LEED investigations the average island size could be determined from an analysis of the line profile.

The substrate oxide film has been prepared according to well known recipes. Rh was evaporated from a rod using a commercial evaporator based on electron bombardment. The evaporation rate was monitored using a quartz microbalance.

3. Results and Discussion

Fig.1 shows a set of C1s spectra taken on a Rh deposit of 4 Å as monitored by the quartz balance. In the SPA-LEED pattern such a deposit exhibits clear indications for island growths of the metal on the surface. As has been discussed in detail for Pt and Pd deposits, the (0,0) reflex shows a pronounced shoulder [8,13]. The profile analysis yields an average isand size of 20 Å for a film of 4 Å thickness of Rh. At low temperature, i.e. 90 K , a single C1s signal with binding energy 286.5 eV is measured. This binding energy is typical for molecularly bound CO. Upon heating the surface to room temperature the single line narrows and shifts to lower binding energy by 0.3 eV. The situation changes when the sample is heated to 400

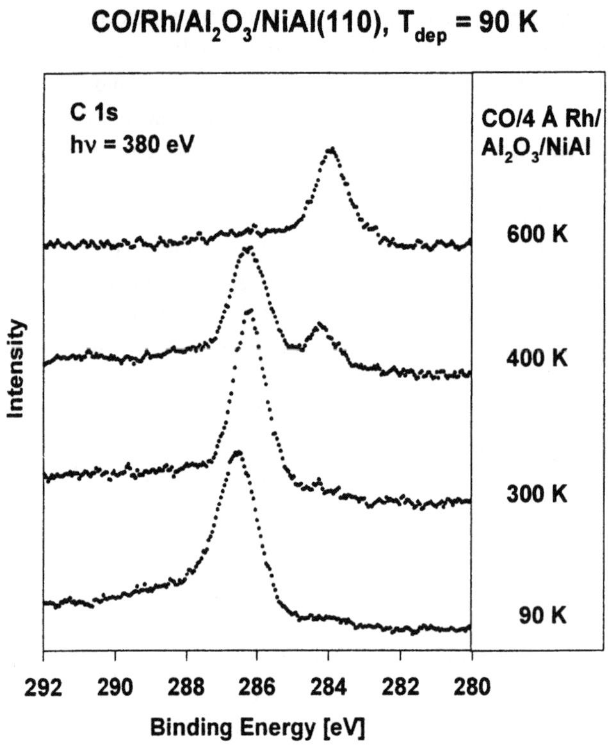

Fig. 1 C1s photoelectron spectra of CO on $Rh/Al_2O_3/NiAl(110)$ after heating to various surface temperatures. The Rh particles have been deposited at T_{dep} = 90 K.

K. At the expense of the molecular species a peak at 284 eV binding energy grows in, and at 600 K surface temperature the molecular peak has completely disappeared. By comparison with literature data it is quite obvious that the peak at 284 eV correponds to atomic carbon being formed on the surface. In Fig.2 the initial situation at low temperature and the final situation at 600 K are plotted together for the various Rh depositions investigated. The amount of deposited Rh is given as determined by the quartz balance. For these metal coverages the average particle size as determined from the SPA-LEED analysis varies by almost an order of magnitude, namely between about 5 Å and about 30 Å. The spectra at the smallest particle size must be taken with care because of the rather limited signal to noise ratio. Several trends may be observed. Firstly the C1s ionization of the molecular species shifts from higher to lower binding energy as the particle size increases. The corresponding shift of the C1s ionization of atomic carbon is considerably smaller but in the same direction. Secondly, the line widths increase systematically for decreasing particle size. Thirdly, and probably most importantly, for the present study, the amount of atomic carbon found on the surface varies with particle size. The last point, which we shall adress first, is highlighted in Fig.3a. In this figure the relative C1s intensity is decomposed into the three components,

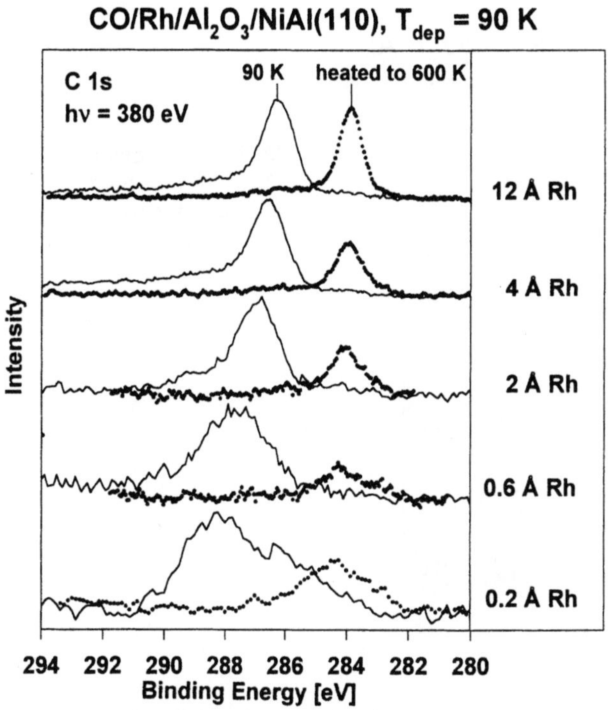

Fig. 2 C1s photoelectron spectra of CO on $Rh/Al_2O_3/NiAl(1\bar{1}0)$ as a function of the amount of deposited Rh metal (T_{dep} = 90 K). For each deposition the full line corresponds to the spectra taken after exposure to CO at 90 K. The dotted line represents the spectrum taken after heating the CO adsorbate to 600 K.

namely carbon stemming from CO dissociation after heating to 400 K, loss of signal due to CO desorption the amount of adsorbed CO, and a carbon contaminant being present on the surface as the CO is dosed. We do not use the data after heating to 600 K for this analysis because at that temperature CO starts to desorb via recombination with oxygen. [24,14,15]. Again, the data for the smallest particle sizes have to be taken with care. We shall therefore base the consecutive conclusions mainly on the data for coverages excluding those for the 0.2 Å deposit. The general result, independent of the temperature, once the dissociation temperature is crossed, is that the amount of dissociated CO increases with increasing Rh particle size. We note, however, that the analysis of the data after heating to 600 K yield the same result.

In order to discuss these findings we adress the three main observations as mentioned above: The first observation was the shift of the C1s ionization towards lower binding energies. This result for Rh clusters corroborates a similar result obtained for Pd clusters on the same substrate. This shift is accompanied by the decrease in width of the C1s ionization line, which was the second observation

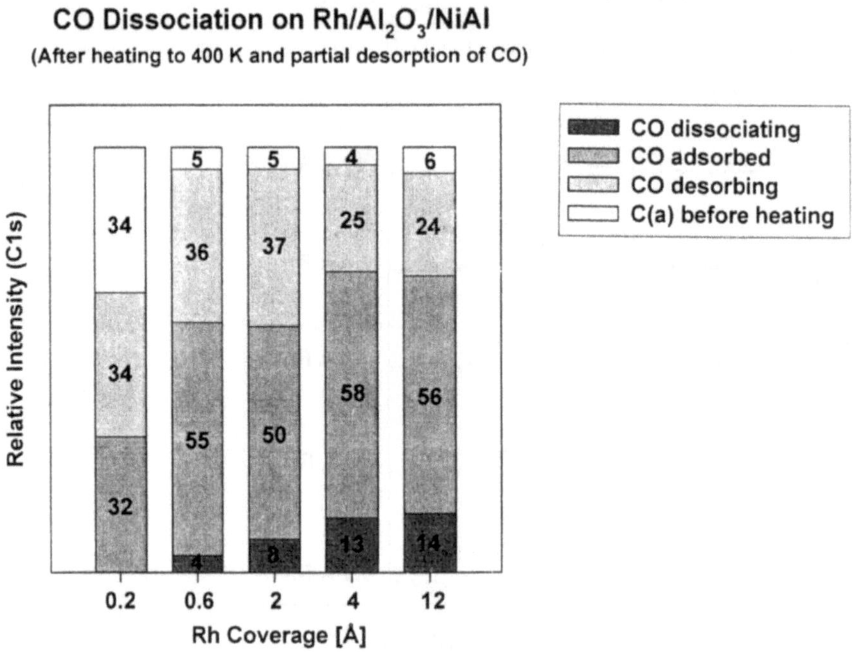

Fig. 3 Relative amounts of various sources of carbon on $Rh/Al_2O_3/NiAl(110)$ as a function of the amount of deposited Rh metal, after heating the system to 400 K.

alluded to above. These two results are interrelated and connected with a change in the interaction strength between the adsorbed molecule and the particle as well as the screening abilities of the metal particle. The latter is strongly influenced by its size as also documented by our studies on Pd particles [12,13]. The presented data are compatible with the idea that the smaller CO-covered particles behave more like a molecular species than the larger particles. The third observation is the increasing tendency for CO dissociation as the size of the particle deposited increases.

Studies of the particle size dependence of CO dissiciation reported in the literature have come to different conclusions [14,15]. None of these investigations, however, has reported XPS data that directly measure the amount of dissociated CO staying on the particle. Studies based purely on the analysis of temperature programmed desorption spectra might lead to the conclusion that dissociation is indeed favoured by increasing the particle size [14]. The conclusions are somewhat indirect as they are based on the CO_2 production rate, resulting from the reaction adsorbed CO with oxygen produced by CO dissociation. In the present study the measurement of the amount of dissociated CO is direct.

One possiblity to interpret the present result is to assume that in order for dissociation to occur the particle has to have a minimal size. After all dissociated

CO needs at least two well separated sites into which the released carbon and oxygen atoms can be accomodated. Taking into account that our smallest particles only contain 10 or less Rh atoms it is quite conceivable that on the smaller particles there is simply not enough space for dissociation to occur. The results for the larger particles found in the present study are compatible with results reported in the literature for stepped and open face Rh single crystal surfaces. This is a quite expected result.

4. Ackowledgements

This work was funded by: The Deutsche Forschungsgemeinschaft, Ministerium für Wissenschaft und Forschung des Landes Nordrhein-Westfalen, Fonds der Chemischen Industrie, the Swedish Natural Science Research Council (NFR) and the Swedish National Board for Industrial and Technical Development. A.S. wishes to acknowledge NFR for a fellowship and J.L. finally thanks the Studienstiftung des Deutschen Volkes for financial support.

5. References

[1] A.B. Stiles (Ed) „Catalyst Supports and Supported Catalysts" Butterworth, Boston (1987).

[2] B.C. Gates, C. Guli, H. Knözinger (Eds.) „Metal Clusters in Catalysis" in „Studies in Surface Science and Catalysis", Vol. 29 Elsevier Publ. Cong., Amsterdam (1986).

[3] G.A. Somorjai, J. Carazza, Ind. Eng. Chem. Fundam. 25, 63 (1986).

[4] a) H.-J. Freund, Ber. Bunsenges. Phys. Chem. 99, 1261 (1995)
b) H.-J. Freund, B. Dillmann, D. Ehrlich, M. Haßel, R.M. Jaeger, H. Kuhlenbeck, C.A. Ventrice Jr., F. Winkelmann, S. Wohlrab, C. Xu, Th. Bertrams, A. Brodde, H. Neddermeyer, J. Mol. Catal. 82, 143 (1993)

[5] F. Winkelmann, S. Wohlrab, J. Libuda, M. Bäumer, D. Cappus, M. Menges, K. Al-Shamery, H. Kuhlenbeck, and H.-J. Freund Surf. Sci. 307 - 309, 1148 (1994).

[6] J. Libuda, F. Winkelmann, M. Bäumer, H.-J. Freund, Th. Bertrams, H. Neddermeyer, and K. Müller, Surf. Sci. 318, 61 (1994)

[7] R.M. Jaeger, H. Kuhlenbeck, H.-J. Freund, M. Wuttig, W. Hoffmann, R. Franchy, and H. Ibach, Surf. Sci. 259, 253 (1991)

[8] J. Libuda, M. Bäumer, and H.-J. Freund, J. Vac. Sci. Technol. A 12, 2259 (1994)

[9] S. Wohlrab, F. Winkelmann, H. Kuhlenbeck,, and H.-J. Freund, to appear in: Springer Proceedings in Physics, Springer, Berlin 1995

[10] A. Sandell, J. Libuda, P. Brühwiler, S. Andersson, A. Maxwell, M. Bäumer, N. Mårtensson, and H.-J. Freund, to appear in: J. Electron Spectrosc. Relat. Phenom.

[11] A. Sandell, J. Libuda, M. Bäumer, and H.-J. Freund, submitted to Surf. Sci.

[12] A. Sandell, J. Libuda, P.A. Brühwiler, S. Andersson, A.J. Maxwell, M. Bäumer, N. Mårtensson, H.-J. Freund, submitted to J. Vac. Sci. Techol.

[13] M. Bäumer, J. Libuda, A. Sandell, H.-J. Freund, G. Grow, Th. Bertrams, H. Neddermeyer, Ber. Bunsenges. Phys. Chem. 99, 1381 (1995)

[14] V. Nehasil, I. Stavá, V. Matolin, Surf. Sci. 331, 105 (1995)

[15] V. Matolin, K. Masek, M.H. Elyakhloufi, E. Gillet, J. Catal. 143, 492 (1993)

[16] E.I. Altman, R.J. Gorte, Surf. Sci. 195, 392 (1988)

[17] D.N. Belton, S.J. Schmieg, Surf. Sci. 202, 238 (1988)

[18] J.T. Yates, E.D. Williams and W.H. Weinberg, Surf. Sci. 91, 562 (1980)

[19] D.G. Castner, B.A. Sextonand G.A. Somorjai, Surf. Sci. 71, 519 (1978)

[20] P.A. Thiel, E.D. Williams, J.T. Yates and W.H. Weinberg, Surf. Sci. 84, 54 (1979)

[21] Y. Kim, H.C. Peebles and J.M. White, Surf. Sci. 114, 363 (1982)

[22] R.A. Marbrow, R.M. Lambert, 67, 489 (1977)

[23] D.G. Castner, G.A. Somorjai, Surf. Sci. 83, 60 (1979)

[24] M. Rebholz, R. Prins, and N. Kruse, Surf. Sci. Lett. 259, L 797 (1991)

[25] M. Rebholz, R. Prins, and N. Kruse, Surf. Sci. Lett. 269/270, 293 (1992)

Effects of Alloying with Ge on the Pt Electronic Structure and Molecular Adsorption

K. Fukutani[1,2], T.T. Magkoev[3,*], Y. Murata[3], and K. Terakura[4]

[1]Institute of Industrial Science, University of Tokyo, Roppongi, Minato-ku, Tokyo 106, Japan
[2]PRESTO, Research Development Corporation of Japan (JRDC)
[3]Institute for Solid State Physics, University of Tokyo, Roppongi, Minato-ku, Tokyo 106, Japan
[4]JRCAT, National Institute for Advanced Interdisciplinary Research, Higashi, Tsukuba, Ibaraki 305, Japan
*Permanent address: University of North Ossetia, Kesaev 121-83, Vladikavkaz 362020, Russia

Abstract. Effects of alloying with a few percents of Ge on the Pt electronic structure and adsorption of NO molecules are investigated by photoemission spectroscopy, reflection-absorption infrared spectroscopy, thermal desorption spectroscopy, and photo-stimulated desorption. The NO molecule exclusively occupies the on-top site on the Pt(111)-Ge surface alloy in contrast to the bridge and the on-top adsorption on Pt(111). The adsorption energy of NO is dramatically reduced compared with that on Pt(111). The Pt *d*-band seems to be downward shifted in energy suggesting the *d*-band filling. The change in chemical properties of Pt(111) for molecular adsorption is discussed in the framework of the rigid-band model.

1. Introduction

Solid surfaces interacting with gas molecules are systems of interest and importance in fundamental science and technological applications. A basic concept included in these systems is, as discussed in the Newns–Anderson model [1], interaction of localized electrons in a molecule with delocalized electrons in a substrate surface. Microscopic understanding of the molecule-surface interaction provides a key to controlling various surface processes including catalytic reactions, surface modification and thin-film fabrication. For this purpose, not only the static properties such as atomic and electronic structures in the ground state, but also dynamical properties need to be clarified.

In catalytic reactions on metal surfaces, a widely accepted idea is that *d*-electrons of a transition metal with narrow band width play a crucial role rather than *sp*-electrons with wide band width. As far as the rigid-band model is fulfilled, two main factors characterizing the *d*-band are the band width and the band filling. Controlling these factors may allow us to control chemical properties of surfaces and thereby design new catalytic materials. One way to achieve this is alloying of metals, which is frequently used to improve the activity and selectivity in catalytic reactions [2,3]. Effects of alloying are twofold; local and non-local effects. The local effect is in principle the geometric effect that the modifying atom geometrically blocks molecular adsorption or accommodation of chemical products. The non-local effect is the electronic effect that alloying significantly modifies the global electronic structure of a host material altering the chemical reactivity of the surface [2–5].

Springer Series in Solid-State Sciences, Volume 121
Elementary Processes in Excitations and Reactions on Solid Surfaces
Eds.: A. Okiji, H. Kasai and K. Makoshi

Adsorption of NO molecules on a Pt(111) surface has been extensively studied by conventional surface science techniques [6–8]. According to these studies, the NO molecule chemisorbs either at bridge or on-top sites depending on the coverage. Photoinduced processes of NO on Pt(111) were investigated by a state-selective detection technique and reflection absorption infrared spectroscopy (RAIRS) [9–12]. It was found that NO at the on-top site is desorbed by photon irradiation, while NO at the bridge site suffers photodissociation on the surface. The photodesorption process induced by valence electron excitation involves the delocalized electron of the substrate and localized unoccupied states originating from the adsorbed molecule. Photo-processes of NO are also investigated on Ni(111) [13,14], being discussed in terms of the *d*-band width [12]. On the other hand, the effect of the *d*-band filling has been discussed on the Pt(111)-Ge surface alloy [15].

In this article, we present experimental results of NO adsorption on a Pt(111) surface alloyed with a small amount of Ge. We also report photoinduced processes of adsorbates on this surface alloy. Alloying of a transition metal with a simple element without *d* electrons in its valence region such as Ge was theoretically investigated by Terakura, and is concluded to fill the *d*-band of the transition metal due to *s*-*d* hybridization [16]. The motivation behind the present study is to examine the relationship between surface chemistry including static and dynamical properties of molecules on the surface and *d*-band filling. Both static and dynamic nature of NO on a metal surface is discussed in connection with the band structure of the substrate.

2. Experiment

The details of the experimental apparatus and procedure except photoemission spectroscopy (PES) measurements have been described elsewhere [10,15]. Briefly, the Pt(111)-Ge surface alloy was prepared in an ultrahigh vacuum chamber (base pressure: 1×10^{-10} Torr) by repeated cycles of deposition of a few monolayer (ML) Ge on Pt(111) and subsequent annealing at 1100 °C until a Ge LMM Auger electron signal corresponding to 0.1 ML was stably detected. Since this surface is stable even after annealing at 1100 °C and exposure to atmosphere, Ge atoms are considered to be distributed into several layers in the surface region. Therefore, we assume that Ge concentration in the top layer of Pt(111) is a few percents and call this surface Pt(111)-Ge surface alloy hereafter. LEED revealing a 1×1 pattern with higher background for this sample, the Ge atoms are considered to be randomly distributed in the surface layer. The sample was cooled to 80–90 K, at which temperature RAIRS and resonance-enhanced multiphoton ionization (REMPI) measurements of desorbed molecules were carried out.

The RAIRS measurements were performed using a conventional Fourier-transform infrared spectrometer with an HgCdTe detector. The absorption spectrum was typically averaged over 60 scans at a resolution of 2 cm^{-1}. The thermal desorption spectrum (TDS) was obtained by quadrupole mass spectrometer at a typical temperature rise rate of 6 K/s. For the photodesorption experiments, ultraviolet light at the wavelength of 193 nm generated from an ArF excimer laser was used to induce desorption of adsorbed molecules. A YAG pumped dye laser was frequency-doubled producing uv light (λ=223–227 nm) to state-selectively ionize NO desorbed from the surface via the $A^2\Sigma^+ \leftarrow X^2\Pi$ transition. From analysis of the REMPI spectrum, the internal-energy distribution was obtained, while the translational-energy distribution was obtained by measuring the time of flight (TOF) for the distance between the sample and the probe beam.

The PES experiments were performed at the synchrotron facility at Institute for Solid State Physics, the University of Tokyo. The linearly polarized light generated from the

SOR ring at photon energies of 40 and 70 eV irradiates the sample at an incidence angle of 45°. The emitted electrons were analyzed at a sample temperature of room temperature by a double-pass cylindrical mirror analyzer facing the sample 90° off from the light direction, where angle-integrated photoemission spectra are obtained.

3. Results

Fig. 1(a) shows RAIR spectra for NO on a Pt(111) surface at a sample temperature of 80 K. An absorption band at 1480 cm^{-1} is observed at a low coverage. As the coverage is increased, this band intensity is reduced, and another absorption peak appears around 1710 cm^{-1}. At saturation coverage, as shown in the curve (2) of Fig. 1(a), only the high-frequency band at 1717 cm^{-1} is clearly seen. In analogy with the structures of nitrosyl complexes, the low-frequency band is assigned to the stretching vibration of NO bound at the bridge site, and the high-frequency band is assigned to that at the on-top site [6,7]. Although the site assignment on the basis of vibrational studies is questioned by a LEED analysis [17], these two species are referred as bridge and on-top ones in this article. Fig. 1(b) shows RAIR spectra for NO on the Pt(111)-Ge surface alloy at the sample temperature of 90 K. At a low coverage after an NO exposure of 0.5 L, an absorption band is seen at 1678 cm^{-1}. This absorption band is increased in intensity with increasing coverage, the peak position being shifted to 1716 cm^{-1} at saturation as shown in the curve (2) of Fig. 1(b). From the spectra for the two surfaces at a low NO coverage, it is concluded that NO is adsorbed at the bridge site on Pt(111) while it is adsorbed at the on-top site on the Pt(111)-Ge surface alloy. The spectra for the two surfaces at saturation, on the other hand, are very similar except for the difference in intensity and peak width. In the case of saturation, LEED exhibits no extra spots except 1×1 on the Pt(111)-Ge surface alloy while a 2×2 pattern is observed on the clean surface.

A thermal desorption spectrum of NO from the Pt(111) surface saturated with NO is shown in Fig. 2(a). The spectrum reveals a predominant peak at 330 K, which is roughly consistent with the literature data [6,8]. Fig. 2(b) shows TDS of NO from the Pt(111)-Ge surface alloy at saturation coverage exhibiting a strong desorption peak at 220 K, which is significantly lower than that on the clean surface suggesting a decrease of the NO adsorption energy on the substrate.

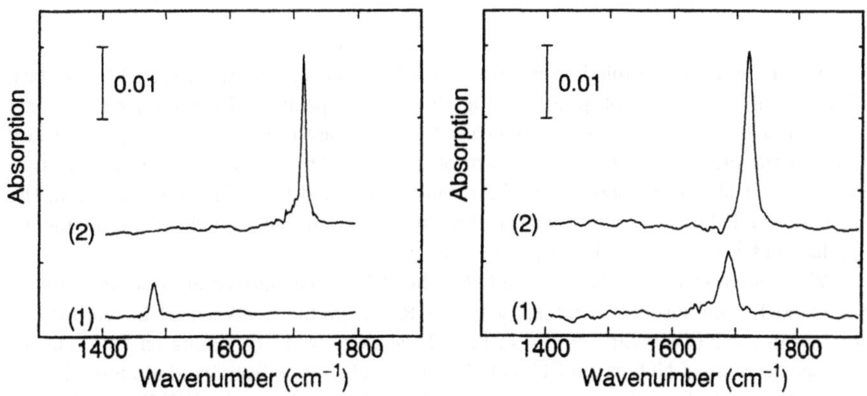

Fig. 1 RAIR spectra for NO adsorbed on (a) the Pt(111) surface at 80 K and (b) Pt(111)-Ge surface alloy at 90 K with low and high NO coverages. NO doses; (a) (1) 0.35, and (2) 2.28 L. (b)(1) 0.6, and (2) 8.0 L.

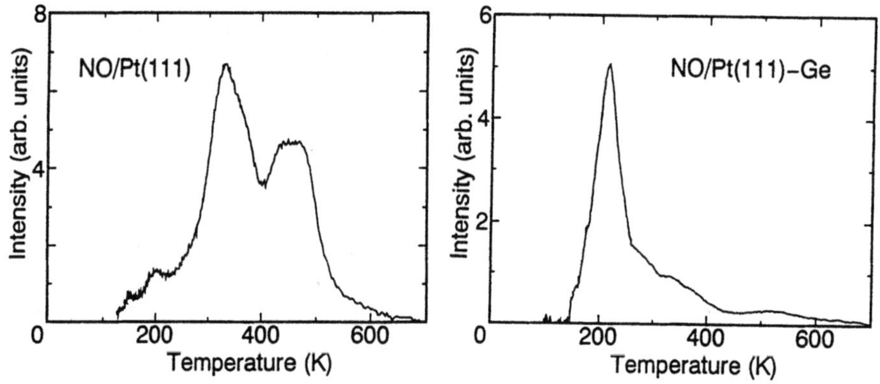

Fig. 2 TDS of NO from (a) the Pt(111) surface and (b) Pt(111)-Ge surface alloy at saturation coverage. Temperature rise rate is about 6 K/s.

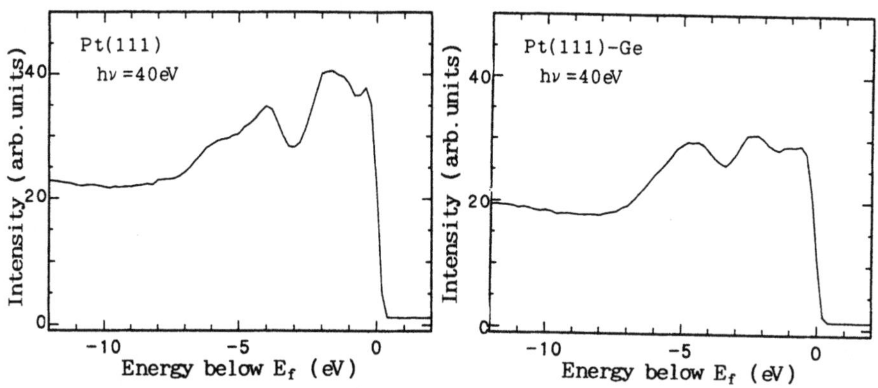

Fig. 3 Angle-integrated photoemission spectra taken at room temperature for (a) Pt(111) and (b) Pt(111)-Ge surface alloy with the photon energy of 40 eV.

Fig. 3 shows photoemission spectra taken for Pt(111) and the Pt(111)-Ge surface alloy at a photon energy of 40 eV. In Fig. 3(a), three peaks and a shoulder are clearly observed at 0.4, 1.6, 4.0, and 5.8 eV below the Fermi level. On the Pt(111)-Ge surface alloy, on the other hand, three peaks are observed at 0.6, 2.4, and 4.6 eV as shown in Fig. 3(b). It should be noted that the shoulder observed for Pt(111) is missing in the spectrum for Pt(111)-Ge, and that the three peaks are shifted to higher binding energy. Similar tendency is observed at a photon energy of 70 eV.

When the laser light of 6.4 eV irradiates the Pt(111)-Ge surface alloy saturated with NO at 90 K, the absorption intensity in RAIRS is gradually decreased indicating the decrease in the NO coverage on the surface. In analogy with the results on Pt(111), the coverage decrease of NO on the Pt(111)-Ge surface alloy upon laser irradiation originates from either photodesorption or photodissociation. By applying the REMPI method, it is found that the laser irradiation causes photodesorption of adsorbate. Figs. 4(a) and 4(b) show the TOF spectra of NO desorbed from the Pt(111) and the Pt(111)-Ge surface alloy, respectively, measured at a distance of 3 mm from the sample. The solid curves

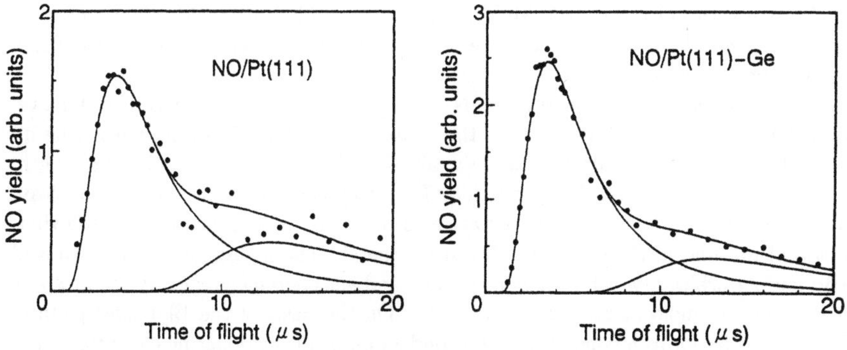

Fig. 4 Time of flight spectra of NO desorbed from (a) Pt(111) and (b) the Pt(111)-Ge surface alloy at 90 K with the laser wavelength of 193 nm. NO molecules in the vibrationally ground, low-rotational and spin-orbit of 1/2 states are detected at 3 mm from the sample .

are fits to a sum of two shifted-maxwellian distributions. The mean translational energy ($<E_t>$) is numerically calculated by the flux-weighted distribution. The mean translational temperatures ($T_t = <E_t>/2k_B$) obtained for the fast and slow components are 800 and 50 K for Pt(111) and 820 and 50 K for the Pt(111)-Ge surface alloy, respectively. It is noted that these values are identical to each other within our experimental accuracy.

4. Discussion

The first point we discuss in this section is the decrease of the binding energy due to alloying. The TDS results indicate that the binding energy of NO on the Pt(111)-Ge surface alloy is substantially lower than that on the clean surface. The RAIR spectra suggest that NO is exclusively adsorbed at the on-top site of Pt on both surfaces at saturation coverage. Since the concentration of Ge at the surface is as low as a few percents as mentioned above, most of the alloy surface exposed are Pt atoms. Therefore, the local adsorption structure is identical for both surfaces in that NO is bound to a single Pt atom. Nevertheless, the binding energy is noticeably affected by the Ge atoms randomly distributed in the surface layer, which definitely indicates that NO adsorption on a Pt surface reflects the non-local feature of the substrate. The significant decrease in the binding energy can be discussed in terms of the change of the band structure. According to a theoretical calculation, the pure Pt crystal treated relativistically has a *d*-electron number of 8.25 per atom meaning that 1.75 of *d*-hole in the *d*-band [18]. As mentioned above, the three peaks observed in the photoemission spectra for Pt(111) are shifted to lower in energy suggesting that the band as a whole is downward shifted. Since the dip position in a band structure is hardly changed by alloy formation, the two band structures shown in Fig. 3 can be compared to each other with respect to the second dip at 3–3.5 eV. Assuming that the dip position in the spectra is at the same energy, it is noticed that the Fermi level for the Pt(111)-Ge surface alloy is higher in energy than that for Pt(111). This result indicates that alloy formation with Ge is well interpreted in the framework of the rigid-band model filling the *d*-hole of the Pt band. This conclusion is in good agreement with the theoretical analysis for the *d*-band structure of a transition metal alloyed with simple *sp* elements, which clearly showed *d*-band filling due to *s*-*d* hybridization with s electrons [16]. Since the presence of d-hole is considered to make the Pt surface chemically active, the decrease of the d-hole density leads to reduction of the NO chemisorption energy.

The second point we note here is the N–O stretching vibrational frequency. A naive idea is that the N–O bond is weaker as the substrate–NO bond is stronger resulting in lower N–O vibrational frequency. Despite weakening of the substrate–NO bond strength for Pt(111)-Ge, however, the N–O stretching frequency is the same as that on the clean surface as shown in Fig. 1. It is informative to recall the orbital character of NO that is involved in bond formation with a transition metal. The molecular orbitals involved are 5σ and 2π orbitals, which have non-bonding and antibonding characters with respect to the N–O intramolecular bond, respectively. On the basis of the Blyholder picture, fully occupied 5σ forms a bond with metal conduction bands by donating electrons to the metal, while partially occupied 2π interacts with metal conduction band by backdonating electrons to NO [19]. Since the 5σ orbital is nonbonding, the degree of 5σ donation (the strength of the 5σ interaction) hardly affects the N–O bond strength. In contrast to this, the 2π backdonation should sensitively affect the N–O bond because of its antibonding character. The experimental results of RAIRS suggest that the d-band filling due to alloy formation reduces the 5σ donation and hardly changes the 2π backdonation. This conclusion is intuitively reasonable considering the d-hole and the d-electron density: Although decrease of the d-hole as small as 1.75 per Pt atom effectively reduces the 5σ donation, increase of the d-electron as much as 8.25 hardly affects the 2π backdonation.

This discussion is consistent with the photodesorption results reflecting the excited states related to the desorption dynamics. As mentioned above, the translational energy distribution characterizing the desorption dynamics is quite similar on both Pt(111) and Pt(111)-Ge surface alloy suggesting the same mechanism in spite of the difference of the binding energy. According to the previous studies, the mechanism of the initial electronic excitation driving NO desorption involves the unoccupied 2π state of NO [10]. Assuming that the same mechanism is operative for the photodesorption of NO on the Pt(111)-Ge surface alloy, the nature of the excited state relevant to the 2π state of NO is unchanged due to alloying with Ge, which is in agreement with the above discussion that the 2π backdonation is unchanged as a result of alloying.

The last point we emphasize in this article is alloying effect on the NO adsorption site, the loss of the bridge species. It should be noted that the Ge concentration in the surface layer of a few percents is not sufficient to geometrically block all the bridge site adsorption. Therefore, blocking of bridge site adsorption originates from the global change of the substrate electronic structure. In the following, a speculative discussion is presented in terms of the symmetry of the Pt d-band and its filling.

The loss of the bridge species indicates that the adsorption energy, the energy gain due to bond formation, at the bridge site is more decreased than that at the on-top site. As discussed above, the chemisorption bond between NO and Pt is mainly attributed to the 5σ donation. Hence, strength of the σ-bond between the molecule 5σ orbital and the substrate d-orbitals is a determining factor behind adsorption-site selection. The five d-orbitals in the O_h space group are generally classified into e_g ($d_{x^2-y^2}$ and $d_{3z^2-r^2}$) and t_{2g} (d_{xy}, d_{yz}, and d_{zx}) orbitals with respect to the cubic crystal axes of x, y, and z. Since Ge atoms are considered to occupy the substitutional sites in the surface alloy, only the t_{2g} state is upward-shifted while the e_g state remains at the original position due to hybridization between the Ge s-orbital and the Pt d-orbital. To maintain charge neutrality, the Fermi level is also upward-shifted resulting in preferential filling of the e_g state. On a (111) surface, new axes of x', y', and z' are taken as $[\bar{1}\bar{1}2]$, $[1\bar{1}0]$, and $[111]$, respectively. The d-orbitals of $d_{3z'^2-r^2}$ and $d_{y'z'}$ are involved in σ–bond formation with on-top and bridge NO, respectively, because the σ–bond for the on-top NO is in the z'

direction, while that for the bridge species is in the $y' - z'$ plane. The d orbitals of $d_{3z'^2-r^2}$ and $d_{y'z'}$ are expressed as [20]

$$d_{3z'^2-r^2} = (d_{xy} + d_{yz} + d_{zx})/\sqrt{3}$$
$$d_{y'z'} = (2d_{x^2-y^2} - d_{yz} + d_{zx})/\sqrt{6}.$$

The e_g states being more filled than the t_{2g} states by alloying with Ge, $d_{y'z'}$ that forms a σ-bond with bridge bound NO is more filled than $d_{3z'^2-r^2}$ that makes a σ-bond with on-top NO. Therefore, the d-band filling by Ge acts to reduce the 5σ donation to the $d_{y'z'}$ orbital leading to suppression of bridge-site adsorption. Nevertheless, full understanding of the d-band filling and its effect on the NO adsorption site requires detailed investigation of the band structure and symmetry, and also theoretical analysis of the electronic structure.

5. Conclusion

Alloying effects with Ge on the Pt electronic structure and NO adsorption are discussed with experimental investigation by PES, RAIRS, TDS, and REMPI. The adsorption energy of NO is substantially decreased compared with that on the Pt(111) surface. Of the two adsorption sites of on-top and bridge observed on Pt(111), the NO molecule exclusively occupies the on-top site on the Pt(111)-Ge surface alloy. Similarity of the translational temperature of photodesorbed NO between Pt(111) and the Pt(111)-Ge surface alloy indicates that the same mechanism is operative for the NO photodesorption process on the two surfaces. Alloy formation with a few percents of Ge changes the global electronic structure of Pt reducing the d-hole density, which causes reduction of the adsorption energy and suppression of bridge site adsorption.

References

1. D.M. Newns, Phys. Rev. **178**, 1123 (1969).
2. F.J.C.M. Toolenaar, F. Stoop, and V. Ponec, J. Catal. **82**, 1 (1983).
3. V. Ponec, Advan. Catal. **32**, 149 (1983).
4. M.T. Paffett, S.C. Gebhard, R.G. Windham, and B.E. Koel, J. Phys. Chem. **94**, 6831 (1990).
5. U. Schneider, G.R. Castro, and K. Wandelt, Surf. Sci. **287/288**, 146 (1993).
6. J. L. Gland and B. A. Sexton, Surf. Sci. **94**, 355 (1980).
7. B.E. Hayden, Surf. Sci. **131**, 419 (1983).
8. R.J. Gorte, L.D. Schmidt, and J.L. Gland Surf. Sci. **109**, 367 (1981).
9. S.A. Buntin, L.J. Richter, D.S. King, and R.R. Cavanagh, J. Chem. Phys. **91**, 6429 (1989).
10. K. Fukutani, Y. Murata, R. Schwarzwald, and T.J. Chuang, Surf. Sci. **311**, 247 (1994).

11. M.-B. Song, M. Suguri, K. Fukutani, F. Komori, and Y. Murata, Appl. Surf. Sci. **79/80**, 25 (1994).

12. K. Fukutani, M.-B. Song, and Y. Murata, Faraday Discuss. Chem. Soc. **96**, 105 (1993).

13. F. Budde, A. V. Hamza, P. M. Ferm, G. Ertl, D. Weide, P. Andresen, and H.-J. Freund, Phys. Rev. Lett. **60**, 1518 (1988).

14. T.T. Magkoev, M.-B. Song, K. Fukutani, and Y. Murata, Surf. Sci. Lett. **330**, L669 (1995).

15. K. Fukutani, T.T. Magkoev, Y. Murata, and K. Terakura, Surf. Sci. in press.

16. K. Terakura, J. Phys. F **7**, 1773 (1977).

17. N. Materer, A. Barbieri, D. Gardin, U. Starke, J.D. Batteas, M.A. Van Hove, and G.A. Somorjai, Phys. Rev. B **48**, 2859 (1993).

18. S. Takizawa, K. Terakura, and T. Mohri, Phys. Rev. B **39**, 5792 (1989).

19. G. Blyholder, J. Phys. Chem. **68**, 2772 (1964).

20. K. Terakura, T. Oguchi, A.R. Williams, and J. Kübler, Phys. Rev. B **30**, 4734 (1984).

Novel Structure Model for the Type C Defect on Si(001) Surface

T. Uda

Joint Research Center for Atom Technology, 1-1-4 Higashi,
Tsukuba, Ibaraki 305, Japan

Abstract. A novel structure model for the type C defect on Si(001) is proposed. The model is based on the assumption that the type C defect is originated from a removal of only one atom in the second layer and not a removal of two atoms on the surface. Optimized structure obtained from the first principles calculation reproduces STM images both at negative and positive sample biases.

1. Introduction

The Si(001) surface has been most extensively investigated since Schlier and Farnthworth first reported evidence for a (2x1) reconstruction [1]. The tight-binding calculation showed that the surface energy can be decreased by relaxation of pairs of surface atoms into buckled dimers [2]. The recent first principles total energy minimization calculations have revealed that the c(4x2) is the lowest energy structure among other possible reconstruction patterns derived from the buckled dimers as building blocks [3-5]. This result was confirmed recently by low temperature STM measurement by Wolkow [6]. The long-standing puzzle that the room temperature STM images suggest apparent symmetric dimer in contradiction to the theoretical result is understood by the thermal flip-flop motion of buckled dimers. Moreover, the phase transition from the c(4x2) surface to the disordered (2x1) surface has been investigated on a model spin system by using renormalization technique [7], and more recently, by using Monte Carlo method with interaction parameters determined from the first principles calculations [8]. The result reproduces fairly well the variation of observed LEED patterns upon temperature [9]. Thus it may safely be said that almost all problems on ideal Si(001) surface have been solved.

This surface, however, contains various defects; point defects, steps, kinks and their complexes. Three typical point defects are the type A, B and C defects after Hammers and Kohler [10]. For the type A defect, the missing dimer model proposed by Pandy [11] has been verified both from theoretical and experimental studies. The double missing dimer is now widely accepted as a credible model for the type B defect [12,13]. On the other hand, structure of the type C defect is yet unknown. The recent STM study revealed that the type C defect is most sensitive to oxidation among other defects [14-16]. This makes it also of practical importance to clarify the structural and electronic properties of the type C defect. The STM image of the type C defect appears as a dimer defect in which two adjacent Si atoms along the dimer rows are missing as shown in Fig.1(a) [10,17]. The calculated

Springer Series in Solid-State Sciences, Volume 121
Elementary Processes in Excitations and Reactions on Solid Surfaces
Eds.: A. Okiji, H. Kasai and K. Makoshi

optimized structure starting from this configuration, however, is far from the expected structure.

In this article, we propose a novel structure, in which only one Si atom on the second layer is missing as shown in Fig.1(b). The optimized structure obtained from the first principles calculation can reproduce characteristic feature of the STM image of the type C defect and can be a possible structure model for that defect.

2. Method

The calculations were carried out using the plane-wave pseudopotential technique within local density approximation (LDA) for the exchange-correlation energy. The wave functions were expanded in a plain wave basis set with an energy cut-off of 9 Ry and 2 k-points were sampled in the irreducible part of the Brillouin zone. The slab model was used and the inversion symmetry was imposed. The size of the unit cell was 4x7 with 14 layers of Si atoms and a vacuum region equivalent to 8 layers.

3. Results and discussion

The removal of a Si atom on the second layer creates four dangling bonds (DBs) as shown in Fig.1(b). Two DBs of the surface atoms 2 and 4 are eliminated by lattice relaxation as shown in Fig. 2. The remaining two DBs of the atoms 5 and 6 in the third layer are also eliminated by dimer formation mainly due to the shift of atom 6 toward the atom 5. By achieving this, however, the bond between the atoms 6 and 7 are broken leading to the creation of two new DBs on them. Since the number of DBs does not change and the strain energy increases evidently, the energy gain due to such bond switching is a rather subtle problem. We will discuss the stabilization mechanism of this rebonding later.

As is clear from Fig. 2, the type C defect bears a close resemblance to a phase break in a buckled dimer row. The heights of upper atoms 1 and 3, and the heights of

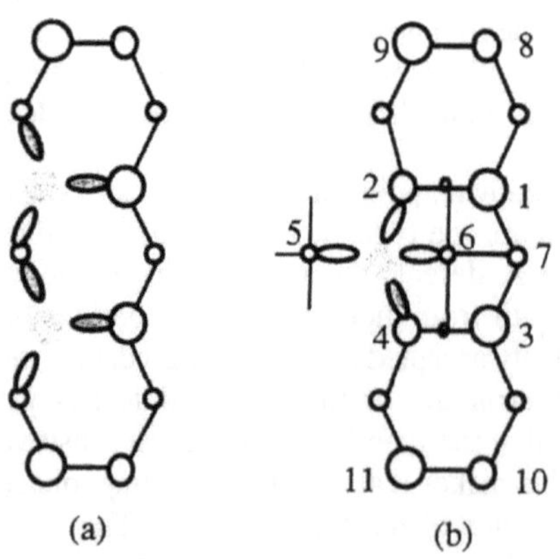

Fig. 1. Structure models for the type C defect (a) from reference [10], and (b) present model. DBs are emphasized by hatching.

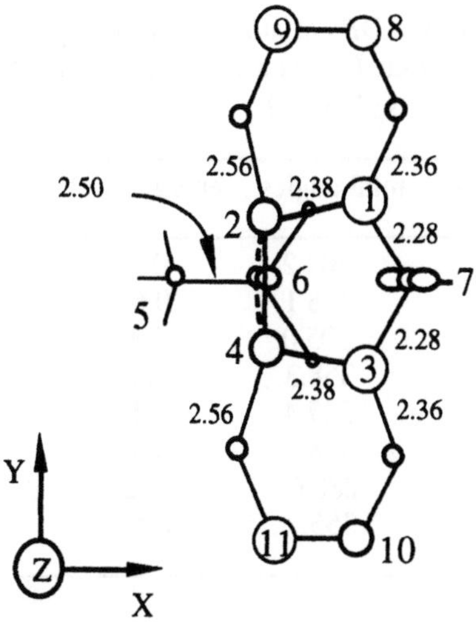

Fig. 2. Optimized structure from Fig.1 (b). Atoms are labelled by large figures and small figures denote bond distances in Å units. Larger circles represent atoms closer to the surface.

the lower atoms 2 and 4 in the type C defect are found to be almost equal to those of up dimer atoms and those of down dimer atoms, respectively. Such phase break may occur on the ideal surface due to flip-flop motion of dimers at finite temperature [8]. The main difference between the two structures is that the phase break moves rather freely on the ideal surface with an activation energy of about 0.1 eV, while the C defect is 'frozen' near the point where the second layer atom is missing. Thus, even at finite temperature the protruded atoms 1, 3, 9, and 11 in Fig. 2 are expected to be observed distinctly by STM measurement, roughly reproducing the observed image of the type C defect. But this is not the whole story. The STM image intensities of atoms 1 and 3 are weaker at negative sample bias and stronger at positive sample bias than those of up dimer atoms [10,16]. Since the heights of both groups of atoms are almost equal each other, this must be explained by an electronic effect. To this end, we have estimated the energies and characters of the DBs on atoms related with the type C defect. The tight-binding calculation [2] showed that the energy and character of a surface DB is strongly related with other three folded σ–bond angles. For example, since the bond angles of a down dimer atom are roughly increased to 120°, the DB is made almost of pure p orbital, and the energy level is shifted upward from the undistorted DB (sp^3 hybridized orbital). On the other hand, the bond angles of an up dimer atom are nearly 90°, the mixing ratio of s-character is increased to more than 30% compared with 25.0 % for the sp^3 hybridized orbital, and the energy is decreased. As is well known this is the reason why occupied surface states are mainly composed of up dimer atom DBs and the unoccupied surface states are made of down dimer atom DBs. We found by using the similar tight-binding estimation that these observations can be extended to general cases, where three bond angles are irregular as is the atoms of the type C defect. More specifically, the remaining DB energy and character are determined mainly by the sum of three bond

TABLE I. Bond angles (degree) and energies of the DBs of atoms near the type C defect. Those values of up and down atoms of a buckled dimer are also listed for the reference. The energies are measured from that of the sp^3 orbital.

atoms	angles			total angles	energy (ev)
1	93.8	108.1	105.3	307.2	-0.3
2	104.2	100.4	153.5	358.1	0.6
3	93.5	106.6	104.9	305.0	-0.3
4	104.2	100.4	153.7	358.3	0.6
6	94.6	111.5	111.3	317.4	-0.18
7	107.0	126.1	125.8	358.9	0.6
up	91.4	90.3	98.9	280.6	-0.65
down	116.8	118.2	120.7	355.7	0.55

angles and depend least on the individual bond angles. The bond angles and energies measured from that of the sp^3 hybridized orbital of three folded atoms 1, 2, 3, 4, 6 and 7 together with those of up and down atoms in a buckled dimers are listed in Table I. While the total bond angle of the lower atom 2 (4) of the C defect is 358° and is almost equal to that of a down dimer atom, the total angle of the upper atom 1 (3) is 307° (305°) which is considerably larger than the value of 281° for an up dimer atom. The energy difference of 0.9 eV between the DBs on atoms 1 (3) and 2 (4) is smaller than that of 1.2 eV between up and down dimer atoms. Therefore the mixing between the DBs on atoms 1 (3) and 2 (4) becomes greater than that between atoms in a buckled dimer. This effect is clearly demonstrated in Fig. 3, where the local densities of states (LDOS) using isotropic Gaussian functions located at 3.5 Å above the atom 1 and the up dimer atom as projection functions, are plotted by the solid and the dashed lines, respectively. In contrast to the familiar result for the up dimer atom LDOS defined as above that the component in the unoccupied region is very small, the LDOS of the atom 1 has considerable amount of component in the unoccupied region at the cost of the decrease of component in the occupied region. This explains why the STM intensity of the C defect is greater than that of up dimer atoms at positive sample bias and is smaller than that at negative sample bias.

The increase of total bond angle of the atom 1 (3) is brought about from the shift of the atom 7 toward positive x- and z- directions. As a result, three bonds of the atom 7 lie almost on a plane (see Table I), and the DB on that atom consists almost of pure p state. On the other hand, the DB on atom 6 contains considerable s-character, as is clear from Table I. Thus the electron transfer occurs from the atom 7 to the atom 6, leading to positively charged atom 7 and negatively charged atom 6. Although the negative charge on the atom 1 (3) and the positive charge on the atom 2 (4) are reduced compared with those of up and down dimer atoms, the energy gain due to coulomb interaction could be one of the driving forces that brings about the bond formation between atoms 5 and 6 and the bond breaking between atoms 6 and 7. Another possible mechanism that favors bond switching among atoms 5, 6 and 7 is that the distances between the atoms 6 and 7 and that between atoms 6 and 4 become extraordinarily small values of about 2.5 Å, leading to the extra energy gain due to formation of metallic-like bonding rather than an isolated DB. We found,

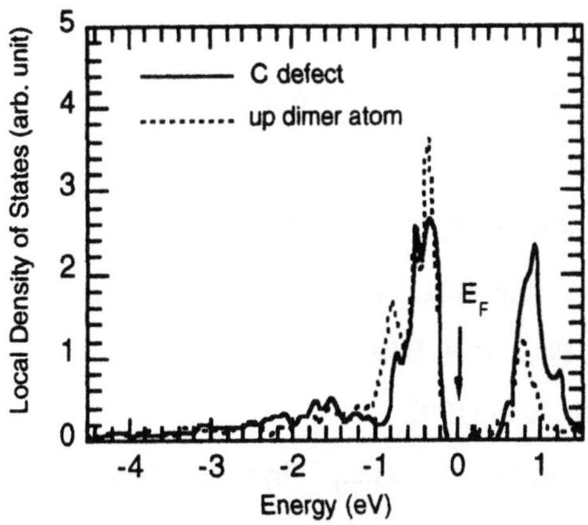

Fig. 3. Local density of states of atom 1 of the C defect (solid line) and of the up dimer atom (dashed line0 as a reference. Projection functions are isotropic Gaussians located 3.5 Å above each atom.

however, the energy gain from this mechanism is rather small at absolute zero temperature. The significance of this specific configuration for a silicon system may manifest itself at finite temperature. As can be seen in Fig. 3, the calculated C defect is not metallic, but the gap opens of about 0.3 eV in contradiction to the experimental results [10]. At finite temperature the thermal vibration of the atom 2 parallel to the dimer row gives rise to the asymmetry in the distances between the atoms 2 and 6, and between the atoms 4 and 6, changing bonding character from metallic-like bonding to highly distorted covalent bonding. This may modulate the energy structure, especially near the Fermi level, smearing out the mini-gap in a time average. In this respect, we strongly desire STS measurement at very low temperature. We are also expecting the detection of lower atoms 2 and 4 of the C defect with high resolution STM measurement.

4. Summary

We have proposed a novel structure model for the type C defect, which is derived from a removal of only one atom in the second layer. The model reproduces characteristic features of STM images both at positive and negative biases. Yet the result of STS at finite temperature remains to be investigated by detailed calculations. Very recently Miyazaki, Uda and Terakura have proposed another structure model for the type C defect with inclusion of H atom [18]. The adsorption of foreign atoms is another problem to be considered.

Acknowledgements

We thank T. Yamasaki and T. Miyazaki for helpful discussion. The numerical calculations were performed on a VPP/500 vector-parallel computer at JRCAT in Tsukuba. This work was partly supported by New Energy and Industrial Technology Development Organization (NEDO).

References

[1] R. E. Schlier and H. E. Farnsworth, J. Chem. Phys. **30**, 917 (1959).

[2] D. J. Chadi, Phys. Rev. Lett. **43**, 43 (1979).

[3] M. T. Yin and M. L. Cohen, Phys. Rev. B**24**, 2303 (1981).

[4] J. Ihm and J. D. Joannopoulos, Phys. Rev. B**24**, 4191 (1981).

[5] Z. Zhu, N. Shima and M. Tsukada, Phys. Rev. B**40**, 11868 (1989).

[6] R. A. Wolkow, Phys. Rev.Lett., **68**, 2636 (1992).

[7] J. Ihm, D. H. Lee, J. D. Joannopoulos, and J. J. Xiong, Phys. Rev. Lett. **51**, 1872 (1983).

[8] K. Inoue, Y. Morikawa, K. Terakura and M. Nakayama, Phys. Rev. B**49**, 14774 (1994).

[9] M. Kubota and M. Murata, Phys. Rev. B**49**, 4810 (1994).

[10] R. J. Hamers and U. K. Kohler, J. Vac. Sci. Technol. A**7** 2854 (1989).

[11] K. C. Pandy, Proc. 7 th Int.conf. on the physics of semiconductors, ed. D. J. Chadi and W. A. Harrison (Springer Verlag, New York,1985).

[12] N. Roberts and R. J. Needs, Surf. Sci. **236**, 112 (1990).

[13] J. Wang, T. A. Arias, and J. D. Joannopoulos, Phys. Rev. B**49**, 10497 (1993).

[14] Ph. Avouris and D. Cahill, Ultramicroscopy **42-44**, 838 (1992).

[15] Yolotsuka and I. Sumita, Ultramicroscopy **42-44** 946 (1992).

[16] M. Udagawa, Y. Umetani, H. Tanaka, M. Itoh, T. Uchiyama, Y. Watanabe, T.

[17] H. Tochihara, T. Amakusa and M. Iwatsuki, Phys. rev. B**50** 12262 (1994).

[18] T. Miyazaki, T. Uda, and K. Terakura, EMRS, Spring Meeting 1995.

Adsorbed PF_3 on Ru(001) – Quantum Gear Problem?

K. Kakitani, H. Kaji, and A. Yoshimori

Department of Electronic Engineering, Okayama University of Science, 1-1 Ridaicho, Okayama 700, Japan

Abstract. According to the recent ESDIAD experiment PF_3 molecules adsorbed on Ru(001) show interesting coverage- and temperature-dependent rotational hindrance. A model is constructed that the molecules sit on the atop site, having short ranged interactions. The analysis is made by using Monte Carlo simulation. On the obtained distributions, rotational motions of isolated adsorbates and adsorbate clusters are discussed quantum-mechanically. A possible quantum gear like situation is found for the adsorbate clusters.

1. Introduction

A PF_3 molecule is adsorbed on Ru(001) in such a way that the P atom forms a bond with a single surface Ru atom with the 3 F atoms upside [1-3]. A very interesting experiment by ESDIAD and LEED [4] indicates that PF_3 molecules adsorbed on Ru(001) could form quantum gears. The experiment has been made with varying coverage, θ, from 0.003 to 0.33 (θ is defined as number of adsorbates per surface atom) and varying temperature, T, from 20 K to 270 K. Important conclusions drawn from their experiment are as follows; (1) the ordered structure only observed is a $(\sqrt{3}\times\sqrt{3})R30^\circ$ one at the saturation coverage θ=0.33 in the measured temperature range, (2) the nearest neighbor adsorbate pair never forms due to geometrical inhibition, (3) the distribution of adsorbates is random except (2), and (4) the azimuthal orientation of PF_3 at low coverages is different by 30° from that at the saturation coverage. Using classical arguments, we have tried to analyze the coverage- and temperature-dependence of the ESDIAD intensity distribution [5-6]. Some of assumption made in the analyses have to be examined quantum mechanically or a totally quantum mechanical treatment may be necessary to explain the ESDIAD results, depending on values of parameters [7].

The other ESDIAD experiment on PF_3 on Ni(111) has been made also, but of one sample at lower coverage and one at saturation coverage [8] with a quantum mechanical analysis on the rotational state of isolated adsorbates. Earlier, a theoretical analysis of classical gears on a square lattice was made [9].

In this paper, referring to the PF_3 on Ru(001) system, we make general analyses of so to speak quantum gears, PF_3 adsorbates. In the next section, a model for an

Springer Series in Solid-State Sciences, Volume 121
Elementary Processes in Excitations and Reactions on Solid Surfaces
Eds.: A. Okiji, H. Kasai and K. Makoshi

isolated adsorbate is introduced. The rotational states are discussed of a pair of PF_3 adsorbates in section 3. The rotational states of larger clusters are discussed in section 4. Discussions in section 5 conclude the paper.

2. Isolated adsorbate

We assume that the adsorbate distribution is random on the triangular lattice of Ru(001) surface except the prohibited nearest neighbor pair and a repulsive interaction between the next nearest neighbor pairs. We use the Monte Carlo simulation to find out probability of an isolated adsorbate and adsorbate clusters if necessary. The geometry of an isolated adsorbate and a next nearest neighbor pair of adsorbates are shown in Fig.1. As you can see in Fig.1, the next nearest neighbor pair of adsorbates may be in a strong coupling situation as far as rotational motions of the adsorbates are concerned. The projected distance of the PF bond on the (001) surface is that of the free molecule, but it may not be so different in the adsorbed state.

An isolated adsorbate may see a crystalline potential for the rotational motion due to the surface atoms and the potential seems to have the six fold symmetry in the PF_3 on Ru(001) system. Then the Hamiltonian for the isolated PF_3 is given by

$$H = -\frac{\hbar^2}{2I}\frac{d^2}{d\varphi^2} + V_0 \cos 6\varphi, \tag{1}$$

where I is the moment of inertia of the PF_3 molecule, φ is the angle of PF bond from the vector $\boldsymbol{a}$ as in Fig.1 and V_0 defines magnitude of the six fold potential. The Schrödinger equation is written as

$$H\Psi(\varphi) = E\Psi(\varphi), \tag{2}$$

where $\Psi(\varphi)$ is the eigenfunction and E the eigenvalue.

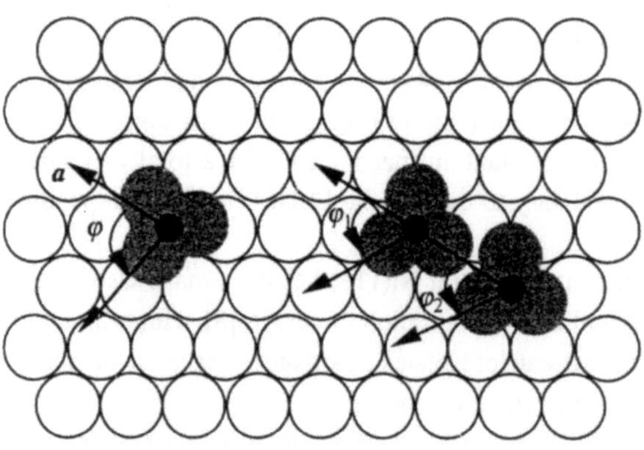

Fig.1 : Isolated PF_3 and the next nearest neighbor pair of PF_3. Rotational angles are defined as indicated.

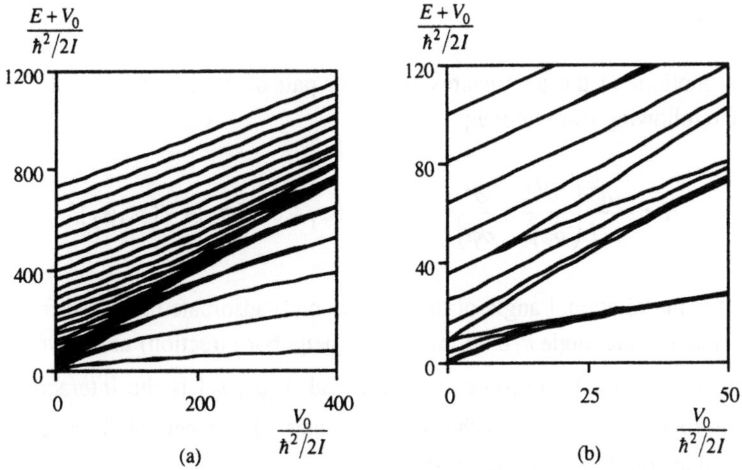

Fig. 2: (a) Energy levels of rotational motion of isolated PF_3 as a function of $V_0(>0)$. (b) The enlarged portion near the origin of (a).

The solutions of (2) are already discussed with the boundary condition, $\Psi(\varphi)=\Psi(\varphi+2\pi)$ [8]. Energy eigenvalues are shown as functions of $V_0/(\hbar^2/2I)$ in Fig.2 by numerical calculation. In the limit of $V_0=0$, (2) can be solved easily to give,

$$E=\left(\hbar^2/2I\right)n^2, \tag{3}$$

$$\Psi(\varphi)=(2\pi)^{-1/2}\exp(in\varphi), \tag{4}$$

where n is integer and $0<\varphi<2\pi$. Naturally, these states are two fold degenerate except the ground state. In the limit of large V_0, the solution becomes to be vibrational around the potential minima. Since there are six minima, the solutions are six fold degenerate;

$$E=\hbar\omega\left(n+\frac{1}{2}\right), \tag{5}$$

$$\Psi(\varphi)=\exp\left[-\frac{I\omega}{2\hbar}\left(\varphi-\frac{(2j+1)\pi}{6}\right)^2\right], \tag{6}$$

where $\omega=6\sqrt{-V_0/I}$, $n=0,1,2,\cdots$ and $j=0,\cdots,5$. When tunneling becomes appreciable with decreasing V_0, the six fold degeneracy will be lifted to a set of four energy levels with separation ratio; 1:2:1, both middle two levels of which are two fold degenerate like the π states of a benzene molecule.

3. Next neighbor pair

The rotational motions of the next nearest neighbor pair of the adsorbates may be subjected to the following Hamiltonian;

$$H=-\frac{\hbar^2}{2I}\left(\frac{\partial^2}{\partial\varphi_1^2}+\frac{\partial^2}{\partial\varphi_2^2}\right)-V_0\left(\cos 6\varphi_1+\cos 6\varphi_2\right)+V(\varphi_1,\varphi_2), \quad (7)$$

where φ_1 (φ_2) is the rotational angle of the first (second) adsorbate measured from the pair axis (making the angle $\pi/6$ to the nearest neighbor direction) as shown in Fig.1 so that the sign of V_0 becomes negative and $V(\varphi_1,\varphi_2)$ is the interaction potential of the rotational motions of the two adsorbates. The potential, $V(\varphi_1,\varphi_2)$, can be expanded in double Fourier series as

$$V(\varphi_1,\varphi_2)=\sum_{m_1,m_2}A_{m_1m_2}\exp\left[-i\left(m_1\varphi_1+m_2\varphi_2\right)\right]. \quad (8)$$

The symmetries of the pair,

$$V(\varphi_1,\varphi_2)=V\left(\varphi_1+(2\pi/3),\varphi_2\right)=V\left(\varphi_1,\varphi_2+(2\pi/3)\right), \quad (9)$$

$$V(\varphi_1,\varphi_2)=V(\varphi_2+\pi,\varphi_1+\pi)=V^*(\varphi_1,\varphi_2), \quad (10)$$

$$V(\varphi_1,\varphi_2)=V(-\varphi_1,-\varphi_2), \quad (11)$$

leave only the following terms in (8),

$$V(\varphi_1,\varphi_2)=2A_{30}\left(\cos 3\varphi_1-\cos 3\varphi_2\right) \\ +2A_{33}\cos 3(\varphi_1+\varphi_2)+2A_{3-3}\cos 3(\varphi_1-\varphi_2)+\cdots. \quad (12)$$

The condition (11) means the mirror symmetry of $V(\varphi_1,\varphi_2)$ with respect to the pair axis. A_{30}, A_{33} and A_{3-3} are real constants.

Since we do not know details of the potential $V(\varphi_1,\varphi_2)$, we take only those three terms of lower order shown in (12) in the following analysis. We make variable transformation,

$$\varphi_1+\varphi_2=2\psi, \quad \varphi_1-\varphi_2=2\varphi, \quad (13)$$

then Hamiltonian (7) is written as

$$H=H_\psi+H_\varphi+H_{\psi,\varphi} \quad (14)$$

$$H_\psi = -\frac{\hbar^2}{2(2I)}\frac{\partial^2}{\partial\psi^2} + 2A_{33}\cos 6\psi, \tag{15}$$

$$H_\varphi = -\frac{\hbar^2}{2(2I)}\frac{\partial^2}{\partial\varphi^2} + 2A_{3-3}\cos 6\varphi, \tag{16}$$

$$\begin{aligned} H_{\psi,\varphi} = &-V_0\left(\cos 6(\psi+\varphi) + \cos 6(\psi-\varphi)\right) \\ &+2A_{30}\left(\cos 3(\psi+\varphi) - \cos 3(\psi-\varphi)\right). \end{aligned} \tag{17}$$

Now we assume that $|A_{33}| >> |A_{3-3}|, |A_{30}|, |V_0|, \hbar^2/2I$. This may be called the gear condition. Under this condition, H_ψ is a dominant term, namely the unperturbed Hamiltonian and the others can be regarded as perturbation.

The ground state solution of the Schrödinger equation

$$H_\psi \mathrm{X} = E\mathrm{X}, \tag{18}$$

for the unperturbed part is given by

$$\begin{aligned} &\mathrm{X}_j(\psi) = \left(\frac{\pi\hbar}{I\omega}\right)^{1/4} \exp\left[-\frac{I\omega}{2\hbar}\left(\psi - \frac{j\pi}{3}\right)^2\right] \\ &\omega = 6\sqrt{-A_{33}/I}, \\ &E = \hbar\omega/2, \end{aligned} \tag{19}$$

where $j = 0, 1, \cdots, 5$. That is, the motion of the freedom, ψ, is vibrational with the potential expanded up to $(\psi - (j\pi/3))^2$ at $\psi = (j\pi/3)$. The solution is six fold degenerate corresponding to j values and if $|A_{33}|$ becomes smaller they split into the four levels with the eigenstates as is shown in the isolated case,

$$\Psi_k = \sum_j \exp\left(i\frac{kj\pi}{3}\right)\mathrm{X}_j. \tag{20}$$

where $k = 0, \pm 1, \pm 2, \pm 3 \ldots$ (the states with the double signs are degenerate). This would correspond to a gear tunneling or a quantum gear slipping.

Then remaining part of Hamiltonian turns out to be in the first order perturbation

$$\begin{aligned} H_\varphi^{(\mathrm{eff})} &= H_\varphi + \left\langle H_{\psi,\varphi}\right\rangle_\psi \\ &= -\frac{\hbar^2}{2(2I)}\frac{\partial^2}{\partial\varphi^2} + \left(2A_{3-3} + 2V_0\left\langle\cos 6\psi\right\rangle_\psi\right)\cos 6\varphi, \end{aligned} \tag{21}$$

where $\langle\cdots\rangle_\psi$ means the expectation value with Ψ_0 or canonical density matrix of Ψ_k. The effective Hamiltonian (21) has a similar form to the one for an isolated adsorbate so that, if the coefficient, $\left(2A_{3-3}+2V_0\langle\cos 6\psi\rangle_\psi\right)$, is small, that is, $\left|\left(2A_{3-3}+2V_0\langle\cos 6\psi\rangle_\psi\right)\right| << \hbar^2/2I$, the motion will be rotational. The factor, $\langle\cos 6\psi\rangle_\psi$, will be the unity for the strong gear limit and, in the moderate gear condition, will become smaller due to quantum gear slipping. The variable, φ, describes a gear motion of the next nearest neighbor pair of the adsorbates, which is a rotation of the two antiferromagnetically coupled rotors.

4. Adsorbate cluster

When we define adsorbate clusters, which consist of adsorbates connected by the next nearest neighbor relation, the analysis of the motion of the next nearest neighbor pair indicates the following. There must be irrotational clusters where all the rotational freedom of adsorbates turn out to be vibrational and rotational clusters where one coupled rotational freedom can be rotational while all the others are vibrational. Those two categories of clusters are shown in Fig.3. As one can see in Fig.3, the minimum unit of irrotational cluster is the triplet connected by the next nearest neighbor pairs. In general, since the gear coupling is essentially antiferromagnetic, every loop cluster of adsorbates of odd number cannot rotate. Then, any clusters containing such a loop of adsorbates of odd number cannot rotate and we call them irrotational clusters.

On the other hand, every loop cluster of adsorbates of even number can have one rotational degree of freedom as is seen easily in Fig.3 every cluster having no loop can rotate in the same sense. That is, all clusters having no odd number loop can rotate. This property is very general one of gears on the two dimensional lattice which cannot decompose into such two sublattices that the nearest neighbors of each sublattice point belong to the other sublattice [9]. The triangular lattice does not have this property while the honeycomb lattice does. The next nearest neighbor sites of the triangular lattice form also a triangular lattice.

For rotational clusters of n adsorbates, one degree of rotaional freedom, φ, which may be called a gear rotation angle, is special to correspond to a rotation as a whole. Hamiltonian for this rotational freedom is given by

$$H=-\frac{\hbar^2}{2nI}\frac{d^2}{d\varphi^2}+\left[2(n-1)A_{3-3}+2nV_0\right]\cos 6\varphi, \qquad (22)$$

for the strong gear coupling limit ($\langle\cos 6\psi\rangle_\psi=-1$ in the same limit for the next nearest neighbor pair). This result seems physically reasonable that the moment of

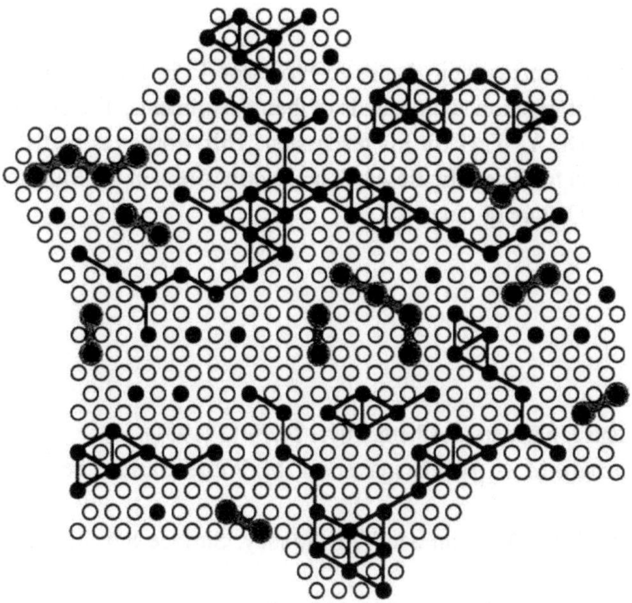

Fig. 3: Snapshot of PF_3 distribution by the Monte Carlo simulation for the $\theta = 0.2$ system. Clusters indicated by solid lines are irrotational. Clusters indicated by thicker gray lines are rotational.

inertia and rotational barrier become n times. Then, equation (22) has the same form of Hamiltonian but the effective barrier height becomes rapidly large as n^2 with increasing size of cluster.

5. Discussions

We have discussed the general quantum properties on rotational degree of freedom of PF_3 adsorbates on the triangular lattice with emphasis of gear like behavior.

We have solved the Schrödinger equations with boundary condition, $\Psi(\varphi) = \Psi(\varphi + 2\pi)$. However, since 100% of F nuclei are ^{19}F of spin 1/2, we have to take into account the symmetry of the nuclear spin state like the para and ortho states of the H_2 molecule with respect to its rotational states. For example, because the rotation of $2\pi/3$ gives a cyclic permutation of three F nuclei, that is, even permutation of Fermions, we believe for the quartet ($S = 3/2$, symmetric with respect to any exchange of spins) that the wavefunction for the isolated adsorbate must be $\Psi(\varphi) = \Psi(\varphi + (2\pi/3))$. In the limit of $V_0 = 0$, this is satisfied only by $n = 3m$ in (3) and (4). The complete solution of his problem will be given in a separate paper.

Acknowledgments

The authors would like to express their thanks to Professor T. E. Madey for his valuable discussions and information about details of their experiments. The authors would also like to thank Dr. Y. Yagi for his discussion. This work was partially supported by a Grant-in-Aid for Scientific Research on Priority Areas of the Ministry of Education, Science and Culture of Japan.

References

[1] F. Nitsché, G. Ertl and J. Küppers, J. Chem. Phys. **74**, 5911 (1981).
[2] Y. Zhou, G. E. Mitchell, M. A. Henderson, and J. M. White, Surface Sci. **214**, 209 (1989).
[3] H.-S. Tao, U. Diebold, N. D. Shinn and T. E. Madey, Surface Sci. **312**, 323 (1994).
[4] N. J. Sack and T. E. Madey, Surface Sci. 347, 367 (1996).
[5] H. Kaji, K. Kakitani, Y. Yagi and A. Yoshimori, Surface Sci. to be published.
[6] H. Kaji, K. Kakitani and A. Yoshimori, to be submitted to Surface Sci.
[7] K. Kakitani, H. Kaji and A. Yoshimori, be submitted to Surface Sci.
[8] M. D. Alvey, J. T. Yates, Jr. and K. J. Uram, J. Chem. Phys. **87**, 7221 (1987).
[9] R. Zwanzig, J. Chem. Phys. **87**, 4870 (1987).

Brief Summary

W. Brenig

This has been an outstanding meeting in many respects. The organizing committee had invited many world leaders in the field of surface processes. The presentations of the talks were excellent. The small size of the conference and the intimate setting of the lecture hall initiated intense discussions.

Let me try to regroup the various sections of the meeting into five larger subjects (partly according to my own biased point of view) as:

1. **Classics.** Since surface science (in contrast to solid state physics) is by no means mature and is still expanding one can find many surprises in the consideration of classical substances, methods, and problems.

H. Ishida, for instance, gave the first microscopic theory of surface resistance starting from the Kubo formula; *K.W. Jacobsen* presented an extended theory of growth patterns observed in metal deposition and growth going all the way from the density functional theory of potential surfaces, via the various growth modes, to a simulation of nonlinear growth. *M. Kawai* showed that for a classic adsorbate system such as CO on Ni and Pt we do not understand the ratio of terminal to bridge site adsorption. *A. Kleyn* gave a new interpretation of his high-energy molecular beam data for O on Ag in terms of a novel state of oxygen. *Y. Murata* presented an interesting model to describe the effect of excited-state life time on the rotational state distribution of NO molecules coming off Pt in laser-induced desorption. The model certainly deserves some decent dynamical calculation. *K. Shobatake* presented evidence for a third component in the time-of-flight spectra of inelastic molecular beam scattering. If this component survives further tests in the future it will pose an interesting problem for theory. *A. Yoshimori* applied quantum theory to the bizarre and amusing adsorbate PF_3/Ru(001). It would be interesting to investigate the collective modes of various adsorbed "coupled quantum gear" clusters experimentally (by inelastic He-scattering?).

2. **Electronic structure, potential-energy surfaces.** With ever increasing computer power, ab initio calculations of potential surfaces become increasingly reliable. But also model calculations are still very useful in giving simple interpretations of large-scale ab initio calculations.

K.W. Jacobsen showed that ab initio calculations for the interaction of hydrogen with transition metals can be understood quite well in terms of a kind of surface molecule cluster with localized d-orbitals and (renormalized) localized σ-orbitals. *H. Nakatsuji* applied his "dipped adcluster" model to various adsorbate systems and gave simple results for charge transfer etc.

S. Tsuneyuki presented ab initio results for excited states as they occur in photo-stimulated desorption of CO and NO from metals. *T. Uda* used ab initio calculations of potential-energy surfaces to determine the structure of defects on silicon surfaces.

3. **Optics.** SHG (second harmonics generation) and SNOM (scanning near-field optical microscopy) have become important tools of surface analysis.

K. Cho presented a nonlocal response theory and applied it to resonant SHG and SNOM. An interesting byproduct is the occurrence of a sizeable imaginary part of quantum electrodynamical radiative corrections in mesoscopic systems. *W. Hübner* gave an introduction to nonlinear magneto-optics, in particular the nonlinear Kerr effect. This is a very sensitive probe of surface magnetism with many future applications, scientific as well as technical.

4. **Anderson-Newns model, Kondo and mixed valence effects, and deviations from the Born-Oppenheimer approximation.** The Newns-Anderson model is the "classic" on the theoretical side and is still going strong.

H. Kasai presented results indicating a gradual breakdown of BOA when the kinetic energy of beam particles hitting a surface is above about 100 eV and discussed orientational effects (including the so-called "steering" effects) in molecular hydrogen adsorption and scattering. *D. Langreth* showed the importance of Kondo and mixed valency effects in charge transfer and energy transfer processes. Such effects should be observable in principle. *K. Makoshi* discussed an interesting interpolation formula for the electron self-energy occurring in Auger processes and resonant tunneling (interpolating between the limits of fast and slow ionic motion). *D. Newns* discussed another situation with strong deviations from BOA: hot electrons occurring in femtosecond pulse laser desorption. The intuitive DIMET description of this process is obtained as the high-T limit of the theory. *A. Okiji* demonstrated that a Fermi-surface-related singularity in charge exchange scattering is (unfortunately) wiped out by quantum recoil effects (though a large peak still remains).

5. **New mechanisms, substances, and probes.** At first sight this may look as if this were the future of surface physics, and in some ways it is. But let me point out again that there are lots of future problems in connection with the more "classic" subjects described above.

W. Brenig presented experimental and theoretical results on the reaction dynamics of H_2, D_2/Si (100),(111). They can be described in terms of a novel type of potential surface with unusually strong coupling between adsorbates and lattice vibrations. *H.-J. Freund* described many new experimental and theoretical results on the structure and dynamics of metal oxide surfaces: CO and NO on

NiO and Cr_2O_3. EELS data can be used as input to a theory of electronic structure and laser-induced desorption. *K. Fukutani* had interesting results concerning the adsorption and desorption of NO and CO on the alloys of Ge with Pt. Surprisingly enough the rigid band model can explain many of the observed data. *M. Persson* discussed the Eley-Rideal reaction as a mechanism which had been proposed a long time ago but has only recently, after the observed asymmetry in the isotope effect (H/DCu $\neq$ D/HCu), become a well-established new mechanism. Certainly many new types of adsorbates can now be investigated using this kind of process. *E. Torikai* presented promising preliminary results from using spin-polarized atomic beams as a new probe of surface spin correlations.

Index of Contributors

Springer-Verlag and the Environment

We at Springer-Verlag firmly believe that an international science publisher has a special obligation to the environment, and our corporate policies consistently reflect this conviction.

We also expect our business partners – paper mills, printers, packaging manufacturers, etc. – to commit themselves to using environmentally friendly materials and production processes.

The paper in this book is made from low- or no-chlorine pulp and is acid free, in conformance with international standards for paper permanency.

Springer Series in Solid-State Sciences

Editors: M. Cardona P. Fulde K. von Klitzing H.-J. Queisser

1 **Principles of Magnetic Resonance** 3rd Edition By C. P. Slichter
2 **Introduction to Solid-State Theory** By O. Madelung
3 **Dynamical Scattering of X-Rays in Crystals** By Z. G. Pinsker
4 **Inelastic Electron Tunneling Spectroscopy** Editor: T. Wolfram
5 **Fundamentals of Crystal Growth I** Macroscopic Equilibrium and Transport Concepts By F. E. Rosenberger
6 **Magnetic Flux Structures in Superconductors** By R. P. Huebener
7 **Green's Functions in Quantum Physics** 2nd Edition By E. N. Economou
8 **Solitons and Condensed Matter Physics** Editors: A. R. Bishop and T. Schneider
9 **Photoferroelectrics** By V. M. Fridkin
10 **Phonon Dispersion Relations in Insulators** By H. Bilz and W. Kress
11 **Electron Transport in Compound Semiconductors** By B. R. Nag
12 **The Physics of Elementary Excitations** By S. Nakajima, Y. Toyozawa, and R. Abe
13 **The Physics of Selenium and Tellurium** Editors: E. Gerlach and P. Grosse
14 **Magnetic Bubble Technology** 2nd Edition By A. H. Eschenfelder
15 **Modern Crystallography I** Symmetry of Crystals. Methods of Structural Crystallography 2nd Edition By B. K. Vainshtein
16 **Organic Molecular Crystals** Their Electronic States By E. A. Silinsh
17 **The Theory of Magnetism I** Statics and Dynamics By D. C. Mattis
18 **Relaxation of Elementary Excitations** Editors: R. Kubo and E. Hanamura
19 **Solitons** Mathematical Methods for Physicists By. G. Eilenberger
20 **Theory of Nonlinear Lattices** 2nd Edition By M. Toda
21 **Modern Crystallography II** Structure of Crystals 2nd Edition By B. K. Vainshtein, V. L. Indenbom, and V. M. Fridkin
22 **Point Defects in Semiconductors I** Theoretical Aspects By M. Lannoo and J. Bourgoin
23 **Physics in One Dimension** Editors: J. Bernasconi and T. Schneider
24 **Physics in High Magnetics Fields** Editors: S. Chikazumi and N. Miura
25 **Fundamental Physics of Amorphous Semiconductors** Editor: F. Yonezawa
26 **Elastic Media with Microstructure I** One-Dimensional Models By I. A. Kunin
27 **Superconductivity of Transition Metals** Their Alloys and Compounds By S. V. Vonsovsky, Yu. A. Izyumov, and E. Z. Kurmaev
28 **The Structure and Properties of Matter** Editor: T. Matsubara
29 **Electron Correlation and Magnetism in Narrow-Band Systems** Editor: T. Moriya
30 **Statistical Physics I** Equilibrium Statistical Mechanics 2nd Edition By M. Toda, R. Kubo, N. Saito
31 **Statistical Physics II** Nonequilibrium Statistical Mechanics 2nd Edition By R. Kubo, M. Toda, N. Hashitsume
32 **Quantum Theory of Magnetism** 2nd Edition By R. M. White
33 **Mixed Crystals** By A. I. Kitaigorodsky
34 **Phonons: Theory and Experiments I** Lattice Dynamics and Models of Interatomic Forces By P. Brüesch
35 **Point Defects in Semiconductors II** Experimental Aspects By J. Bourgoin and M. Lannoo
36 **Modern Crystallography III** Crystal Growth By A. A. Chernov
37 **Modern Chrystallography IV** Physical Properties of Crystals Editor: L. A. Shuvalov
38 **Physics of Intercalation Compounds** Editors: L. Pietronero and E. Tosatti
39 **Anderson Localization** Editors: Y. Nagaoka and H. Fukuyama
40 **Semiconductor Physics** An Introduction 5th Edition By K. Seeger
41 **The LMTO Method** Muffin-Tin Orbitals and Electronic Structure By H. L. Skriver
42 **Crystal Optics with Spatial Dispersion, and Excitons** 2nd Edition By V. M. Agranovich and V. L. Ginzburg
43 **Structure Analysis of Point Defects in Solids** An Introduction to Multiple Magnetic Resonance Spectroscopy By J.-M. Spaeth, J. R. Niklas, and R. H. Bartram
44 **Elastic Media with Microstructure II** Three-Dimensional Models By I. A. Kunin
45 **Electronic Properties of Doped Semiconductors** By B. I. Shklovskii and A. L. Efros
46 **Topological Disorder in Condensed Matter** Editors: F. Yonezawa and T. Ninomiya
47 **Statics and Dynamics of Nonlinear Systems** Editors: G. Benedek, H. Bilz, and R. Zeyher

Springer Series in Solid-State Sciences

Editors: M. Cardona P. Fulde K. von Klitzing H.-J. Queisser

48 **Magnetic Phase Transitions**
Editors: M. Ausloos and R. J. Elliott

49 **Organic Molecular Aggregates**
Electronic Excitation and Interaction Processes
Editors: P. Reineker, H. Haken, and H. C. Wolf

50 **Multiple Diffraction of X-Rays in Crystals**
By Shih-Lin Chang

51 **Phonon Scattering in Condensed Matter**
Editors: W. Eisenmenger, K. Laßmann, and S. Döttinger

52 **Superconductivity in Magnetic and Exotic Materials** Editors: T. Matsubara and A. Kotani

53 **Two-Dimensional Systems, Heterostructures, and Superlattices**
Editors: G. Bauer, F. Kuchar, and H. Heinrich

54 **Magnetic Excitations and Fluctuations**
Editors: S. W. Lovesey, U. Balucani, F. Borsa, and V. Tognetti

55 **The Theory of Magnetism II** Thermodynamics and Statistical Mechanics By D. C. Mattis

56 **Spin Fluctuations in Itinerant Electron Magnetism** By T. Moriya

57 **Polycrystalline Semiconductors**
Physical Properties and Applications
Editor: G. Harbeke

58 **The Recursion Method and Its Applications**
Editors: D. G. Pettifor and D. L. Weaire

59 **Dynamical Processes and Ordering on Solid Surfaces** Editors: A. Yoshimori and M. Tsukada

60 **Excitonic Processes in Solids**
By M. Ueta, H. Kanzaki, K. Kobayashi, Y. Toyozawa, and E. Hanamura

61 **Localization, Interaction, and Transport Phenomena** Editors: B. Kramer, G. Bergmann, and Y. Bruynseraede

62 **Theory of Heavy Fermions and Valence Fluctuations** Editors: T. Kasuya and T. Saso

63 **Electronic Properties of Polymers and Related Compounds**
Editors: H. Kuzmany, M. Mehring, and S. Roth

64 **Symmetries in Physics** Group Theory Applied to Physical Problems 2nd Edition
By W. Ludwig and C. Falter

65 **Phonons: Theory and Experiments II**
Experiments and Interpretation of Experimental Results By P. Brüesch

66 **Phonons: Theory and Experiments III**
Phenomena Related to Phonons
By P. Brüesch

67 **Two-Dimensional Systems: Physics and New Devices**
Editors: G. Bauer, F. Kuchar, and H. Heinrich

68 **Phonon Scattering in Condensed Matter V**
Editors: A. C. Anderson and J. P. Wolfe

69 **Nonlinearity in Condensed Matter**
Editors: A. R. Bishop, D. K. Campbell, P. Kumar, and S. E. Trullinger

70 **From Hamiltonians to Phase Diagrams**
The Electronic and Statistical-Mechanical Theory of sp-Bonded Metals and Alloys By J. Hafner

71 **High Magnetic Fields in Semiconductor Physics**
Editor: G. Landwehr

72 **One-Dimensional Conductors**
By S. Kagoshima, H. Nagasawa, and T. Sambongi

73 **Quantum Solid-State Physics**
Editors: S. V. Vonsovsky and M. I. Katsnelson

74 **Quantum Monte Carlo Methods in Equilibrium and Nonequilibrium Systems** Editor: M. Suzuki

75 **Electronic Structure and Optical Properties of Semiconductors** 2nd Edition
By M. L. Cohen and J. R. Chelikowsky

76 **Electronic Properties of Conjugated Polymers**
Editors: H. Kuzmany, M. Mehring, and S. Roth

77 **Fermi Surface Effects**
Editors: J. Kondo and A. Yoshimori

78 **Group Theory and Its Applications in Physics**
2nd Edition
By T. Inui, Y. Tanabe, and Y. Onodera

79 **Elementary Excitations in Quantum Fluids**
Editors: K. Ohbayashi and M. Watabe

80 **Monte Carlo Simulation in Statistical Physics**
An Introduction 2nd Edition
By K. Binder and D. W. Heermann

81 **Core-Level Spectroscopy in Condensed Systems**
Editors: J. Kanamori and A. Kotani

82 **Photoelectron Spectroscopy**
Principle and Applications 2nd Edition
By S. Hüfner

83 **Physics and Technology of Submicron Structures**
Editors: H. Heinrich, G. Bauer, and F. Kuchar

84 **Beyond the Crystalline State** An Emerging Perspective By G. Venkataraman, D. Sahoo, and V. Balakrishnan

85 **The Quantum Hall Effects**
Integral and Fractional 2nd Edition
By T. Chakraborty and P. Pietiläinen

86 **The Quantum Statistics of Dynamic Processes**
By E. Fick and G. Sauermann

87 **High Magnetic Fields in Semiconductor Physics II**
Transport and Optics Editor: G. Landwehr

88 **Organic Superconductors**
By T. Ishiguro and K. Yamaji

89 **Strong Correlation and Superconductivity**
Editors: H. Fukuyama, S. Maekawa, and A. P. Malozemoff

MIX
Papier aus verantwortungsvollen Quellen
Paper from responsible sources
FSC® C105338

If you have any concerns about our products, you can contact us on
ProductSafety@springernature.com

In case Publisher is established outside the EU, the EU authorized representative is:
Springer Nature Customer Service Center GmbH
Europaplatz 3, 69115 Heidelberg, Germany

Printed by Libri Plureos GmbH
in Hamburg, Germany